INTRODUCTION TO THE THEORY OF FOURIER'S SERIES AND INTEGRALS

BY

H. S. CARSLAW, Sc.D., LL.D., F.R.S.E.

PROFESSOR OF MATHEMATICS IN THE UNIVERSITY OF SYDNEY
FELLOW OF EMMANUEL COLLEGE, CAMBRIDGE
AND FORMERLY LECTURER IN MATHEMATICS IN THE UNIVERSITY OF GLASGOW

THIRD EDITION, REVISED AND ENLARGED

DOVER PUBLICATIONS, INC.
NEW YORK

Published in Canada by General Publishing Company, Ltd., 30 Lesmill Road, Don Mills, Toronto, Ontario.

Published in the United Kingdom by Constable and Company, Ltd., 10 Orange Street, London WC 2.

This Dover edition, first published in 1950, is an unabridged and unaltered republication of the third edition, published by Macmillan and Company, Ltd., in 1930.

Standard Book Number: 486-60048-3
Library of Congress Catalog Card Number: 50-1367

Manufactured in the United States of America
Dover Publications, Inc.
180 Varick Street
New York, N. Y. 10014

PREFACE TO THE THIRD EDITION

In preparing this new edition the text has been carefully revised, and considerable new matter has been introduced without altering the numbering of the sections or the character and aim of the book.

The use of the upper and lower limits of indetermination simplifies some of the proofs of the earlier edition. Additional tests for uniform convergence of series are included. Term by term integration and the Second Theorem of Mean Value are treated more fully. The sets of examples on Infinite Series and Integrals have been enlarged by the insertion of questions drawn from recent Cambridge Scholarship and Intercollegiate Examinations, as well as from the papers set in the Mathematical Tripos. The introduction of functions of bounded variation extends the class of functions to which the elementary discussion of Fourier's Series given in the text applies.

In the chapters dealing particularly with Fourier's Series space has been found for the Riemann-Lebesgue Theorem and its consequences, and for Parseval's Theorem under fairly general conditions.

For all ordinary purposes the discussion of the properties of Fourier's Series and Fourier's Constants given in the text will, it is hoped, be found both sufficient and satisfactory.

For the specialist who wishes to go further a treatment of the Lebesgue Definite Integral is given in a new Appendix, which takes the place of the former Appendix containing a detailed bibliography of Trigonometrical and Fourier's Series. In this Appendix I have tried to show, in as simple a way as possible, what the Lebesgue Integral is, and in what respects the rules to which it is subject differ from, and are superior to, those for the classical Riemann Integral.

So many papers are being written on Trigonometrical Series, and on Fourier's Series, Fourier's Constants, and Fourier's Integrals, that a mere list of their titles, brought up to date, would cover many pages. And it is doubtful if such a list is of much value to the student. In any case he has now at his disposal other works from which bibliographical information of this kind can be obtained. It is hoped that the lists of books and memoirs given at the ends of the chapters and of Appendix II will make up for the omission of the detailed bibliography.

In the revision of part of the proofs I am fortunate in having had the assistance of Mr. George Walker of the University of Sydney, and now at Emmanuel College, Cambridge. His criticism and suggestions have been of great service to me.

H. S. CARSLAW.

EMMANUEL COLLEGE,
CAMBRIDGE, *24th September*, 1929.

PREFACE TO THE SECOND EDITION

THIS book forms the first volume of the new edition of my book on *Fourier's Series and Integrals and the Mathematical Theory of the Conduction of Heat*, published in 1906, and now for some time out of print. Since 1906 so much advance has been made in the Theory of Fourier's Series and Integrals, as well as in the mathematical discussion of Heat Conduction, that it has seemed advisable to write a completely new work, and to issue the same in two volumes. The first volume, which now appears, is concerned with the Theory of Infinite Series and Integrals, with special reference to Fourier's Series and Integrals. The second volume will be devoted to the Mathematical Theory of the Conduction of Heat.

No one can properly understand Fourier's Series and Integrals without a knowledge of what is involved in the convergence of infinite series and integrals. With these questions is bound up the development of the idea of a limit and a function, and both are founded upon the modern theory of real numbers. The first three chapters deal with these matters. In Chapter IV the Definite Integral is treated from Riemann's point of view, and special attention is given to the question of the convergence of infinite integrals. The theory of series whose terms are functions of a single variable, and the theory of integrals which contain an arbitrary parameter are discussed in Chapters V and VI. It will be seen that the two theories are closely related, and can be developed on similar lines.

The treatment of Fourier's Series in Chapter VII depends on Dirichlet's Integrals. There, and elsewhere throughout the book, the Second Theorem of Mean Value will be found an essential part of the argument. In the same chapter the work of Poisson is adapted to modern standards, and a prominent place is given to Fejér's work, both in the proof of the fundamental theorem and in the discussion of the convergence of Fourier's Series. Chapter IX is devoted to Gibb's Phenomenon, and the last chapter to Fourier's Integrals. In this chapter the work of Pringsheim, who has greatly extended the class of functions to which Fourier's Integral Theorem applies, has been used.

Two appendices are added. The first deals with *Practical Harmonic Analysis* and *Periodogram Analysis*. In the second a bibliography of the subject is given.

The functions treated in this book are "ordinary" functions. An interval (a, b) for which $f(x)$ is defined can be broken up into a finite number of open partial intervals, in each of which the function is monotonic. If infinities occur in the range, they are isolated and finite in number. Such functions will satisfy the demands of the Applied Mathematician.

The modern theory of integration, associated chiefly with the name of Lebesgue, has introduced into the Theory of Fourier's Series and Integrals functions of a far more complicated nature. Various writers, notably W. H. Young, are engaged in building up a theory of these and applied series much more advanced than anything treated in this book. These developments are in the meantime chiefly interesting to the Pure Mathematician specialising in the Theory of Functions of a Real Variable. My purpose has been to remove some of the difficulties of the Applied Mathematician.

The preparation of this book has occupied some time, and much of it has been given as a final course in the Infinitesimal Calculus to my students. To them it owes much. For assistance in the revision of the proofs and for many valuable suggestions, I am much indebted to Mr. E. M. Wellish, Mr. R. J. Lyons, and Mr. H. H. Thorne of the Department of Mathematics in the University of Sydney.

H. S. CARSLAW.

EMMANUEL COLLEGE,
CAMBRIDGE, *Jan.* 1921.

CONTENTS

CHAPTER IV

THE DEFINITE INTEGRAL

CHAPTER V

THE THEORY OF INFINITE SERIES, WHOSE TERMS ARE FUNCTIONS OF A SINGLE VARIABLE

CHAPTER VI

DEFINITE INTEGRALS CONTAINING AN ARBITRARY PARAMETER

CHAPTER VII

FOURIER'S SERIES

CHAPTER VIII

THE NATURE OF THE CONVERGENCE OF FOURIER'S SERIES AND SOME PROPERTIES OF FOURIER'S CONSTANTS

CHAPTER IX

THE APPROXIMATION CURVES AND THE GIBBS PHENOMENON IN FOURIER'S SERIES

CHAPTER X

FOURIER'S INTEGRALS

HISTORICAL INTRODUCTION

A trigonometrical series

$$a_0 + (a_1 \cos x + b_1 \sin x) + (a_2 \cos 2x + b_2 \sin 2x) + \ldots$$

is said to be a Fourier's Series, if the constants $a_0, a_1, b_1, \ldots$ satisfy the equations

$$a_0 = \frac{1}{2\pi}\int_{-\pi}^{\pi} f(x)\,dx,$$

$$\left.\begin{aligned} a_n &= \frac{1}{\pi}\int_{-\pi}^{\pi} f(x)\cos nx\,dx, \\ b_n &= \frac{1}{\pi}\int_{-\pi}^{\pi} f(x)\sin nx\,dx \end{aligned}\right\} n \geqq 1,$$

and the Fourier's Series is said to correspond to the function $f(x)$.*

In many important cases the sum of the Fourier's Series which corresponds to $f(x)$ is equal to $f(x)$; but if the function is arbitrary, there is no *a priori* reason that the series should converge at all in the interval $(-\pi, \pi)$, nor, if it does converge at a point, is there any *a priori* reason that its sum for that value of x should be $f(x)$.

Fourier in his *Théorie analytique de la Chaleur* (1822) was the first to assert that an arbitrary function, given in the interval $(-\pi, \pi)$, could be expressed in this way. He proved quite rigorously that the expansion is true for certain simple functions, which he needed in the problems of the conduction of heat; and, though he did not develop his proof for the general case with the precision the importance of the theorem demanded, the substantial accuracy of his method must be admitted. That the expansion was possible

---

*This correspondence is sometimes denoted by

$$f(x) \sim a_0 + \sum_{1}^{\infty}(a_n \cos nx + b_n \sin nx),$$

the notation being due to Hurwitz, *Math. Annalen*, 57 (1903), 427.

in the case of an arbitrary function, as such was understood at that time, was assumed to be true from the date at which his work became known. Since then these series have been freely used in the solution of the differential equations of mathematical physics. For this reason they are now called Fourier's Series—or the Fourier's Series corresponding to the function $f(x)$—and the coefficients in the series,

$$\frac{1}{2\pi}\int_{-\pi}^{\pi} f(x)\,dx,\quad \frac{1}{\pi}\int_{-\pi}^{\pi} f(x)\frac{\sin}{\cos}nx\,dx,$$

are called Fourier's Coefficients, or Fourier's Constants, for that function.

The Theory of Fourier's Series has had—and still is having—an immense influence on the development of the theory of functions of a real variable, and the influence and importance of these series in this field are comparable with those of the power series in the general theory of functions.

FIRST PERIOD [1750–1850]

The question of the possibility of the expansion of an arbitrary function of x in a trigonometrical series of sines and cosines of multiples of x arose in the middle of the eighteenth century in connection with the problem of the vibration of strings.

The theory of these vibrations reduces to the solution of the differential equation

$$\frac{\partial^2 y}{\partial t^2}=a^2\frac{\partial^2 y}{\partial x^2},$$

and the earliest attempts at its solution were made by d'Alembert,* Euler,† and D. Bernoulli.‡ Both d'Alembert and Euler obtained the solution in the functional form

$$y=\phi(x+at)+\psi(x-at).$$

The principal difference between them lay in the fact that d'Alembert supposed the initial form of the string to be given by a single analytical expression, while Euler regarded it as lying along any arbitrary continuous curve, different parts of which might be given by different analytical expressions. Bernoulli, on

---

* *Mém. de l'Académie de Berlin*, **3** (1747), 214.

† *loc. cit.*, **4** (1748), 69. ‡ *loc. cit.*, **9** (1753), 173.

the other hand, gave the solution, when the string starts from rest, in the form of a trigonometrical series

$$y = A_1 \sin x \cos at + A_2 \sin 2x \cos 2at + \dots ,$$

and he asserted that this solution, being perfectly general, must contain those given by Euler and d'Alembert. The importance of his discovery was immediately recognised, and Euler pointed out that if this statement of the solution were correct, an arbitrary function of a single variable must be developable in an infinite series of sines of multiples of the variable. This he held to be obviously impossible, since a series of sines is both periodic and odd, and he argued that if the arbitrary function had not both of these properties it could not be expanded in such a series.

While the debate was at this stage a memoir appeared in 1759* by Lagrange, then a young and unknown mathematician, in which the problem was examined from a totally different point of view. While he accepted Euler's solution as the most general, he objected to the mode of demonstration, and he proposed to obtain a satisfactory solution by first considering the case of a finite number of particles stretched on a weightless string. From the solution of this problem he deduced that of a continuous string by making the number of particles infinite.† In this way he showed that when the initial displacement of the string of unit length is given by $f(x)$ and the initial velocity by $F(x)$, the displacement at time t is given by

$$y = 2\int_0^1 \sum_1^\infty (\sin n\pi x' \sin n\pi x \cos n\pi at) f(x')dx'$$

$$+ \frac{2}{a\pi}\int_0^1 \sum_1^\infty \frac{1}{n} (\sin n\pi x' \sin n\pi x \sin n\pi at) F(x')dx'.$$

This result and the discussion of the problem which Lagrange gave in this and other memoirs have prompted some mathematicians to deny the importance of Fourier's discoveries, and to attribute to Lagrange the priority in the proof of the development of an arbitrary function in trigonometrical series. It is true that in the formula quoted above it is only necessary to change the order of summation and integration, and to put $t=0$, in order

*Cf. Lagrange, *Œuvres*, 1 (Paris, 1867), 37.

†*loc. cit.*, § 37.

that we may obtain the development of the function $f(x)$ in a series of sines, and that the coefficients shall take the definite integral forms with which we are now familiar. Still Lagrange did not take this step, and, as Burkhardt remarks,* the fact that he did not do so is a very instructive example of the ease with which an author omits to draw an almost obvious conclusion from his results, when his investigation has been undertaken with another end in view. Lagrange's purpose was to demonstrate the truth of Euler's solution and to defend its general conclusions against d'Alembert's attacks. When he had obtained his solution, he therefore proceeded to transform it into the functional form given by Euler. Having succeeded in this, he held his demonstration to be complete.

The further development of the theory of these series was due to the astronomical problem of the expansion of the reciprocal of the distance between two planets in a series of cosines of multiples of the angle between the radii. As early as 1749 and 1754 d'Alembert and Euler had published discussions of this question in which the idea of the definite integral expressions for the coefficients in Fourier's Series may be traced, and Clairaut, in 1757,† gave his results in a form which practically contained these coefficients. Again, Euler,‡ in a paper written in 1777 and published in 1793, actually employed the method of multiplying both sides of the equation

$$f(x)=a_0+2a_1 \cos x+2a_2 \cos 2x+\ldots+2a_n \cos nx+\ldots$$

by $\cos nx$ and integrating the series term by term between the limits 0 and π. In this way he found that

$$a_n=\frac{1}{\pi}\int_0^\pi f(x) \cos nx\, dx.$$

It is curious that these papers seem to have had no effect upon the discussion of the problem of the Vibrations of Strings in which, as we have seen, d'Alembert, Euler, Bernoulli, and Lagrange were about the same time engaged. The explanation is probably to be found in the fact that these results were not accepted with

---

*Burkhardt, "Entwicklungen nach oscillirenden Functionen," *Jahresber. d. Math. Ver., Leipzig*, 10, Heft II (1901), 32.

†*Paris, Hist. Acad. Sci.* (1754 [59]), Art. iv. (July 1757).

‡*Petrop. N. Acta*, 11 (1793 [98]), p. 94 (May 1777).

confidence, and that they were only used in determining the coefficients of expansions whose existence could be demonstrated by other means.

It was left to Fourier to place our knowledge of the theory of trigonometrical series on a firmer foundation.

The methods he adopted were suggested by the problems he met in the Mathematical Theory of the Conduction of Heat. He discussed the subject in various memoirs, the most important having been presented to the Paris Academy in 1811, although it was not printed till 1824-6. These memoirs are practically contained in his book, *Théorie analytique de la Chaleur* (1822). In a number of special cases he verified that a function $f(x)$, given in the interval $(-\pi, \pi)$, can be expressed as the sum of the series

$$a_0+(a_1 \cos x+b_1 \sin x)+(a_2 \cos 2x+b_2 \sin 2x)+\ldots$$

where $$a_0=\frac{1}{2\pi}\int_{-\pi}^{\pi} f(x)dx, \quad a_n=\frac{1}{\pi}\int_{-\pi}^{\pi} f(x) \cos nx\, dx,$$

$$b_n=\int_{-\pi}^{\pi} f(x) \sin nx\, dx, \quad (n \geqq 1).$$

Some of the proofs he gave for the general case of an arbitrary function are far from rigorous. One is the same as that given by Euler. But in his final discussion of the general case (Cf. §§ 415, 416 and 423), the method he employs is perfectly sound, and not unlike that which Dirichlet used later in his classical memoir. However, this discussion is little more than a sketch of a proof, and it contains no reference to the conditions which the arbitrary function must satisfy.

Fourier made no claim to the discovery of the values of the coefficients

$$a_0=\frac{1}{2\pi}\int_{-\pi}^{\pi} f(x)dx,$$

$$\left.\begin{aligned} a_n&=\frac{1}{\pi}\int_{-\pi}^{\pi} f(x) \cos nx\, dx, \\ b_n&=\frac{1}{\pi}\int_{-\pi}^{\pi} f(x) \sin nx\, dx, \end{aligned}\right\} n \geqq 1.$$

We have already seen that they were employed both by Clairaut and Euler before this time. Still there is an important difference between Fourier's interpretation of these integrals and that which was current among the mathematicians of the eighteenth century.

The earlier writers by whom they were employed (with the possible exception of Clairaut) applied them to the determination of the coefficients of series whose existence had been demonstrated by other means. Fourier was the first to apply them to the representation of an entirely arbitrary function, in the sense in which that sum was then understood. In this he made a distinct advance upon his predecessors. Indeed Riemann* asserts that when Fourier, in his first paper to the Paris Academy in 1807, stated that a completely arbitrary function could be expressed in such a series, his statement so surprised Lagrange that he denied the possibility in the most definite terms. It should also be noted that he was the first to allow that the arbitrary function might be given by different analytical expressions in different parts of the interval; also that he asserted that the sine series could be used for other functions than odd ones, and the cosine series for other functions than even ones. Further, he was the first to see that, when a function is defined for a given range of the variable, its value outside that range is in no way determined, and it follows that no one before him can have properly understood the representation of an arbitrary function by a trigonometrical series.

The treatment which his work received from the Paris Academy is evidence of the doubt with which his contemporaries viewed his arguments and results. His first paper upon the Theory of Heat was presented in 1807. The Academy, wishing to encourage the author to extend and improve his theory, made the question of the propagation of heat the subject of the *grand prix de mathématiques* for 1812. Fourier submitted his *Mémoire sur la propagation de la Chaleur* at the end of 1811 as a candidate for the prize. The memoir was referred to Laplace, Lagrange, Legendre, and the other adjudicators; but, while awarding him the prize, they qualified their praise with criticisms of the rigour of his analysis and methods,† and the paper was not published at

---

*Cf. Riemann, "Über die Darstellbarkeit einer Function durch eine trigonometrische Reihe," *Göttingen, Abh. Ges. Wiss.*, 13 (1867), § 2, and *Mathematische Werke* (2 Aufl., 1892), p. 232.

†Their report is quoted by Darboux in his Introduction (p. vii) to *Œuvres de Fourier*, T. I:—"Cette pièce renferme les véritables équations différentielles de la transmission de la chaleur, soit à l'intérieur des corps, soit à leur surface; et la nouveauté du sujet, jointe à son importance, a déterminé la Classe à couronner cet Ouvrage, en observant cependant que la manière dont l'Auteur parvient à ses

the time in the *Mémoires de l'Academie des Sciences.* Fourier always resented the treatment he had received. When publishing his treatise in 1822, he incorporated in it, practically without change, the first part of this memoir; and two years later, having become Secretary of the Academy on the death of Delambre, he caused his original paper, in the form in which it had been communicated in 1811, to be published in these *Mémoires.*‡ Probably this step was taken to secure to himself the priority in his discoveries, in consequence of the attention the subject was receiving at the hands of other mathematicians. It is also possible that he wished to show the injustice of the criticisms which had been passed upon his work. After the publication of his treatise, when the results of his different memoirs had become known, it was recognised that real advance had been made by him in the discussion of the subject and the substantial accuracy of his reasoning was admitted.§

équations n'est pas exempte de difficultés, et que son analyse, pour les intégrer, laisse encore quelque chose à désirer, soit relativement à la généralité, soit même du côté de la rigueur."

‡*Mémoires de l'Acad. des Sc.*, **4**, p. 185, and **5**, p. 153.

§It is interesting to note the following references to his work in the writings of modern mathematicians:

Kelvin, *Coll. Works*, Vol. III, p. 192 (Article on "Heat," *Enc. Brit.*, 1878).

"Returning to the question of the Conduction of Heat, we have first of all to say that the theory of it was discovered by Fourier, and given to the world through the French Academy in his *Théorie analytique de la Chaleur*, with solutions of problems naturally arising from it, of which it is difficult to say whether their uniquely original quality, or their transcendently intense mathematical interest, or their perennially important instructiveness for physical science, is most to be praised."

Darboux, Introduction, *Œuvres de Fourier*, **1** (1888), p. v.

"Par l'importance de ses découvertes, par l'influence décisive qu'il a exercée sur le développement de la Physique mathématique, Fourier méritait l'hommage qui est rendu aujourd'hui à ses travaux et à sa mémoire. Son nom figurera dignement à côté des noms, illustres entre tous, dont la liste, destinée à s'accroître avec les années, constitue dès à présent un véritable titre d'honneur pour notre pays. La *Théorie analytique de la Chaleur* . . . , que l'on peut placer sans injustice à côté des écrits scientifiques les plus parfaits de tous les temps, se recommande par une exposition intéressante et originale des principes fondamentaux; il éclaire de la lumière la plus vive et la plus pénétrante toutes les idées essentielles que nous devons à Fourier et sur lesquelles doit reposer désormais la Philosophie naturelle; mais il contient, nous devons le reconnaître, beaucoup de négligences, des erreurs de calcul et de détail que Fourier a su éviter dans d'autres écrits."

Poincaré, *Théorie analytique de la propagation de la Chaleur* (1891), p. 1, § 1.

"La théorie de la chaleur de Fourier est un des premiers exemples de l'application de l'analyse à la physique; en partant d'hypothèses simples qui ne sont

The next writer upon the Theory of Heat was Poisson. He employed an altogether different method in his discussion of the question of the representation of an arbitrary function by a trigonometrical series in his papers from 1820 onwards, which are practically contained in his books, *Traité de Mécanique* (2e éd., 1833) and *Théorie mathématique de la Chaleur* (1835). He began with the equation

$$\frac{1-h^2}{1-2h\cos(x'-x)+h^2}=1+2\sum_1^\infty h^n \cos n(x'-x),$$

h being numerically less than unity, and he obtained, by integration,

$$\int_{-\pi}^{\pi} \frac{(1-h^2)f(x')\,dx'}{1-2h\cos(x'-x)+h^2}$$
$$=\int_{-\pi}^{\pi} f(x')dx' + 2\sum_1^\infty h^n \int_{-\pi}^{\pi} f(x')\cos n(x'-x)\,dx'.$$

While it is true that by proceeding to the limit we may deduce that at a point of continuity or ordinary discontinuity

$$f(x) \text{ or } \tfrac{1}{2}[f(x+0)+f(x-0)]$$

is equal to

$$\lim_{h\to 1}\left[\frac{1}{2\pi}\int_{-\pi}^{\pi} f(x')\,dx' + \frac{1}{\pi}\sum_1^\infty h^n \int_{-\pi}^{\pi} f(x')\cos n(x'-x)\,dx'\right],$$

we are not entitled to assert that this holds *for the value* $h=1$, unless we have already proved that the series converges for this value. This is the real difficulty in the theory of Fourier's Series, and this limitation on Poisson's discussion has been lost sight of in

---

autre chose que des faits expérimentaux généralisés, Fourier en a déduit une série de conséquences dont l'ensemble constitue une théorie complète et cohérente. Les résultats qu'il a obtenus sont certes intéressants par eux-mêmes, mais ce qui l'est plus encore est la méthode qu'il a employée pour y parvenir et qui servira toujours de modèle à tous ceux qui voudront cultiver une branche quelconque de la physique mathématique. J'ajouterai que le livre de Fourier a une importance capitale dans l'histoire des mathématiques et que l'analyse pure lui doit peut-être plus encore que l'analyse appliquée."

Boussinesq, *Théorie analytique de la Chaleur*, **1** (1901), 4.

"Les admirables applications qu'il fit de cette méthode (*i.e.* his method of integrating the equations of Conduction of Heat) sont, à la fois, assez simples et assez générales, pour avoir servi de modèle aux géomètres de la première moitié de ce siècle; et elles leur ont été d'autant plus utiles, qu'elles ont pu, avec de légères modifications tout au plus, être transportées dans d'autres branches de la Physique mathématique, notamment dans l'Hydrodynamique et dans la Théorie de l'élasticité."

some presentations of Fourier's Series. There are, however, other directions in which Poisson's method has led to most notable results. The importance of his work cannot be exaggerated.*

After Poisson, Cauchy attacked the subject in different memoirs published from 1826 onwards using his method of residues, but his treatment did not attract so much attention as that given about the same time by Dirichlet, to which we now turn.

Dirichlet's investigation is contained in two memoirs which appeared in 1829† and 1837.‡ The method which he employed we have already referred to in speaking of Fourier's work. He based his proof upon a careful discussion of the limiting values of the integrals

$$\int_0^a f(x) \frac{\sin \mu x}{\sin x}\, dx \ldots, \; a > 0,$$

$$\int_a^b f(x) \frac{\sin \mu x}{\sin x}\, dx \ldots, \; b > a > 0,$$

as μ increases indefinitely. By this means he showed that the sum of the Fourier's Series for $f(x)$ is $\frac{1}{2}(f(x+0)+f(x-0))$ at every point between $-\pi$ and π, and $\frac{1}{2}(f(-\pi+0)+f(\pi-0))$ at $x = \pm\pi$, provided that $f(x)$ has only a finite number of ordinary discontinuities and turning points, and that it does not become infinite in $(-\pi, \pi)$. In a later paper,§ in which he discussed the expansion in Spherical Harmonics, he showed that the restriction that $f(x)$ must remain finite is not necessary, provided that $\int_{-\pi}^{\pi} f(x)dx$ converges absolutely.

SECOND PERIOD [1850–1905]

The principal names in the First Period are those of Fourier and Dirichlet, and the position as left by Dirichlet was that, when the function $f(x)$ is bounded in the interval $(-\pi, \pi)$, and this interval can be broken up into a finite number of partial intervals in each of which $f(x)$ is monotonic, the Fourier's Series converges at every point within the interval to $\frac{1}{2}[f(x+0)+f(x-0)]$, and at the end-points to $\frac{1}{2}[f(-\pi+0)+f(\pi-0)]$. These sufficient conditions—and

*For a full treatment of Poisson's method, reference may be made to Bôcher's paper, "Introduction to the Theory of Fourier's Series," *Ann. of Math.* (2), **7** (1906).

†*Journal für Math.*, **4** (1829).

‡*Dove's Repertorium der Physik*, **1** (1837), 152.

§*Journal für Math.*, **17** (1837).

their extension to the unbounded function—cover most of the cases that are likely to be required in the applications of Fourier's Series to the solution of the differential equations of mathematical physics.

In the Second Period we pass more definitely into the domain of the pure mathematician, and the first name we meet is that of Riemann. His memoir* *Über die Darstellbarkeit einer Function durch eine trigonometrische Reihe* formed his *Habilitationsschrift* at Göttingen in 1854, but it was not published till 1867, after his death. It led to most important developments in mathematical analysis, as well as to the discovery of many striking properties of trigonometrical series, in general, and of Fourier's Series, in particular. His aim was to find a necessary and sufficient condition which the arbitrary function must satisfy so that, at a point x in the interval, the corresponding Fourier's Series shall converge to $f(x)$. Dirichlet had shown that certain conditions were sufficient. The question Riemann set himself to answer has not yet been solved. It is quite probable that it is not solvable. But in the consideration of the problem he realised that the concept of the definite integral should be widened. And the Riemann Integral we owe to the study of Fourier's Series.

Cauchy in 1823† had defined the definite integral of a continuous function as the limit of a sum, much in the way it is still treated in elementary text-books. He divided the interval of integration into partial intervals by the points

$$a=a_0,\ a_1 \dots a_{n-1},\ a_n=b.$$

The sum S was given by the equation

$$S=(a_1-a_0)f(x_1)+(a_2-a_1)f(x_2)+\dots+(a_n-a_{n-1})f(x_n),$$

where x_r is any point in (a_{r-1}, a_r).

He showed that, when the number of points of section tends to infinity and the length of the largest partial interval tends to zero, the sums S tend to a limit. The definite integral $\int_a^b f(x)dx$ he defined to be this limit.

If the function is continuous in (a, b) except at the point c, in

---

*See note on p. 6.

†Cf. Cauchy, *Résumé des leçons données a l'École roy. Polytechnique sur le calcul infinitésimal*, 1 (Paris, 1823), pp. 81–84, and *Œuvres* (2), 4, p. 122–29.

the neighbourhood of which it may be bounded or not, the integral is taken to be the sum of the limits

$$\lim_{h\to 0}\int_a^{c-h} f(x)dx \quad \text{and} \quad \lim_{h\to 0}\int_{c+h}^{b} f(x)dx,$$

when these limits exist. And if $f(x)$ is discontinuous at a finite number of points $c_1, c_2, \dots c_m$, the interval is divided into parts each of which contains only one of these points. To each of these parts the preceding definition is applied, when this is possible; and then the sum of the numbers so obtained is taken as the integral from a to b.

In dealing with the bounded function, Riemann did not assume that it was continuous in the interval, or had only a finite number of discontinuities therein. But he used the sum S as before, and the integral $\int_a^b f(x)dx$ was defined as the limit of these sums S, provided this limit existed. He obtained a necessary and sufficient condition for the existence of the limit, and placed the definite integral on a wider and purely arithmetical basis.

With Riemann's definition of the integral $\int_a^b f(x)dx$, for a bounded function—given in the text in a slightly modified form—functions that were previously without an integral became integrable. A striking example* due to him was the sum of the series,

$$(x)+\frac{(2x)}{2^2}+\frac{(3x)}{3^2}+\dots,$$

where (nx) stands for the positive or negative difference between nx and the nearest integer, unless it lies midway between two consecutive integers, when (nx) is to be taken as zero. This function is discontinuous for every rational number of the form $p/2n$, where p is an odd number, prime to n; and there are an infinite number of points of discontinuity in every interval, however small.

A fundamental theorem proved by Riemann deals with the Fourier's Constants $\frac{1}{\pi}\int_{-\pi}^{\pi} f(x)\,\frac{\sin}{\cos}\,nx\,dx$. He showed that for any bounded and integrable function $f(x)$ these constants tend to zero as n tends to infinity. And this holds also for the integral

$$\int_a^b f(x)\,\frac{\sin}{\cos}\,nx\,dx.$$

*Cf. *loc. cit.* § 6.

This theorem shows at once that, if $f(x)$ is bounded and integrable in $(-\pi, \pi)$, the convergence of its Fourier's Series at a point in $(-\pi, \pi)$ depends only on the behaviour of $f(x)$ in the neighbourhood of that point.

Riemann was also led to examine the theory of trigonometrical series of the type

$$a_0 + (a_1 \cos x + b_1 \sin x) + (a_2 \cos 2x + b_2 \sin 2x) + \dots,$$

when the coefficients are not Fourier's Constants. He obtained many of the properties of such series. The most important question to be answered was whether a function could be represented by more than one such series in an interval $(-\pi, \pi)$. This reduces to the question whether the sum of a trigonometrical series in which the coefficients do not all vanish can be zero right through the interval. The discussion of this and similar problems was carried on, chiefly by Heine and G. Cantor, from 1870 onwards; in these papers Cantor laid the foundation of the Theory of Sets of Points, another example of the remarkable influence the theory of Fourier's Series has had upon the development of mathematics. It will be sufficient in this place to state that Cantor showed in 1872 that all the coefficients of the trigonometrical series must vanish, if its sum is zero at all points of $(-\pi, \pi)$, with the exception of the points of a set of the nth order.*

In 1875 P. du Bois-Reymond proved† that if a trigonometrical series converges in $(-\pi, \pi)$ to $f(x)$, where $f(x)$ is integrable, the series must be the Fourier's Series for $f(x)$. He also settled the question as to whether the Fourier's Series for a continuous function always has $f(x)$ for its sum; for he gave not only an example of a function, continuous in $(-\pi, \pi)$, whose Fourier's Series did not converge at a particular point, but he also constructed another, whose Fourier's Series fails to converge at the points of an everywhere dense set. Many years later Fejér gave several much simpler examples.‡

The nature of the convergence of Fourier's Series received attention, especially after the introduction by Stokes (1847) and

---

* *Math. Annalen*, 5 (1872), 123.

† *Abh. d. Bay. Akad.*, 12 (1875), p. 117.

‡ Cf. *Journal für Math.*, 137 (1909); 138 (1910), 22. *Rend. Circ. Mat. Palermo*, 28 (1909), 402.

Seidel (1848) of the concept of uniform convergence. It had been known since Dirichlet's time that the series were, in general, only conditionally convergent, if at all; and that their convergence depended upon the presence of positive and negative terms. It was not till 1870 that Heine showed* that if $f(x)$ is bounded and integrable, and otherwise satisfies Dirichlet's Conditions in $(-\pi, \pi)$, its Fourier's Series converges uniformly in any interval (a, b), which contains neither inside it nor at an end any discontinuity of the function.

The importance attached to the question of uniform convergence of the series was due to the impression that term by term integration would only be permissible, if the series converged uniformly. It was not till much later that it was found that a Fourier's Series could be integrated term by term, even if the series itself did not converge.

The sufficient conditions of Dirichlet were succeeded by three conditions, now classical, associated with the names of Dini, Lipschitz and Jordan. Dini† in 1880 showed that *the Fourier's Series for the integrable function* $f(x)$ *has* $\lim_{h\to 0} \frac{1}{2}[f(x+h)+f(x-h)]$ *for its sum at any point in* $(-\pi, \pi)$ *for which this limit exists, provided that there is a positive* δ *such that*

$$\int_0^{\delta} \frac{\left|f(x+t)+f(x-t)-\lim_{h\to 0}[f(x+h)+f(x-h)]\right|}{t}\,dt$$

is a convergent integral.

A special case of Dini's criterion had been given in 1864 by Lipschitz.‡ This can be put in the form:

The Fourier's Series for $f(x)$ *converges at* x *to*

$$\lim_{h\to 0} \tfrac{1}{2}[f(x+h)+f(x-h)],$$

when this limit exists, if there is a positive δ *such that*

$$\left|f(x+t)+f(x-t)-\lim_{h\to 0}[f(x+h)+f(x-h)]\right| < Ct^k,$$

when $$0 < t \leqq \delta,$$

where C *and* k *are positive numbers.*

---

* *Journal für Math.*, **71** (1870), 353.

† Cf. Dini, "Serie di Fourier e altre rappresentazioni analitiche delle funzioni di una variabile reale" (Pisa, 1880), p. 102.

‡ Cf. *Journal für Math.*, **63** (1864), 296.

The treatment of Fourier's Series was simplified by Jordan* by the introduction of his *functions of bounded variation* and his criterion states that the *Fourier's Series for the integrable function $f(x)$ converges to*

$$\tfrac{1}{2}[f(x+0)+f(x-0)]$$

at every point in the neighbourhood of which $f(x)$ is of bounded variation.

During this period the properties of Fourier's Constants were also examined, and among the important results obtained, when the Riemann integral was still used, it is sufficient to cite that usually called Parseval's Theorem,† according to which, when $f(x)$ and $[f(x)]^2$ are integrable in $(-\pi, \pi)$,

$$\frac{1}{\pi}\int_{-\pi}^{\pi}[f(x)]^2dx = 2a_0^2 + \sum_1^\infty (a_n^2 + b_n^2).$$

Also, if $f(x)$ and $g(x)$, as well as their squares, are integrable,

$$\frac{1}{\pi}\int_{-\pi}^{\pi} f(x)g(x)\,dx = 2a_0\alpha_0 + \sum_1^\infty (a_n\alpha_n + b_n\beta_n),$$

where a_n, b_n, and α_n, β_n are the Fourier's Constants for $f(x)$ and $g(x)$ respectively.

If Fourier's Series for $f(x)$ is not convergent, it may converge when one or other of the methods of "summation" applied to divergent series is adopted. Fejér in 1904 discovered the remarkable theorem‡ that, when the series is summed by the method of arithmetical means, its sum is $\tfrac{1}{2}[f(x+0)+f(x-0)]$ at every point in $(-\pi, \pi)$ at which $f(x\pm 0)$ exist, the only condition attached to $f(x)$ being that, if bounded, it shall be integrable in $(-\pi, \pi)$, and, if unbounded, that $\int_{-\pi}^{\pi} f(x)\,dx$ shall be absolutely convergent.

THIRD PERIOD [1905–]

The theory of Fourier's Series, as built up by Dirichlet, Riemann, Cantor, Dini, Jordan and other mathematicians of the nineteenth century, with a fuller understanding of the limiting processes

*Cf. *Comptes Rendus*, **92** (1881), 228, and Jordan, *Cours d'Analyse*, **2** (1e éd., 1882), Ch. V.

†Cf. de la Vallée Poussin, *Ann. Soc. sc. Brux.*, **17B** (1893), 18, and Hurwitz, *Math. Annalen*, **57** (1903), 175.

‡Cf. *Math. Annalen*, **58** (1904), 51.

involved, placed in the hands of applied mathematicians a quite satisfactory instrument. But the properties of the series, which we owe to them, failed in many ways to give a theory with which the pure mathematician could be fully content. Unity, symmetry and completeness were still wanting. In this respect the last twenty-five years have seen a great improvement, due, chiefly, to the new definition of the definite integral put forward in 1902 by Lebesgue in his Paris thesis—*Intégrale, longueur, aire**—and further developed in his *Leçons sur l'intégration et la recherche des fonctions primitives* (1904).†

Lebesgue's integral is founded upon the subtle and rather difficult idea of the measure of a set of points. In the modern theory of functions of a real variable, Lebesgue's integral (or one of the others associated with it) is indispensable. But for practical purposes the Riemann integral will suffice. The progress which we now describe lies in the field of the specialist; and in no department of pure mathematics has greater activity been displayed in recent years than in the theory of trigonometrical series.‡ Most important contributions have been made by Lebesgue himself, Fejér, Hobson, Hardy and Littlewood, de la Vallée Poussin and W. H. Young.

The first point to notice is that, if $f(x)$ is bounded and integrable according to Riemann's definition, it is also integrable with Lebesgue's definition, and the integrals are equal. But a bounded function may be integrable with Lebesgue's definition, and fail to be integrable with Riemann's. It is convenient to say that a function is integrable (L), when it is integrable according to Lebesgue's definition, and that it is integrable (R), when it is integrable according to Riemann's definition. If $f(x)$ is integrable (L), but not bounded in the interval of integration, the Lebesgue integral converges absolutely. Unbounded functions may be integrable (L), but not integrable (R); and conversely.

The fundamental theorems of integration apply to both integrals, but one of the advantages for our present purpose of the Lebesgue integral is that a function integrable (L) need not be continuous

**Annali di Mat.* (3), 7 (1902), 231.

†A revised and enlarged second edition has appeared in 1928.

‡A full account of work in this field is to be found in Hobson's *Theory of Functions of a Real Variable*, 2 (2nd ed., 1926), Ch. VIII.

"almost everywhere"* in the interval of integration, as is the case with a function integrable (R).† Also if $f_n(x)$ is integrable (L), and $\lim_{n\to\infty} f_n(x)$ exists, finite or infinite, this limit is integrable (L). And more important still, with Lebesgue integrals, under much more general conditions,‡ we can make use of the relation:

$$\text{If}\quad f(x)=\lim_{n\to\infty} f_n(x),\quad \text{then}\quad \int_a^b f(x)\,dx=\lim_{n\to\infty}\int_a^b f_n(x)\,dx.$$

Returning to Fourier's Series, we remark first that Riemann's theorem, according to which the Fourier's Constants of a bounded and integrable function $f(x)$ tend to zero when n tends to infinity—or, more generally, that $\lim_{n\to\infty}\int_a^b f(x) \frac{\sin}{\cos} nx\,dx=0$—applies with the Lebesgue integrable to any function, bounded or not, integrable (L). This is now usually referred to as the Riemann-Lebesgue Theorem—or Fundamental Lemma—and may be stated as follows:

If $f(x)$ is integrable (L) in (a, b), then

$$\lim_{n\to\infty}\int_a^b f(x) \frac{\sin}{\cos} nx\,dx=0.$$

This was proved by Lebesgue in 1903.§

Now the sum $s_n(x)$ of the terms up to those in $\cos nx$ and $\sin nx$ of the Fourier's Series for $f(x)$ integrable (L) can be written

$$s_n(x)=\frac{1}{\pi}\int_0^{\frac{1}{2}\pi}[f(x+2\alpha)+f(x-2\alpha)]\frac{\sin(2n+1)\alpha}{\sin\alpha}\,d\alpha,$$

and we find from this that

$$\lim_{n\to\infty} s_n(x)-f(x)=\frac{1}{\pi}\lim_{n\to\infty}\int_0^{\frac{1}{2}\pi}\phi(\alpha)\frac{\sin(2n+1)\alpha}{\alpha}\,d\alpha,$$

where $\qquad \phi(\alpha)=f(x+2\alpha)+f(x-2\alpha)-2f(x).$

Hence, by the Riemann-Lebesgue Theorem,

$$\lim_{n\to\infty} s_n(x)=f(x),\ \text{if}\ \lim_{n\to\infty}\int_0^h \phi(\alpha)\frac{\sin(2n+1)\alpha}{\alpha}\,d\alpha$$

for some positive h.

---

*A property is said to hold *almost everywhere* in an interval, if it holds for all points except those forming a set of measure zero.

†Cf. Appendix II, § 10.

‡See Appendix II, §§ 15, 18.

§*Annales Sci. de l'École Normale* (3), **20** (1903), 453. Also see Lebesgue, *Leçons sur les séries trigonométriques* (Paris, 1906), 61.

Also the question of the convergence of the Fourier's Series for $f(x)$ at a point in the interval $(-\pi, \pi)$ depends only on the behaviour of $f(x)$ in the neighbourhood of that point.

In 1905 Lebesgue gave a new sufficient condition for the convergence of the Fourier's Series for $f(x)$, which included all the previously known conditions.*

Another point to notice is that the question of term by term integration of Fourier's Series does not depend, as used to be thought to be the case, on the uniform convergence of the series. Indeed, with the usual notation, we have†

$$\int_{-\pi}^{x} f(x)\,dx = a_0(x+\pi) + \sum_{1}^{\infty} \frac{1}{n}(a_n \sin nx + b_n(\cos n\pi - \cos nx)),$$

where x is any point in $(-\pi, \pi)$, for any function integrable (L), whether the Fourier's Series converges or not. And the new series converges uniformly to $\int_{-\pi}^{x} f(x)\,dx$ in the interval $(-\pi, \pi)$. Term by term integration can then be continued indefinitely.

This result can be used as a test in determining whether a trigonometrical series is a Fourier's Series. If, on integrating the series term by term, it fails to converge in the range $(-\pi, \pi)$, it cannot be a Fourier's Series. In this way it can be seen that $\sum_{n=2}^{\infty} \frac{\sin nx}{\log n}$ is not a Fourier's Series, as the integrated series diverges at $x=0$.‡

Again Parseval's Theorem, that

$$\frac{1}{\pi}\int_{-\pi}^{\pi} [f(x)]^2 dx = 2a_0^2 + \sum_{1}^{\infty} (a_n^2 + b_n^2),$$

holds for any function $f(x)$, whose square is integrable (L) in $(-\pi, \pi)$, and a similar remark applies to the relation

$$\frac{1}{\pi}\int_{-\pi}^{\pi} f(x)\,g(x)\,dx = 2a_0\alpha_0 + \sum_{1}^{\infty} (a_n\alpha_n + b_n\beta_n),$$

where a_n, b_n and α_n, β_n are the Fourier's Constants for the functions $f(x)$ and $g(x)$, whose squares are integrable (L) in $(-\pi, \pi)$.§

---

*Cf. *Math. Annalen*, **61** (1905), 82, and Lebesgue, *Leçons sur les séries trigonométriques*, p. 59.

†Cf. Lebesgue, *Leçons sur les séries trigonométriques*, p. 102.

‡This example is due to Fatou, *Comptes Rendus*, **142** (1906), p. 765. Other examples are given by Perron, *Math. Annalen*, **87** (1922), 84.

§Cf. Lebesgue, *Leçons sur les séries trigonométriques*, p. 100.
Fatou, *Acta Math.*, **30** (1906), 352.

As $g(x)$ can be put equal to zero in the partial intervals $(-\pi, \alpha)$ and (β, π), it follows that when $f(x)$ and $g(x)$ are functions whose squares are integrable (L) in $(-\pi, \pi)$ and (α, β) respectively, the integral $\int_\alpha^\beta f(x)g(x)\,dx$ may be obtained by substituting for $f(x)$ its Fourier's Series and applying term by term integration.

But one of the most remarkable results which follow from the use of the Lebesgue integral in the theory of Fourier's Series is the converse of Parseval's Theorem, known from its discoverers as the Riesz-Fischer Theorem:*

Any trigonometrical series for which $\sum_1^\infty (a_n^2 + b_n^2)$ converges is the Fourier's Series of a function whose square is integrable (L) in $(-\pi, \pi)$.

Reference has already been made to the application of summation by Fejér's arithmetical means to Fourier's Series. This method is a special case $(C, 1)$ of the general Cesàro sum, usually denoted by (C, r). A great deal of work has been done in the investigation of sufficient conditions that Fourier's Series be summable (C, r) at a point in $(-\pi, \pi)$. The results obtained by this method, when r is fractional, have thrown light on ordinary convergence and Cesàro summation, when r is integral.

Another field in which much progress has been made is the investigation of the behaviour and properties of Fourier's Constants when Lebesgue integrals are used. The Parseval and Riesz-Fischer Theorems belong to this class, and extensions of both have been made, when the condition that $f(x)$ and $g(x)$ shall be functions whose squares are integrable (L) is replaced by a more general condition.

The convergence problem for Fourier's Series is still unsolved. There is no property of the arbitrary function $f(x)$, integrable (L) in $(-\pi, \pi)$, which is known to be both necessary and sufficient for the convergence of Fourier's Series. There are simple sufficient conditions, which are known not to be necessary, and the necessary conditions obtained are known not to be sufficient; and the same remark applies to summation by an assigned Cesàro mean.

---

*Cf. F. Riesz, *Comptes Rendus*, **144** (1907), 615–619, 734–736.
Fischer, *Comptes Rendus*, **144** (1907), 1022.
Young, W. H. and Grace Chisholm, *Quarterly J. of Math.*, **44** (1912), 49.

REFERENCES.

BURKHARDT, "Entwicklungen nach oscillirenden Functionen," *Jahresber. d. Math. Ver.*, 10 (1901).

"Trigonometrische Reihen und Integrale bis etwa 1850," *Enc. d. math. Wiss.*, Bd. II, Tl. I, p. 819 *et seq.* (Leipzig, 1914).

GIBSON, "On the History of the Fourier Series," *Proc. Edinburgh Math. Soc.*, 11 (1893).

HILB U. M. RIESZ, "Neuere Untersuchungen über trigonometrische Reihen," *Enc. d. math. Wiss.*, Bd. II, Tl. III, p. 1189 *et seq.* (Leipzig, 1922).

PLANCHEREL, "Le développement de la théorie des séries trigonométriques dans le dernier quart de siècle," *L'ens. math.*, 24 (1924–5).

PLESSNER, "Trigonometrische Reihen" in Pascal's *Repertorium der höheren Mathematik* (2 Aufl. Leipzig, 1929), Bd. I, Tl. III, Kap. XXV.

RIEMANN, *loc. cit.*

SACHSE, "Versuch einer Geschichte der Darstellung willkürlicher Functionen einer Variabeln durch trigonometrische Reihen," *Schlömilch's Zeitschrift*, 25 (1880), and *Bull. des sciences math.* (2), 4 (1880), 43.

TONELLI, *Serie trigonometriche* (1928, Bologna).

CHAPTER I

RATIONAL AND IRRATIONAL NUMBERS THE SYSTEM OF REAL NUMBERS

1. Rational Numbers. The question of the convergence of Infinite Series is only capable of satisfactory treatment when the difficulties underlying the conception of irrational numbers have been overcome. For this reason we shall first of all give a short discussion of that subject.

The idea of number is formed by a series of generalisations. We begin with the positive integers. The operations of addition and multiplication upon these numbers are always possible; but if a and b are two positive integers, we cannot determine positive integers x and y, so that the equations $a=b+x$ and $a=by$ are satisfied, unless, in the first case, a is greater than b, and, in the second case, a is a multiple of b. To overcome this difficulty fractional and negative numbers are introduced, and the system of *rational numbers* placed at our disposal.*

The system of rational numbers is *ordered*, *i.e.* if we have two different numbers a and b of this system, one of them is greater than the other. Also, if $a>b$ and $b>c$, then $a>c$, when a, b and c are numbers of the system.

Further, if two different rational numbers a and b are given, we can always find another rational number greater than the

*The reader who wishes an extended treatment of the system of rational numbers is referred to Stolz und Gmeiner, *Theoretische Arithmetik* (Leipzig, 1900–1902) and Pringsheim, *Vorlesungen über Zahlen- und Funktionenlehre* (Leipzig, 1916).

one and less than the other. It follows from this that between any two different rational numbers there are an infinite number of rational numbers.*

2. The introduction of fractional and negative rational numbers may be justified from two points of view. The fractional numbers are necessary for the representation of the subdivision of a unit magnitude into several equal parts, and the negative numbers form a valuable instrument for the measurement of magnitudes which may be counted in opposite directions. This may be taken as the argument of the applied mathematician. On the other hand there is the argument of the pure mathematician, with whom the notion of number, positive and negative, integral and fractional, rests upon a foundation independent of measurable magnitude, and in whose eyes analysis is a scheme which deals with numbers only, and has no concern *per se* with measurable quantity. It is possible to found mathematical analysis upon the notion of positive integral number. Thereafter the successive definitions of the different kinds of number, of equality and inequality among these numbers, and of the four fundamental operations, may be presented abstractly.†

3. Irrational Numbers. The extension of the idea of number from the rational to the irrational is as natural, if not as easy, as is that from the positive integers to the fractional and negative rational numbers.

Let a and b be any two positive integers. The equation $x^b=a$ cannot be solved in terms of positive integers unless a is a perfect b^{th} power. To make the solution possible in general the irrational numbers are introduced. But it will be seen below that the

---

*When we say that a set of things has a finite number of members, we mean that there is a positive integer n, such that the total number of members of the set is less than n.

When we say that it has an infinite number of members, we mean that it has not a finite number. In other words, however large n may be, there are more members of the set than n.

A set is said to be countably infinite (or enumerable) when its members can be represented by a sequence $u_1, u_2, u_3, \ldots$.

In this case there is a one-one correspondence between the members of the set and the set of positive integers 1, 2, 3, ...

†Cf. Hobson, *Proc. London Math. Soc.* (1), 35 (1913), 126; also the same author's *Theory of Functions of a Real Variable*, 1 (3rd. ed., 1927), 11.

system of irrational numbers is not confined to numbers which arise as the roots of algebraical equations whose coefficients are integers.

So much for the desirability of the extension from the abstract side. From the concrete the need for the extension is also evident. We have only to consider the measurement of any quantity to which the property of unlimited divisibility is assigned, *e.g.* a straight line L produced indefinitely. Take any segment of this line as unit of length, a definite point of the line as origin or zero point, and the directions of right and left for the positive and negative senses. To every rational number corresponds a

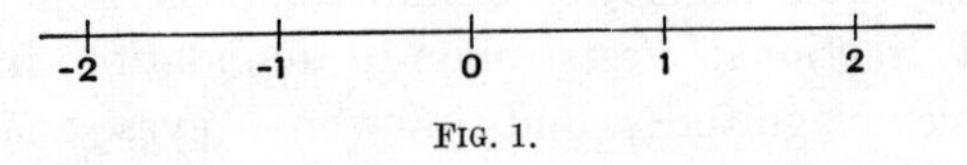

FIG. 1.

definite point on the line. If the number is an integer, the point is obtained by taking the required number of unit segments one after the other in the proper direction. If it is a fraction $\pm p/q$, it is obtained by dividing the unit of length into q equal parts and taking p of these to the right or left according as the sign is positive or negative. These numbers are called the *measures* of the corresponding segments, and the segments are said to be *commensurable* with the unit of length. The points corresponding to rational numbers may be called *rational points.*

There are, however, an infinite number of points on the line L which are not rational points. Although we may approach them as nearly as we please by choosing more and more rational points on the line, we can never quite reach them in this way. The simplest example is the case of the points coinciding with one end of the diagonal of a square, the sides of which are the unit of length, when the diagonal lies along the line L and its other end coincides with any rational point.

Thus, without considering any other case of incommensurability, we see that *the line L is infinitely richer in points than the system of rational numbers in numbers.*

Hence it is clear that if we desire to follow arithmetically all the properties of the straight line, the rational numbers are insufficient, and it will be necessary to extend this system by the creation of other numbers.

4. Returning to the point of view of the pure mathematician, we shall now describe Dedekind's method of introducing the irrational number, in its most general form, into analysis.*

Let us suppose that by some method or other we have divided all the rational numbers into two classes, a lower class A and an upper class B, such that every number α of the lower class is less than every number β of the upper class.

When this division has been made, if a number α belongs to the class A, every number less than α does so also; and if a number β belongs to the class B, every number greater than β does so also.

Three different cases can arise :

(I) *The lower class can have a greatest number and the upper class no smallest number.*

This would occur, if, for example, we put the number 5 and every number less than 5 in the lower class, and if we put in the upper class all the numbers greater than 5.

(II) *The upper class can have a smallest number and the lower class no greatest number.*

This would occur if, for example, we put the number 5 and all the numbers greater than 5 in the upper class, while in the lower class we put all the numbers less than 5.

It is impossible that the lower class can have a greatest number m, and the upper class a smallest number n, in the same division of the rational numbers; for between the rational numbers m and n there are rational numbers, so that our hypothesis that the two classes contain all the rational numbers is contradicted.

But a third case can arise :

(III) *The lower class can have no greatest number and the upper class no smallest number.*

For example, let us arrange the positive integers and their squares in two rows, so that the squares are underneath the numbers to which they correspond. Since the square of a fraction in its lowest terms is a fraction whose numerator and

*Dedekind (1831–1916) published his theory in *Stetigkeit und irrationale Zahlen* (Braunschweig, 1872); English translation in Dedekind's *Essays on Number* (Chicago, 1901).

denominator are perfect squares,* we see that there are not rational numbers whose squares are 2, 3, 5, 6, 7, 8, 10,11, ... ,

1 2 3 4 ...
1 2 3 **4** 5 6 7 8 **9** 10 11 12 13 14 15 **16** .... .

However there are rational numbers whose squares are as near these numbers as we please. For instance, the numbers

$$2,\ 1{\cdot}5,\ 1{\cdot}42,\ 1{\cdot}415,\ 1{\cdot}4143,\ \ldots,$$
$$1,\ 1{\cdot}4,\ 1{\cdot}41,\ 1{\cdot}414,\ 1{\cdot}4142,\ \ldots,$$

form an upper and a lower set in which the squares of the terms in the lower are less than 2, and the squares of the terms in the upper are greater than 2. We can find a number in the upper set and a number in the lower set such that their squares differ from 2 by as little as we please.†

Now form a lower class, as described above, containing all negative rational numbers, zero and all the positive rational numbers whose squares are less than 2; and an upper class containing all the positive rational numbers whose squares are greater than 2. Then every rational number belongs to one class or the other. Also every number in the lower class is less than every number in the upper. The lower class has no greatest number and the upper class has no smallest number.

5. When by any means we have obtained a division of all the rational numbers into two classes of this kind, the lower class having no greatest number and the upper class no smallest number, we create a new number defined by this division. We call it an *irrational number*, and we say that it is greater than all the rational numbers of its lower class, and less than all the rational numbers of its upper class.

Such divisions are usually called *sections*.‡ The irrational number $\sqrt{2}$ is defined by the section of the rational numbers described above. Similar sections would define the irrational numbers $\sqrt[3]{3}$, $\sqrt[4]{5}$, etc. The system of irrational numbers is given by all the possible divisions of the rational numbers into a lower class A and an upper class B, such that every rational

---

*If a formal proof of this statement is needed, see Dedekind, *loc. cit.*, English translation, p. 14, or Hardy, *Course of Pure Mathematics* (5th ed., 1928), 6.

†Cf. Hardy, *loc. cit.*, p. 8.

‡French, *coupure*; German, *Schnitt*.

number is in one class or the other, the numbers of the lower class being less than the numbers of the upper class, while the lower class has no greatest number, and the upper class no smallest number.

In other words, every irrational number is defined by its section (A, B). It may be said to "correspond" to this section.

The system of *rational* numbers and *irrational* numbers together make up the system of *real* numbers.

The rational numbers themselves "correspond" to divisions of rational numbers.

For instance, take the rational number m. In the lower class A put all the rational numbers less than m, and m itself. In the upper class B put all the rational numbers greater than m. Then m corresponds to this division of the rational numbers.

Extending the meaning of the term *section*, as used above in the definition of the irrational number, to divisions in which the lower and upper classes have greatest or smallest numbers, we may say that the rational number m corresponds to a *rational section* (A, B),* and that the irrational numbers correspond to *irrational sections*. When the rational and irrational numbers are defined in this way, and together form the system of real numbers, the real number which corresponds to the rational number m (to save confusion it is sometimes called the *rational-real* number) is conceptually distinct from m. However, the relations of magnitude, and the fundamental operations for the real numbers, are defined in such a way that this rational-real number has no properties distinct from those of m, and it is usually denoted by the same symbol.

6. Relations of Magnitude for Real Numbers. We have extended our conception of number. We must now arrange the system of real numbers in order; *i.e.* we must say when two numbers are equal or unequal to, greater or less than, each other.

In this place we need only deal with cases where at least one of the numbers is irrational.

An irrational number is never *equal* to a rational number. They are always *different or unequal.*

Next, in § 5, we have seen that the *irrational* number given by the section (A, B) is said to be *greater* than the *rational* number m, when m is a member of the lower class A, and that the *rational* number m is said to be *greater* than the *irrational* number given by the section (A, B), when m is a member of the upper class B.

---

*The rational number m could correspond to two sections: the one named in the text, and that in which the lower class A contains all the rational numbers less than m, and the upper class B, m and all the rational numbers greater than m. To save ambiguity, one of these sections only must be chosen.

Two irrational numbers are *equal*, when they are both given by the same section. They are *different* or *unequal*, when they are given by different sections.

The *irrational* number a given by the section (A, B) is *greater* than the *irrational* number a' given by the section (A′, B′), when the class A contains numbers of the class B′. Now the class A has no greatest number. But if a certain number of the class A belongs to the class B′, all the numbers of A greater than this number also belong to B′. The class A thus contains an infinite number of members of the class B′, when $a > a'$.

If a real number a is *greater* than another real number a', then a' is *less* than a.

It will be observed that the notation $>$, $=$, $<$ is used in dealing with real numbers as in dealing with rational numbers.

The real number β is said to *lie between* the real numbers a and γ, when one of them is greater than β and the other less.

With these definitions the system of real numbers is *ordered*. If we have two different real numbers, one of them is greater than the other; and if we have three real numbers such that $a > \beta$ and $\beta > \gamma$, then $a > \gamma$.

These definitions can be simplified when the rational numbers themselves are given by sections, as explained at the end of § 5.

7. Between any two different rational numbers there is an infinite number of rational numbers. A similar property holds for the system of real numbers, as will now be shown:

(I) *Between any two different real numbers a, a' there are an infinite number of rational numbers.*

If a and a' are rational, the property is known.

If a is rational and a' irrational, let us assume $a > a'$. Let a' be given by the section (A′, B′). Then the rational number a is a member of the upper class B′, and B′ has no least number. Therefore an infinite number of members of the class B′ are less than a. It follows from the definitions of § 5 that there are an infinite number of rational numbers greater than a' and less than a.

A similar proof applies to the case when the irrational number a' is greater than the rational number a.

There remains the case when a and a' are both irrational. Let a be given by the section (A, B) and a' by the section (A′, B′). Also let $a > a'$.

Then the class A of a contains an infinite number of members of the class B′ of a'; and these numbers are less than a and greater than a'.

A similar proof applies to the case when $a < a'$.

The result which has just been proved can be made more general:

(II) *Between any two different real numbers there are an infinite number of irrational numbers.*

Let a, a' be the two given numbers, and suppose $a < a'$.

Take any two rational numbers β and β', such that $a < \beta < \beta' < a'$. If we can show that between β and β' there must be an irrational number, the theorem is established.

Let i be an irrational number. If this does not lie between β and β', by adding to it a suitable rational number we can make it do so. For we can find two rational numbers m, n, such that $m < i < n$ and $(n - m)$ is less than $(\beta' - \beta)$. The number $\beta - m + i$ is irrational, and lies between β and β'.

8. Dedekind's Theorem. We shall now prove a very important property of the system of *real** numbers, which will be used frequently in the pages which follow.

If the system of real numbers is divided into two classes A and B, in such a way that

(i) *each class contains at least one number,*

(ii) *every number belongs to one class or the other,*

(iii) *every number in the lower class A is less than every number in the upper class B;*

then there is a number a such that

every number less than a belongs to the lower class A, and

every number greater than a belongs to the upper class B.

The separating number a itself may belong to either class.

Consider the rational numbers in A and B.

These form two classes—*e.g.* A′ and B′—such that every rational number is in one class or the other, and the numbers in the lower class A′ are all less than the numbers in the upper class B′.

As we have seen in § 4, three cases, and only three, can arise.

(i) *The lower class A′ can have a greatest number m and the upper class B′ no smallest number.*

The rational number m is the number a of the theorem. For it is clear that every real number a less than m belongs to the class A, since m is a member of this class. Also every real number b, greater than m, belongs to the class B. This is evident if b is rational, since b then belongs to the class B′, and B′ is part of B. If b is irrational, we can take a rational number n between m and b. Then n belongs to B, and therefore b does so also.

(ii) *The upper class B′ can have a smallest number m and the lower class A′ no greatest number.*

It follows, as above, that the rational number m is the number a of our theorem.

(iii) *The lower class A′ can have no greatest number and the upper class B′ no smallest number.*

---

*It will be observed that the system of *rational* numbers does not possess this property.

Let m be the irrational number defined by this section (A′, B′). Every rational number less than m belongs to the class A, and every rational number greater than m belongs to the class B.

We have yet to show that every irrational number less than m belongs to the class A, and every irrational number greater than m to the class B.

But this follows at once from § 6. For if m' is an irrational number less than m, we know that there are rational numbers between m and m'. These belong to the class A, and therefore m' does so also.

A similar argument applies to the case when $m' > m$.

In the above discussion the *separating* number a belongs to the lower class, and is rational, in case (i); it belongs to the upper class, and is again rational, in case (ii); it is irrational, and may belong to either class, in case (iii).

9. The Linear Continuum. Dedekind's Axiom. We return now to the straight line L of § 3, in which a definite point O has been taken as origin and a definite segment as the unit of length.

We have seen how to effect a correspondence between the rational numbers and the "rational points" of this line. The "rational points" are the ends of segments obtained by marking

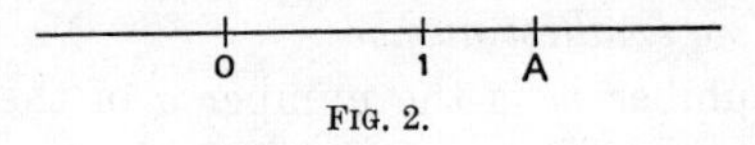

FIG. 2.

off from O on the line lengths equal to multiples or sub-multiples of the unit segment, and the numbers are the measures of the corresponding segments.

Let OA be a segment incommensurable with the unit segment. The point A divides the rational points of the line into two classes, such that all the points of the lower class are to the left of all the points of the upper class. The lower class has no last point, and the upper class no first point.

We then say that A is an *irrational point* of the line, and that the measure of the segment OA is the irrational number defined by this section of the rational numbers.

Thus to any point of the line L corresponds a real number, and to different points of the line correspond different real numbers.

There remains the question—*To every real number does there correspond a point of the line* ?

For all rational numbers we can answer the question in the affirmative. When we turn to the irrational numbers, the question amounts to this: *If all the rational points of the line are divided into two classes, a lower and an upper, so that the lower class has no last point and the upper class no first point, is there one, and only one, point on the line which brings about this separation* ?

The existence of such a point on the line cannot be proved. The assumption that there is one, and only one, for every section of the rational points is nothing less than an axiom by means of which we assign its continuity to the line.

This assumption is Dedekind's Axiom of Continuity for the line. In adopting it we may now say that *to every point P of the line corresponds a number, rational or irrational, the measure of the segment OP, and that to every real number corresponds a point P of the line, such that the measure of OP is that number.*

The correspondence between the points of the line L (*the linear continuum*) and the system of real numbers (*the arithmetical continuum*) is now perfect. The points can be taken as the images of the numbers, and the numbers as the signs of the points. In consequence of this perfect correspondence, we may, in future, use the terms number and point in this connection as identical.

10. The Development of the System of Real Numbers. It is instructive to see how the idea of the system of real numbers, as we have described it, has grown.* The irrational numbers, belonging as they do in modern arithmetical theory to the realm of arithmetic, arose from the geometrical problems which required their aid. They appeared first as an expression for the ratios of incommensurable pairs of lines. In this sense the Fifth Book of Euclid, in which the general theory of Ratio is developed, and the Tenth Book, which deals with Incommensurable Magnitudes, may be taken as the starting point of the theory. But the irrationalities which Euclid examines are only definite cases of the ratios of incommensurable lines, such as may be obtained with the aid of ruler and compass; that is to say, they depend on square roots

*Cf. Pringsheim, "Irrationzahlen u. Konvergenz unendlicher Prozesse," *Enc. d. math. Wiss.*, Bd. I, Tl. I, p. 49 *et seq.* (Leipzig, 1898).

alone. The idea that the ratio of any two such incommensurable lines determined a definite (irrational) number did not occur to him, nor to any of the mathematicians of that age.

Although there are traces in the writings of at least one of the mathematicians of the sixteenth century of the idea that every irrational number, just as much as every rational number, possesses a determinate and unique place in the ordered sequence of numbers, these irrational numbers were still considered to arise only from certain cases of evolution, a limitation which is partly due to the commanding position of Euclid's methods in Geometry, and partly to the belief that the problem of finding the nth root of an integer, which lies between the nth powers of two consecutive integers, was the only problem whose solution could not be obtained in terms of rational numbers.

The introduction of the methods of Coordinate Geometry by Descartes in 1637, and the discovery of the Infinitesimal Calculus by Leibnitz and Newton in 1684-7, made mathematicians regard this question in another light, since the applicability of number to spatial magnitude is a fundamental postulate of Coordinate Geometry. "The view now prevailed that number and quantity were *the* objects of mathematical investigation, and that the two were so similar as not to require careful separation. Thus number was applied to quantity without any hesitation, and, conversely, where existing numbers were found inadequate to measurement, new ones were created on the sole ground that every quantity must have a numerical measure."*

It was reserved for the mathematicians of the nineteenth century—notably Weierstrass, Cantor, Dedekind and Heine—to establish the theory on a proper basis. Until their writings appeared, a number was looked upon as an expression for the result of the measurement of a line by another which was regarded as the unit of length. To every segment, or, with the natural modification, to every point, of a line corresponded a definite number, which was either rational or irrational; and by the term irrational number was meant a number defined by an infinite set of arithmetical operations (*e.g.* infinite decimals or continued fractions). The justification for regarding such an

*Cf. Russell, *Principles of Mathematics* (1903), Ch. XIX, 417.

unending sequence of rational numbers as a definite number was considered to be the fact that this system was obtained as the equivalent of a given segment by the aid of the same methods of measurement as those which gave a definite rational number for other segments. However it does not in any way follow from this that, conversely, any arbitrarily given arithmetical representation of this kind can be regarded in the above sense as an irrational number; that is to say, that we can consider as evident the existence of a segment which would produce by suitable measurement the given arithmetical representation. Cantor* has the credit of first pointing out that the assumption that a definite segment must correspond to every such sequence is neither self-evident nor does it admit of proof, but involves an actual axiom of Geometry. Almost at the same time Dedekind showed that the axiom in question (or more exactly one which is equivalent to it) gave a meaning, which we can comprehend, to that property which, so far without any sufficient definition, had been spoken of as the continuity of the line.

To make the theory of number independent of any geometrical axiom and to place it upon a basis entirely independent of measurable magnitude was the object of the arithmetical theories associated with the names of Weierstrass, Dedekind and Cantor. The theory of Dedekind has been followed in the previous pages. Those of Weierstrass and Cantor, which regard irrational numbers as the limits of convergent sequences, may be deduced from that of Dedekind. In all these theories irrational numbers appear as new numbers, to each of which a definite place in the domain of rational numbers is assigned, and with which we can operate according to definite rules. The ordinary operations of arithmetic for these numbers are defined in such a way as to be in agreement with the ordinary operations upon the rational numbers. They can be used for the representation of definite quantities, and to them can be ascribed definite quantities, according to the axiom of continuity to which we have already referred.

---

* *Math. Annalen*, 5 (1872), 127.

REFERENCES.

Bromwich, *Theory of Infinite Series* (2nd. ed., 1926, App. I).

Dedekind, *Stetigkeit und irrationale Zahlen* (Braunschweig, 1872); English translation in Dedekind's *Essays on Number* (Chicago, 1901).

De la Vallée Poussin, *Cours d'Analyse*, 1 (5e éd., Paris, 1923), Introduction, § 1.

Dini, *Fondamenti per la Teorica delle Funzioni di Variabili Reali* (Pisa, 1878), §§ 1-9.

Goursat, *Cours d'Analyse*, 1 (4e éd., Paris 1923), Ch. I.

Hobson, *Theory of Functions of a Real Variable*, 1 (3rd ed., 1927), Ch. I.

Knopp, *Theorie und Anwendung der unendlichen Reihen* (2 Aufl., Berlin, 1924; English translation, 1928), Kap. I-II.

Pringsheim, *Vorlesungen über Zahlen- und Funktionenlehre*, 1 (Leipzig, 1916), Absch. I, Kap. I-III.

Russell, *Principles of Mathematics* (1903), Ch. XXXIV.

Stolz u. Gmeiner, *Theoretische Arithmetik* (Leipzig, 1902), Abth. II, Absch. VII.

Tannery, *Introduction à la Théorie des Fonctions*, 1 (2e éd., Paris, 1914), Ch. I.

And

Pringsheim, "Irrationalzahlen u. Konvergenz unendlicher Prozesse," *Enc. d. math. Wiss.*, Bd. I, Tl. I (Leipzig, 1898).

CHAPTER II

INFINITE SEQUENCES AND SERIES

11. Infinite Aggregates. We are accustomed to speak of the positive integral numbers, the prime numbers, the integers which are perfect squares, etc. These are all examples of infinite sets of numbers or sets which have more than a finite number of terms. In mathematical language they are termed aggregates, and the theory of such infinite aggregates forms an important branch of modern pure mathematics.*

The terms of an aggregate are all different. Their number may be finite or infinite. In the latter case the aggregates are usually called infinite aggregates, but sometimes we shall refer to them simply as aggregates. After the discussion in the previous chapter, there will be no confusion if we speak of an aggregate of points on a line instead of an aggregate of numbers. The two notions are identical. We associate with each number the point of which it is the abscissa. It may happen that, however far we go along the line, there are points of the aggregate further on. In this case we say that it extends to infinity. An aggregate is said to be *bounded on the right*, or *bounded above*, when there is

*Cantor may be taken as the founder of this theory, which the Germans call *Mengenlehre.* In a series of papers published from 1870 onward he showed its importance in the Theory of Functions of a Real Variable, and especially in the rigorous discussion of the conditions for the development of an arbitrary function in trigonometric series.

Reference may be made to the standard treatise on the subject by W. H. and Grace Chisholm Young, *Theory of Sets of Points* (1906), and to the earlier chapters of Hobson's *Theory of Functions of a Real Variable*, Vol. I, already cited.

The most recent book on the subject, from the advanced point of view, is *Mengenlehre*, by Hausdorff (2 Aufl., Berlin, 1927).

no point of it to the right of some fixed point. It is said to be *bounded on the left*, or *bounded below*, when there is no point of it to the left of some fixed point. The aggregate of rational numbers greater than zero is bounded on the left. The aggregate of rational numbers less than zero is bounded on the right. The aggregate of real positive numbers less than unity is bounded above and below; in such a case we simply say that it is *bounded.* The aggregate of integral numbers is unbounded.

12. The Upper and Lower Bounds of an Aggregate. *When an aggregate* (E)* *is bounded on the right, there is a number* M *which possesses the following properties:*

no number of (E) *is greater than* M*;*
however small the positive number ϵ *may be, there is a number of* (E) *greater than* $M-\epsilon$.

We can arrange all the real numbers in two classes, A and B, relative to the aggregate. A number x will be put in the class A if one or more numbers of (E) are greater than x. It will be put in the class B if no number of (E) is greater than x. Since the aggregate is bounded on the right, there are members of both classes, and any number of the class A is smaller than any number of the class B.

By Dedekind's Theorem (§ 8) there is a number M separating the two classes, such that every number less than M belongs to the class A, and every number greater than M to the class B. We shall now show that this is the number M of our theorem.

In the first place, there is no number of (E) greater than M. For suppose there is such a number $M+h$ $(h>0)$. Then the number $M+\frac{1}{2}h$, which is also greater than M, would belong to the class A, and M would not separate the two classes A and B.

In the second place, whatever the positive number ϵ may be, the number $M-\epsilon$ belongs to the class A. It follows from the way in which the class A is defined that there is at least one number of (E) greater than $M-\epsilon$.

This number M is called the *upper bound* of the aggregate (E). It may belong to the aggregate. This occurs when the aggregate

---

*This notation is convenient, the letter E being the first letter of the French term *ensemble.*

contains a finite number of terms. But when the aggregate contains an infinite number of terms, the upper bound need not belong to it. For example, consider the rational numbers whose squares are not greater than 2. This aggregate is bounded on the right, its upper bound being the irrational number $\sqrt{2}$, which does not belong to the aggregate. On the other hand, the aggregate of real numbers whose squares are not greater than 2 is also bounded on the right, and has the same upper bound. But $\sqrt{2}$ belongs to this aggregate.

If the upper bound M of the aggregate (E) does not belong to it, there must be an infinite number of terms of the aggregate between M and $M-\epsilon$, however small the positive number ϵ may be. If there were only a finite number of such terms, there would be no term of (E) between the greatest of them and M, which is contrary to our hypothesis.

It can be shown in the same way that *when an aggregate (E) is bounded on the left, there is a number m possessing the following properties:*

no number of (E) is smaller than m;

however small the positive number ϵ may be, there is a number of (E) less than $m+\epsilon$.

The number m defined in this way is called the *lower bound* of the aggregate (E). As above, it may, or may not, belong to the aggregate when it has an infinite number of terms. But when the aggregate has only a finite number of terms it must belong to it.

13. Limiting Points of an Aggregate. Consider the aggregate

$$1, \frac{1}{2}, \frac{1}{3}, \ldots \frac{1}{n}, \ldots .$$

There are an infinite number of points of this aggregate in any interval, however small, extending from the origin to the right. Such a point, round which an infinite number of points of an aggregate cluster, is called a *limiting point** of the aggregate. More definitely, *a will be a limiting point of the aggregate (E) if, however small the positive number ϵ may be, there is in (E) a point other than a whose distance from a is less than ϵ.* If there be one

*French, *point limite*; German, *Häufungspunkt*.

such point within the interval $(a-\epsilon, a+\epsilon)$, there will be an infinite number, since, if there were only n of them, and a_n were the nearest to a, there would not be in (E) a point other than a whose distance from a was less than $|a-a_n|$*. In that case a would not be a limiting point, contrary to our hypothesis.

An aggregate may have more than one limiting point. The rational numbers between zero and unity form an aggregate with an infinite number of limiting points, since every point of the segment (0, 1) is a limiting point. It will be noticed that some of the limiting points of this aggregate belong to it, and some, namely the irrational points of the segment and its end-points, do not.

In the example at the beginning of this section,

$$1, \frac{1}{2}, \frac{1}{3}, \ldots \frac{1}{n}, \ldots,$$

the lower bound, zero, is a limiting point, and does not belong to the aggregate. The upper bound, unity, belongs to the aggregate, and is not a limiting point.

The set of real numbers from 0 to 1, inclusive, is an aggregate which is identical with its limiting points.

14. Weierstrass's Theorem. *An infinite aggregate, bounded above and below, has at least one limiting point.*

Let the infinite aggregate (E) be bounded, and have M and m for its upper and lower bounds.

We can arrange all the real numbers in two classes relative to the aggregate (E). A number x will be said to belong to the class A when an infinite number of terms of (E) are greater than x. It will be said to belong to the class (B) in the contrary case.

Since m belongs to the class A and M to the class B, there are numbers of both classes. Also any number in the class A is less than any number in the class B.

By Dedekind's Theorem, there is a number μ separating the two classes. However small the positive number ϵ may be, $\mu-\epsilon$ belongs to the class A, and $\mu+\epsilon$ to the class B. Thus the interval contains an infinite number of terms of the aggregate.

Hence μ is a limiting point.

---

*It is usual to denote the difference between two real numbers a and b, taken positive, by $|a-b|$, and to call it the *absolute value* or *modulus* of $(a-b)$. With this notation $|x+y| \leqq |x|+|y|$, $|xy|=|x||y|$.

As will be seen from the example of § 13, the bounds M and m may be limiting points.

An infinite aggregate, when unbounded, need not have a limiting point; *e.g.* the set of integers, positive or negative. But if the aggregate has an infinite number of points in an interval of finite length, then it must have at least one limiting point.

15. Convergent Sequences. We speak of an infinite sequence of numbers

$$u_1, \quad u_2, \quad u_3, \ldots u_n, \ldots$$

when some law is given according to which the general term u_n may be written down.

The sequence

$$u_1, \quad u_2, \quad u_3, \ldots$$

is said to be convergent and to have the limit A, when, by indefinitely increasing n, the difference between A and u_n becomes, and thereafter remains, as small as we please.

This property is so fundamental that it is well to put it more precisely, as follows: *The sequence is said to be convergent and to have the limit A, when, any positive number ϵ having been chosen, as small as we please, there is a positive integer ν such that*

$$|A - u_n| < \epsilon, \textit{ provided that } n \geqq \nu.$$

For example, the sequence

$$1, \quad \frac{1}{2}, \quad \frac{1}{3}, \ldots \frac{1}{n}, \ldots$$

has the limit zero, since $1/n$ is less than ϵ for all values of n greater than $1/\epsilon$.

The notation that is employed in this connection is

$$\lim_{n \to \infty} u_n = A,$$

and we say that *as n tends to infinity, u_n has the limit A.**

The letter ϵ is usually employed to denote an arbitrarily small positive number, as in the above definition of convergence to a limit as n tends to infinity. Strictly speaking, the words *as small as we please* are unnecessary in the definition, but they are inserted as making clearer the property that is being defined.

We shall very frequently have to employ the form of words which occurs in this definition, or words analogous to them, and

*The phrase " u_n tends to the limit A as n tends to infinity" is also used.

the beginner is advised to make himself familiar with them by formally testing whether the following sequences are convergent or not :

(a) $1, \frac{1}{2}, \frac{1}{2^2}, \ldots$. (c) $1, 1+\frac{1}{2}, 1+\frac{1}{2}+\frac{1}{2^2}, \ldots$.

(b) $1, -\frac{1}{2}, \frac{1}{3}, \ldots$. (d) $1, -1, 1, -1, \ldots$.

A sequence cannot converge to two distinct limits A and B. If this were possible, let $\epsilon < \frac{1}{2}|A-B|$. Then there are only a finite number of terms of the sequence outside the interval $(A-\epsilon, A+\epsilon)$, since the sequence converges to the value A. This contradicts the statement that the sequence has also the limit B, for we would only have a finite number of terms in the interval of the same length with B as centre.

The application of the test of convergency contained in the definition involves the knowledge of the limit A. Thus it will frequently be impossible to use it. The required criterion for the convergence of a sequence, when we are not simply asked to test whether a given number is or is not the limit, is contained in the fundamental *general principle of convergence*:—*

A necessary and sufficient condition for the existence of a limit to the sequence

$$u_1, \quad u_2, \quad u_3, \ldots$$

is that a positive integer ν exists such that $|u_{n+p}-u_n|$ becomes as small as we please when $n \geqq \nu$, for every positive integer p.

More exactly:

A necessary and sufficient condition for the existence of a limit to the sequence

$$u_1, \quad u_2, \quad u_3, \ldots$$

is that, if any positive number ϵ has been chosen, as small as we please, there shall be a positive integer ν such that

$$|u_{n+p}-u_n| < \epsilon, \text{ when } n \geqq \nu, \text{ for every positive integer } p.$$

We shall first of all show that the condition is *necessary*; *i.e.* if

*This is one of the most important theorems of analysis. In the words of Pringsheim, "Dieser Satz, mit seiner Übertragung auf *beliebige* (z.B. stetige) Zahlenmengen—von *du Bois-Reymond* als das '*allgemeine Convergenzprinzip*' bezeichnet (Allg. Funct.-Theorie, pp., 6, 260)—ist der eigentliche *Fundamentalsatz der gesamten Analysis* und sollte mit genügender Betonung seines fundamentalen Characters an der Spitze jedes rationellen Lehrbuches der Analysis stehen," *loc. cit., Enc. d. math. Wiss.* p. 66.

the sequence converges, this condition is satisfied; secondly that, if this condition is satisfied, the sequence converges; in other words, the condition is *sufficient*.

(i) *The condition is necessary.*

Let the sequence converge to the limit A.

Having chosen the arbitrary positive number ϵ, then take $\frac{1}{2}\epsilon$.

We know that there is a positive integer ν such that

$$|A-u_n|<\tfrac{1}{2}\epsilon, \text{ when } n \geqq \nu.$$

But $$(u_{n+p}-u_n)=(u_{n+p}-A)+(A-u_n).$$

Therefore $$\begin{aligned}|u_{n+p}-u_n| &\leqq |u_{n+p}-A|+|A-u_n| \\ &< \tfrac{1}{2}\epsilon + \tfrac{1}{2}\epsilon, \\ &\qquad \text{if } n \geqq \nu, \text{ for every positive integer } p, \\ &< \epsilon.\end{aligned}$$

(ii) *The condition is sufficient.*

We must examine two cases; first, when the sequence contains an infinite number of terms equal to one another; second, when it does not.

(*a*) Let there be an infinite number of terms equal to A.

Then, if

$$|u_{n+p}-u_n|<\epsilon, \text{ when } n \geqq \nu, \text{ and } p \text{ is any positive integer},$$

we may take $u_{n+p}=A$ for some value of p, and we have

$$|A-u_n|<\epsilon, \text{ when } n \geqq \nu.$$

Therefore the sequence converges, and has A for its limit.

(*b*) Let there be only a finite number of terms equal to one another.

Having chosen the arbitrary positive number ϵ, then take $\frac{1}{2}\epsilon$.

We know that there is a positive integer N such that

$$|u_{n+p}-u_n|<\tfrac{1}{2}\epsilon, \text{ when } n \geqq N, \text{ for every positive integer } p.$$

It follows that we have

$$|u_n-u_N|<\tfrac{1}{2}\epsilon, \text{ when } n \geqq N.$$

Therefore all the terms of the sequence

$$u_{N+1}, \quad u_{N+2}, \quad u_{N+3}, \ldots$$

lie within the interval whose end-points are $u_N-\frac{1}{2}\epsilon$ and $u_N+\frac{1}{2}\epsilon$.

There must be an infinite number of distinct terms in this sequence. Otherwise we would have an infinite number of terms equal to one another.

Consider the infinite aggregate (E) formed by the distinct terms in

$$u_1, \quad u_2, \quad u_3, \ldots .$$

This aggregate is bounded and must have at least one limiting point A within, or at an end of, the above interval. (Cf. § 14.)

There cannot be another limiting point A', for if there were we could choose ϵ equal to $\frac{1}{4}|A - A'|$ say, and the formula

$$|u_{n+p} - u_n| < \epsilon, \text{ when } n \geqq \nu, \text{ for every positive integer } p,$$

shows that all the terms of the sequence

$$u_1, \quad u_2, \quad u_3, \ldots ,$$

except a finite number, would lie within an interval of length $\frac{1}{2}|A - A'|$. This is impossible if A, A' are limiting points of the aggregate.

Thus the aggregate (E) has one and only one limiting point A.

We shall now show that the sequence

$$u_1, \quad u_2, \quad u_3, \ldots$$

converges to A as n tends to ∞.

We have $\qquad u_n - A = (u_n - u_N) + (u_N - A).$

Therefore

$$\begin{aligned} |u_n - A| &\leqq |u_n - u_N| + |u_N - A| \\ &< \tfrac{1}{2}\epsilon + \tfrac{1}{2}\epsilon, \quad \text{when } n \geqq N, \\ &< \epsilon, \quad \text{when } n \geqq N. \end{aligned}$$

Thus the sequence converges, and has A for its limit.

We have therefore proved this theorem:

A necessary and sufficient condition for the convergence of the sequence

$$u_1, \quad u_2, \quad u_3, \ldots$$

is that, to the arbitrary positive number ϵ, there shall correspond a positive integer ν such that

$$|u_{n+p} - u_n| < \epsilon, \textit{ when } n \geqq \nu, \textit{ for every positive integer } p.$$

It is easy to show that the above condition may be replaced by the following:

In order that the sequence

$$u_1, \quad u_2, \quad u_3, \ldots$$

may converge, it is necessary and sufficient that, to the arbitrary positive number ϵ, there shall correspond a positive integer n such that

$$|u_{n+p} - u_n| < \epsilon \textit{ for every positive integer } p.$$

It is clear that if the sequence converges, this condition is satisfied by $n = \nu$.

Further, if this condition is satisfied, and ϵ is an arbitrary positive number, to the number $\frac{1}{2}\epsilon$ there corresponds a positive integer n such that

$$|u_{n+p} - u_n| < \tfrac{1}{2}\epsilon \text{ for every positive integer } p.$$

But
$$|u_{n+p''} - u_{n+p'}| \leqq |u_{n+p''} - u_n| + |u_{n+p'} - u_n|$$
$$< \tfrac{1}{2}\epsilon + \tfrac{1}{2}\epsilon, \text{ when } p', p'' \text{ are any positive integers.}$$

Therefore the condition in the text is also satisfied, and the sequence con verges.

16. Divergent and Oscillatory Sequences.*

When the sequence

$$u_1, \quad u_2, \quad u_3, \ldots$$

does not converge, several different cases arise.

(i) In the first place, the terms may have the property that if any positive number A, however large, is chosen, there is a positive integer ν such that

$$u_n > A, \text{ when } n \geqq \nu.$$

In this case we say that the sequence is *divergent*, and that it *diverges* to $+\infty$, and we write this

$$\lim_{n\to\infty} u_n = +\infty .$$

(ii) In the second place, the terms may have the property that if any negative number $-A$ is chosen, however large A may be, there is a positive integer ν such that

$$u_n < -A, \text{ when } n \geqq \nu.$$

In this case we say that the sequence is *divergent*, and that it *diverges to* $-\infty$, and we write this

$$\lim_{n\to\infty} u_n = -\infty .$$

The terms of a sequence may all be very large in absolute value, when n is very large, yet the sequence may not diverge to $+\infty$ or to $-\infty$. A sufficient illustration of this is given by the sequence whose general term is $(-1)^n n$.

After some value of n the terms must all have the same sign, if the sequence is to diverge to $+\infty$ or to $-\infty$, the sign being positive in the first alternative, and negative in the second.

(iii) *When the sequence does not converge, and does not diverge to* $+\infty$ *or to* $-\infty$, *it is said to* **oscillate.**

---

*In the first edition of this book, the term divergent was used as meaning merely not convergent. In this edition the term is applied only to the case of divergence to $+\infty$ or to $-\infty$, and sequences which oscillate infinitely are placed among the oscillatory sequences.

An oscillatory sequence is said to **oscillate finitely,** *if there is a positive number A such that* $|u_n| < A$, *for all values of n; and it is said to* **oscillate infinitely** *when there is no such number.*

For example, the sequence whose general term is $(-1)^n$ **oscillates finitely**; the sequence whose general term is $(-1)^n n$ **oscillates infinitely.**

We may distinguish between convergent and divergent sequences by saying that *a convergent sequence has a finite limit, i.e.* $\lim_{n\to\infty} u_n = A$, where A is a definite number; *a divergent sequence has an infinite limit, i.e.* $\lim_{n\to\infty} u_n = +\infty$ or $\lim_{n\to\infty} u_n = -\infty$.

But it must be remembered that the symbol ∞, and the terms *infinite, infinity* and *tend to infinity,* have purely conventional meanings. **There is no number infinity.** Phrases in which the term is used have only a meaning for us when we have previously, by definition, attached a meaning to them.

When we say that *n tends to infinity,* we are using a short and convenient phrase to express the fact that n assumes an endless series of values which eventually become and remain greater than any arbitrary (large) positive number. So far we have supposed n, in this connection, to advance through integral values only. This restriction will be removed later.

A similar remark applies to the phrases *divergence to* $+\infty$ or *to* $-\infty$, and *oscillating infinitely,* as well as to our earlier use of the terms *an infinite number, infinite sequence* and *infinite aggregates.* In each case a definite meaning has been attached to the term, and it is employed only with that meaning.

It is true that much of our work might be simplified by the introduction of new numbers $+\infty$, $-\infty$, and by assuming the existence of corresponding points upon the line which we have used as the domain of the numbers. But the creation of these numbers, and the introduction of these points, would be a matter for separate definition.

17. 1. Monotonic Sequences. *If the terms of the sequence*

$$u_1, \quad u_2, \quad u_3, \dots$$

satisfy either of the following relations

$$u_1 \leqq u_2 \leqq u_3 \dots \leqq u_n, \dots$$

or

$$u_1 \geqq u_2 \geqq u_3 \dots \geqq u_n, \dots,$$

the sequence is said to be **monotonic.**

In the first case, the terms never decrease, and the sequence may be called *monotonic increasing*; in the second case, the terms never increase, and the sequence may be called *monotonic decreasing.**

Obviously, when we are concerned with the convergence or divergence of a sequence, the monotonic property, if such exist, need not enter till after a certain stage.

The tests for convergence or divergence are extremely simple in the case of monotonic sequences.

If the sequence $$u_1, \quad u_2, \quad u_3, \ldots$$
is monotonic increasing, and its terms are all less than some fixed number B, the sequence is convergent and has for its limit a number β such that $u_n \leqq \beta \leqq B$ for every positive integer n.

Consider the aggregate formed by distinct terms of the sequence. It is bounded by u_1 on the left and by B on the right. Thus it must have an upper bound β (cf. § 12) equal to or less than B, and, however small the positive number ϵ may be there will be a term of the sequence greater than $\beta - \epsilon$.

Let this term be u_ν. Then all the terms after $u_{\nu-1}$ are to the right of $\beta - \epsilon$ and not to the right of β. If any of them coincide with β, from that stage on the terms must be equal.

Thus we have shown that
$$|\beta - u_n| < \epsilon, \text{ when } n \geqq \nu,$$
and therefore the sequence is convergent and has β for its limit.

The following test may be proved in the same way :

If the sequence $$u_1, \quad u_2, \quad u_3, \ldots$$
is monotonic decreasing, and its terms are all greater than some fixed number A, then the sequence is convergent and has for its limit a number α such that $u_n \geqq \alpha \geqq A$ for every positive integer n.

It is an immediate consequence of these theorems that a monotonic sequence either tends to a limit or diverges to $+\infty$ or to $-\infty$.

17. 2. The Upper and Lower Limits of Indetermination of a Bounded Sequence. Let $u_1, u_2, u_3, \ldots$ be a sequence, bounded above and below.

Let $\quad M_1$ be the upper bound of $u_1, u_2, u_3, u_4, \ldots$,

and $\quad M_2$ be the upper bound of $u_2, u_3, u_4, \ldots$,

and so on.

---

*The words *steadily increasing* and *steadily decreasing* are sometimes employed in this connection, and when none of the terms of the sequence are equal, the words *in the stricter sense* are added.

Similarly let $m_1, m_2, m_3, \ldots$ be the lower bounds of the corresponding sequences.

Then $$M_1 \geqq M_2 \geqq M_3 \geqq \ldots \geqq m_1.$$

Thus $\lim_{n\to\infty} M_n$ exists (§ 17. 1.).

Let this limit be Λ.

To the arbitrary positive number ϵ there corresponds a positive integer ν, such that
$$\Lambda \leqq M_n < \Lambda + \epsilon, \text{ when } n \geqq \nu.$$

But M_ν is the upper bound of $u_\nu, u_{\nu+1}, u_{\nu+2}, \ldots$

Therefore $u_n \leqq M_\nu$, when $n \geqq \nu$.

Thus $u_n < \Lambda + \epsilon$, when $n \geqq \nu$.

Also at least one of the set $u_\nu, u_{\nu+1}, u_{\nu+2}, \ldots$ (say u_N) is greater than $M_\nu - \epsilon$, and thus greater than $\Lambda - \epsilon$.

Take ν' a positive integer $> N$.

Then $$\Lambda \leqq M_{\nu}' \leqq M_\nu.$$

And at least one (say $u_{N'}$) of the set $u_{\nu'}, u_{\nu'+1}, u_{\nu'+2}, \ldots$ is greater than $\Lambda - \epsilon$.

In this way we have the infinite set
$$u_N, \quad u_{N'}, \quad u_{N''}, \ldots$$
all greater than $\Lambda - \epsilon$.

We have thus shown that there is a number Λ associated with the bounded sequence $u_1, u_2, u_3 \ldots$ which has the following properties:

If ϵ is an arbitrary positive number, $u_n < \Lambda + \epsilon$, for all positive integers greater than a definite integer depending on ϵ; and $u_n > \Lambda - \epsilon$, for an infinite number of positive integers.

Similarly for the lower bounds $m_1, m_2, m_3, \ldots$ we see that $\lim_{n\to\infty} m_n$ exists.

Denoting this limit by λ, we have the corresponding result; $u_n > \lambda - \epsilon$, *for all values of n greater than a definite value depending on* ϵ; *and* $u_n < \lambda + \epsilon$, *for an infinite number of values of n.*

The numbers Λ and λ are called the *upper* and *lower limits of indetermination of the sequence,** and we write
$$\Lambda = \overline{\lim_{n\to\infty}}\, u_n, \quad \lambda = \underline{\lim_{n\to\infty}}\, u_n.$$

It is clear that $\Lambda \geqq \lambda$.

17. 3. *Let $u_1, u_2, u_3, \ldots, v_1, v_2, v_3, \ldots$ be sequences of positive terms, the first sequence being bounded above, and in the second $\lim_{n\to\infty} v_n$ being equal to unity.*

Then $\overline{\lim_{n\to\infty}}\,(u_n v_n) = \overline{\lim_{n\to\infty}}\, u_n$.

It is clear that as u_1, u_2, u_3, are positive and the sequence $u_1, u_2, u_3, \ldots$ is bounded above, the sequence of positive terms $u_1v_1, u_2v_2, u_3v_3, \ldots$ is bounded above.

---

*French: *la plus grande limite* and *la plus petite limite*, or *limites d'indetermination.* German: *obere* and *untere Unbestimmtheitsgrenze.* Other terms are used; e.g. *obere (untere) Häufungsgrenze*; *oberes (unteres) Limes*; and *Limes superior (inferior).*

If possible, let $\overline{\lim_{n\to\infty}}(u_n v_n)=\mu'>\mu=\overline{\lim_{n\to\infty}}\, u_n$.

Take $2\epsilon=\mu'-\mu$, and let the upper bound of $u_1, u_2, u_3, \ldots$ be K.

Since $\lim_{n\to\infty} v_n=1$, there is a positive integer ν_1, such that

$$|v_n-1|<\frac{\epsilon}{2K}, \text{ when } n \geqq \nu_1.$$

Thus $|u_n v_n - u_n| = |u_n||v_n-1| < K\times\frac{\epsilon}{2K}<\frac{1}{2}\epsilon$, when $n\geqq\nu_1$.

Therefore $$u_n v_n < u_n+\tfrac{1}{2}\epsilon, \text{ when } n\geqq \nu_1.$$

But since $\overline{\lim_{n\to\infty}}\, u_n=\mu$, there is a positive integer ν_2, such that

$$u_n<\mu+\tfrac{1}{2}\epsilon, \text{ when } n\geqq\nu_2.$$

Therefore $$u_n v_n<\mu+\epsilon, \text{ when } n\geqq\nu,$$

where ν is the larger of the integers ν_1 and ν_2.

But since $\overline{\lim_{n\to\infty}}(u_n v_n)=\mu'$, we know that $u_n v_n>\mu'-\epsilon$, for an infinite number of values of n.

Therefore μ' cannot be greater than μ.

Similarly it can be shown that μ' is not less than μ.

Hence $\mu'=\mu$ and the theorem is proved.

17.4. (i) *If the sequence $u_1, u_2, u_3, \ldots$ converges, then* $\overline{\lim_{n\to\infty}}\, u_n=\underline{\lim_{n\to\infty}}\, u_n=\lim_{n\to\infty} u_n$.

Since $u_1, u_2, u_3, \ldots$ converges, it must be a bounded sequence, and $\overline{\lim_{n\to\infty}}\, u_n$, $\underline{\lim_{n\to\infty}}\, u_n$ both exist.

We have to show that they are equal to $\lim_{n\to\infty} u_n$.

Let $\lim_{n\to\infty} u_n=l$, $\overline{\lim_{n\to\infty}}\, u_n=\Lambda$, and $\underline{\lim_{n\to\infty}}\, u_n=\lambda$.

If possible, let $\Lambda>l$, and take $2\epsilon=\Lambda-l$.

Then there is a positive integer ν, such that

$$u_n<l+\epsilon, \text{ when } n\geqq\nu.$$

But we are given that $u_n>\Lambda-\epsilon$, for an infinite number of values of n.

These two inequalities cannot both be true.

Thus Λ cannot be greater than l.

In a similar way we can show that λ cannot be less than l.

But $$\Lambda\geqq\lambda.$$

Therefore we must have $\Lambda=\lambda=l$.

(ii) Conversely, *if the upper and lower limits of indetermination of the bounded sequence $u_1, u_2, u_3, \ldots$ are equal, then the sequence converges to their common value.*

We are given that

$$\overline{\lim_{n\to\infty}}\, u_n=\Lambda=\lambda=\underline{\lim_{n\to\infty}}\, u_n.$$

Thus, to the arbitrary positive number ϵ, there correspond positive integers ν_1 and ν_2, such that

$$u_n<\Lambda+\epsilon, \text{ when } n\geqq\nu_1,$$
$$u_n>\lambda-\epsilon, \text{ when } n\geqq\nu_2,$$

Hence $$|u_n - \Lambda| < \epsilon, \text{ when } n \geqq \nu,$$
where ν is the larger of the two positive integers ν_1 and ν_2.

Thus $$\overline{\lim_{n\to\infty}}\, u_n = \underline{\lim_{n\to\infty}}\, u_n = \lim_{n\to\infty} u_n.$$

(iii) With this notation it is easy to establish *the general principle of convergence* (p. 38), namely that *a necessary and sufficient condition for the existence of a limit to the sequence $u_1, u_2, u_3 \ldots$ is that, if any positive number ϵ has been chosen, as small as we please, there shall be a positive integer ν, such that $|u_{n+p} - u_n| < \epsilon$, when $n \geqq \nu$ and p is any positive integer.*

There is no difficulty in showing that this condition is necessary for the convergence of the sequence. (Cf. p. 39.)

The difficulty in our former proof was to show that the condition was sufficient.

But, it is clear that with this condition, the sequence $u_1, u_2, u_3, \ldots$ is bounded. Its upper and lower limits of indetermination therefore exist.

With the same notation as before, let Λ and λ be unequal.

Take $$2\epsilon = \Lambda - \lambda.$$

There is a positive integer ν, such that

$$|u_{n+p} - u_n| < \epsilon, \text{ when } n \geqq \nu \text{ and } p = 1, 2, 3 \ldots \quad \text{(1)}$$

But $$u_n < \lambda + \tfrac{1}{2}\epsilon, \text{ for an infinite number of values of } n. \quad \text{(2)}$$

And $$u_n > \Lambda - \tfrac{1}{2}\epsilon, \text{ for an infinite number of values of } n. \quad \text{(3)}$$

Let ν', ν'' be the first positive integers greater than ν which satisfy (2) and (3) respectively.

Then $|u_{\nu''} - u_{\nu'}| > \epsilon$, contrary to (1).

Therefore Λ and λ are equal, and by (ii) $\lim_{n\to\infty} u_n$ exists.

18. *Let $A_1, A_2, A_3, \ldots$ be an infinite set of closed intervals, each lying entirely within the preceding, or lying within it and having with it a common end-point; also let the length of A_n tend to zero as n tends to infinity. Then there is one, and only one, point which belongs to all the intervals, either as an internal point of all, or, from and after a definite stage, as a common end-point of all.*

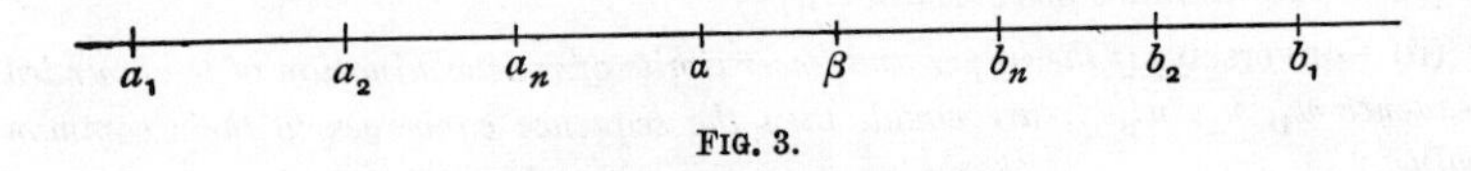

FIG. 3.

Let the representative interval A_n be given by
$$a_n \leqq x \leqq b_n.$$

Then we have $$a_1 \leqq a_2 \leqq a_3 \ldots < b_1,$$
and $$b_1 \geqq b_2 \geqq b_3 \ldots > a_1.$$

Thus the sequence of end-points

$$a_1, \quad a_2, \quad a_3 \ldots \qquad \ldots\ldots\ldots\ldots\ldots\ldots(1)$$

has a limit, say α, and $a_n \leqq \alpha$ for every positive integer n (§ 17. 1).

Also the sequence of end-points

$$b_1, \quad b_2, \quad b_3 \ldots, \qquad \ldots\ldots\ldots\ldots\ldots\ldots(2)$$

has a limit, say β, and $b_n \geqq \beta$ for every positive integer n (§ 17. 1).

Now it is clear that, under the given conditions, β cannot be less than α.

Therefore, for every value of n,

$$b_n - a_n > \beta - \alpha \geqq 0.$$

But
$$\lim_{n\to\infty} (b_n - a_n) = 0.$$

It follows that $\alpha = \beta$.*

Therefore this common limit of the sequences (1) and (2) satisfies the inequalities

$$a_n \leqq \alpha \leqq b_n \text{ for every positive integer } n,$$

and thus belongs to all the intervals.

Further, no other point (*e.g.* γ) can satisfy $a_n \leqq \gamma \leqq b_n$ for all values of n.

Since we would have at the same time

$$\lim_{n\to\infty} a_n \leqq \gamma \quad \text{and} \quad \lim_{n\to\infty} b_n \geqq \gamma,$$

which is impossible unless $\gamma = \alpha$.

19. The Sum of an Infinite Series.

Let
$$u_1, \quad u_2, \quad u_3, \ldots$$
be an infinite sequence, and let the successive sums

$$s_1 = u_1,$$
$$s_2 = u_1 + u_2,$$
$$\ldots\ldots\ldots\ldots$$
$$s_n = u_1 + u_2 + u_3 + \ldots + u_n$$

be formed.

If the sequence
$$s_1, \quad s_2, \quad s_3, \ldots$$
is convergent and has the limit s, then s is called the sum of the infinite series
$$u_1 + u_2 + u_3 + \ldots$$
and this series is said to be convergent.

---

*This result also follows at once from the fact that, if $\lim a_n = \alpha$ and $\lim b_n = \beta$ then $\lim (a_n - b_n) = \alpha - \beta$. (Cf. § 26, Theorem I.)

It must be carefully noted that what we call the sum of the infinite series is *a limit*, the limit of the sum of n terms of

$$u_1+u_2+u_3+\dots,$$

as n tends to infinity. Thus we have no right to assume without proof that familiar properties of finite sums are necessarily true for sums such as s.

When $\lim_{n\to\infty} s_n=+\infty$ or $\lim_{n\to\infty} s_n=-\infty$, we shall say that the infinite series is *divergent*, or *diverges to* $+\infty$ or $-\infty$, as the case may be.

If s_n does not tend to a limit, or to $+\infty$ or to $-\infty$, then it oscillates finitely or infinitely according to the definitions of these terms in § 16. In this case we shall say that the series *oscillates finitely or infinitely.**

The conditions obtained in § 15 for the convergence of a sequence allow us to state the criteria for the convergence of the series in either of the following ways:

(i) *The series converges and has s for its sum, if, any positive number ϵ having been chosen, as small as we please, there is a positive integer ν such that* $|s-s_n|<\epsilon$, *when* $n\geqq\nu$.

(ii) *A necessary and sufficient condition for the convergence of the series is that, if any positive number ϵ has been chosen, as small as we please, there shall be a positive integer ν such that*

$$|s_{n+p}-s_n|<\epsilon, \text{ when } n\geqq\nu, \text{ for every positive integer } p.\dagger$$

It is clear that, *if the series converges*, $\lim_{n\to\infty} u_n=0$. This is contained in the second criterion. It is a *necessary* condition for convergence, but it is not a *sufficient* condition; *e.g.* the series

$$1+\tfrac{1}{2}+\tfrac{1}{3}+\dots$$

is divergent, though $\lim_{n\to\infty} u_n=0$.

If we denote

$$u_{n+1}+u_{n+2}+\dots+u_{n+p}, \quad \text{or} \quad s_{n+p}-s_n, \text{ by } {}_pR_n$$

the above necessary and sufficient condition for convergence of the series may be written

$$|{}_pR_n|<\epsilon, \text{ when } n\geqq\nu, \text{ for every positive integer } p.$$

---

*Cf. footnote, p. 41.

†As remarked in § 15, this condition can be replaced by: *To the arbitrary positive number ϵ there must correspond a positive integer n such that*

$$|s_{n+p}-s_n|<\epsilon \textit{ for every positive integer } p.$$

Again, *if the series* $u_1 + u_2 + u_3 + \ldots$ *converges and has s for its sum, the series*

$$u_{n+1} + u_{n+2} + u_{n+3} + \ldots$$

converges and has $s - s_n$ *for its sum.*

For we have $$s_{n+p} = s_n + {}_pR_n.$$

Also keeping n fixed, it is clear that

$$\lim_{p\to\infty} s_{n+p} = s.$$

Therefore $$\lim_{p\to\infty} ({}_pR_n) = s - s_n.$$

Thus if we write R_n for the sum of the series

$$u_{n+1} + u_{n+2} + \ldots ,$$

we have $$s = s_n + R_n.$$

The first criterion for convergence can now be put in the form

$$|R_n| < \epsilon, \text{ when } n \geqq \nu.$$

R_n is usually called the *remainder* of the series after n terms, and ${}_pR_n$, or $s_{n+p} - s_n$, a *partial remainder.*

20. Series whose Terms are all Positive.

Let $$u_1 + u_2 + u_3 + \ldots$$

be a series whose terms are all positive. The sum of n terms of this series either tends to a limit, or it diverges to $+\infty$.

Since the terms are all positive, the successive sums

$$s_1 = u_1,$$
$$s_2 = u_1 + u_2,$$
$$s_3 = u_1 + u_2 + u_3,$$
$$\ldots\ldots\ldots\ldots\ldots\ldots ,$$

form a monotonic increasing sequence, and the theorem stated above follows from § 17. 1.

When a series whose terms are all positive is convergent, the series we obtain when we take the terms in any order we please is also convergent and has the same sum.

This change of the order of the terms is to be such that there will be a one-one correspondence between the terms of the old series and the new. The term in any assigned place in the one series is to have a definite place in the other.

Let

$$s_1 = u_1,$$
$$s_2 = u_1 + u_2,$$
$$s_3 = u_1 + u_2 + u_3,$$
$$\dots\dots\dots\dots\dots .$$

Then the aggregate (U), which corresponds to the sequence

$$s_1, \quad s_2, \quad s_3, \dots ,$$

is bounded and its upper bound s is the sum of the series.

Let (U') be the corresponding aggregate for the series obtained by taking the terms in any order we please, on the understanding we have explained above. Every number in (U') is less than s. In addition, if A is any number less than s, there must be a number of (U) greater than A, and *a fortiori* a number of (U') greater than A. The aggregate (U') is thus bounded on the right, and its upper bound is s. The sum of the new series is therefore the same as the sum of the old.

It follows that *if the series*

$$u_1 + u_2 + u_3 + \dots ,$$

whose terms are all positive, diverges, the series we obtain by changing the order of the terms must also diverge.

The following theorems may be proved at once by the use of the second condition for convergence (§ 19):

If the series $$u_1 + u_2 + u_3 + \dots$$

is convergent and all its terms are positive, the series we obtain from this, either

(1) *by keeping only a part of its terms,*

or (2) *by replacing certain of its terms by others, either positive or zero, which are respectively equal or inferior to them,*

or (3) *by changing the signs of some of its terms,*

are also convergent.

21. Absolute and Conditional Convergence. The trigonometrical series, whose properties we shall investigate later, belong to the class of series whose convergence is due to the presence of both positive and negative terms, in the sense that the series would diverge if all the terms were taken with the same sign.

A series with positive and negative terms is said to be absolutely convergent, when the series in which all the terms are taken with the same sign converges.

In other words, the series

$$u_1 + u_2 + u_3 + \dots$$

is absolutely convergent when the series of absolute values

$$|u_1| + |u_2| + |u_3| + \dots$$

is convergent.

It is obvious that an absolutely convergent series is also convergent in the ordinary sense, since the absolute values of the partial remainders of the original series cannot be greater than those of the second series. There are, however, convergent series which are not absolutely convergent:

e.g. $1 - \frac{1}{2} + \frac{1}{3} \dots$ is convergent.

$1 + \frac{1}{2} + \frac{1}{3} \dots$ is divergent.

Series in which the convergence depends upon the presence of both positive and negative terms are said to be conditionally convergent.

The reason for this name is that, as we shall now prove, an *absolutely convergent series* remains convergent, and has the same sum, even although we alter the order in which its terms are taken; while a *conditionally convergent series* may converge for one arrangement of the terms and diverge for another. Indeed we shall see that we can make a conditionally convergent series have any sum we please, or be greater than any number we care to name, by changing the order of its terms. There is nothing very extraordinary in this statement. The rearrangement of the terms introduces a new function of n, say s'_n, instead of the old function s_n, as the sum of the first n terms. There is no *a priori* reason why this function s'_n should have a limit as n tends to infinity, or, if it has a limit, that this should be the same as the limit of s_n.*

22. Absolutely Convergent Series. *The sum of an absolutely convergent series remains the same when the order of the terms is changed.*

Let (S) be the given absolutely convergent series; (S') the series composed of the *positive* terms of (S) in the order in which they appear; (S'') the series composed of the absolute values of the *negative* terms of (S), also in the order in which they appear.

If the number of terms either in (S') or (S'') is limited, the theorem requires no proof, since we can change the order of the

*Cf. Osgood, *Introduction to Infinite Series* (1897), 44.

terms in the finite sum, which includes the terms of (S) up to the last of the class which is limited in number, without altering its sum, and we have just seen that when the terms are of the same sign, as in those which follow, the alteration in the order in the convergent series does not affect its sum.

Let σ be the sum of the infinite series formed by the absolute values of the terms of (S).

Let s_n be the sum of the first n terms of (S).

In this sum let n' terms be positive and n'' negative.

Let $s_{n'}$ be the sum of these n' terms.

Let $s_{n''}$ be the sum of the absolute values of these n'' terms, taking in each case these terms in the order in which they appear in (S).

Then
$$s_n = s_{n'} - s_{n''},$$
$$s_{n'} < \sigma,$$
$$s_{n''} < \sigma.$$

Now, as n increases $s_{n'}$, $s_{n''}$ never diminish. Thus, as n increases without limit, the successive values of $s_{n'}$, $s_{n''}$ form two infinite monotonic sequences such as we have examined in § 17. 1, whose terms do not exceed the fixed number σ. These sequences, therefore, tend to fixed limits, say, s' and s''.

Thus
$$\lim_{n\to\infty} (s_n) = s' - s''.$$

Hence the sum of the absolutely convergent series (S) is equal to the difference between the sums of the two infinite series formed one with the positive terms in the order in which they appear, and the other with the absolute values of the negative terms, also in the order in which they appear in (S).

Now any alteration in the order of the terms of (S) does not change the values of s' and s''; since we have seen that in the case of a convergent series whose terms are all positive we do not alter the sum by rearranging the terms. It follows that (S) *remains convergent and has the same sum when the order of its terms is changed in any way we please, provided that a one-one correspondence exists between the terms of the old series and the new.*

We add some other results with regard to absolutely convergent series which admit of simple demonstration:

Any series whose terms are either equal or inferior in absolute value to the corresponding terms of an absolutely convergent series is also absolutely convergent.

An absolutely convergent series remains absolutely convergent when we suppress a certain number of its terms.

If

$$u_1 + u_2 + \dots ,$$

$$v_1 + v_2 + \dots ,$$

are two absolutely convergent series whose sums are U and V, the series

$$(u_1 + v_1) + (u_2 + v_2) + \dots$$

and

$$(u_1 - v_1) + (u_2 - v_2) + \dots$$

are also absolutely convergent and their sums are equal to $U \pm V$ respectively.

23. Conditionally Convergent Series. *The sum of a conditionally convergent series depends essentially on the order of its terms.*

Let (S) be such a series. The positive and negative terms must both be infinite in number, since otherwise the series would converge absolutely.

Further, the series formed by the positive terms in the order in which they occur in (S), and the series formed in the same way by the negative terms, must both be *divergent.*

Both could not *converge*, since in that case our series would be equal to the difference of two absolutely convergent series, some of whose terms might be zero, and therefore would be absolutely convergent (§ 22). Also (S) could not converge, if one of these series converged and the other diverged.

We can therefore take sufficient terms from the positive terms to make their sum exceed any positive number we care to name. In the same way we can take sufficient terms from the negative terms to make the sum of their absolute values exceed any number we care to name.

Let a be any positive number.

First take positive numbers from (S) in the order in which they appear, stopping whenever the sum is greater than a. Then take negative terms from (S), in the order in which they appear, stopping whenever the combined sum is less than a. Then add on as many from the remaining positive terms as will make the sum exceed a, stopping when the sum first exceeds a; and then proceed to the negative terms; and so on.

In this way we form a new series (S') composed of the same terms as (S), in which the sum of n terms is sometimes greater than a and sometimes less than a.

Now the series (S) converges. Let its terms be $u_1, u_2, u_3, \dots$. Then, with the usual notation,

$$|u_n| < \epsilon, \text{ when } n \geqq \nu.$$

Let the points B_ν and A_ν (Fig. 4) correspond to the sums obtained in (S'), as described above, when ν groups of positive terms and ν groups of negative terms have been taken.

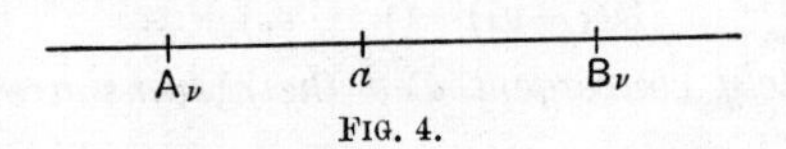

FIG. 4.

Then it is clear that $(a - A_\nu)$ and $(B_\nu - a)$ are each less than ϵ, since each of these groups contains at least one term of (S), and $(a - A_\nu)$, $(B_\nu - a)$ are at most equal to the absolute value of the last term in each group.

Let these 2ν groups contain in all ν' terms.

The term $u'_{n'}$ in (S'), when $n' \geqq \nu'$, is less in absolute value than ϵ. Thus, if we proceed from A_ν, the sums $s'_{n'}$ lie within the interval $(a - \epsilon, a + \epsilon)$, when $n' \geqq \nu'$.

In other words, $|s'_{n'} - a| < \epsilon$, when $n' \geqq \nu'$.

Therefore $$\lim_{n' \to \infty} s'_{n'} = a.$$

A similar argument holds for the case of a negative number, the only difference being that now we begin with the negative terms of the series.

We have thus established the following theorem:

If a conditionally convergent series is given, we can so arrange the order of the terms as to make the sum of the new series converge to any value we care to name.

REFERENCES.

BROMWICH, *loc. cit.*, Ch. I-IV.

DE LA VALLÉE POUSSIN, *loc. cit.*, 1 (5e éd., 1923), Introduction § 2, Ch. XI, §§ 1, 2.

GOURSAT, *loc. cit.*, 1 (4e éd., 1923), Ch. I and VIII.

HARDY, *Course of Pure Mathematics* (5th ed., 1928), Ch. IV and VIII.

KNOPP, *loc. cit.*, English translation, Ch. III, VIII-X.

PRINGSHEIM, *loc. cit.*, Bd. I, Absch. I, Kap. III, V, Absch. II, Kap. I-III.

STOLZ U. GMEINER, *loc. cit.*, Abth. II, Absch. IX.

And

PRINGSHEIM, "Irrationalzahlen u. Konvergenz unendlicher Prozesse," *Enc. d. math. Wiss.*, Bd. I, Tl. I (Leipzig, 1898).

CHAPTER III

FUNCTIONS OF A SINGLE VARIABLE LIMITS AND CONTINUITY

24. The Idea of a Function. In Elementary Mathematics, when we speak of a function of x, we usually mean a real expression obtained by certain operations, *e.g.* x^2, $\sqrt{x}$, $\log x$, $\sin^{-1}x$. In some cases, from the nature of the operations, the range of the variable x is indicated. In the first of the above examples, the range is unlimited; in the second, $x \geqq 0$; in the third $x > 0$; and in the last $|x| \leqq 1$.

In Higher Mathematics the term "function of x" has a much more general meaning. *Let a and b be any two real numbers, where $b > a$. If to every value of x in the interval $a \leqq x \leqq b$ there corresponds a (real) number y, then we say that y is a function of x in the interval (a, b), and we write $y = f(x)$.*

Sometimes the end-points of the interval are excluded from the domain of x, which is then given by $a < x < b$. In this case the interval is said to be *open* at both ends; when both ends are included (*i.e.* $a \leqq x \leqq b$) it is said to be *closed*. An interval may be *open* at one end and *closed* at the other (*e.g.* $a < x \leqq b$).

Unless otherwise stated, when we speak of an interval in the rest of this work, we shall refer to an interval *closed* at both ends. And when we say that x *lies in the interval* (a, b), we mean that $a \leqq x \leqq b$, but when x is to lie between a and b, and not to coincide with either, we shall say that x *lies in the open interval* (a, b).*

Consider the aggregate formed by the values of a function $f(x)$

*In Ch. II, when a point x lies between a and b, and does not coincide with either, we have referred to it as *within* the interval (a, b). This form of words is convenient, and not likely to give rise to confusion.

given in an interval (a, b). If this aggregate is bounded (cf. § 11), we say that the function $f(x)$ is *bounded* in the interval. The numbers M and m, the *upper* and *lower bounds* of the aggregate (cf. § 12), are called the *upper* and *lower bounds* of the function in the interval. And a function can have an upper bound and no lower bound, and *vice versa*.

The difference $(M - m)$ is called the *oscillation of the function in the interval.**

It should be noticed that a function may be determinate in an interval, and yet not bounded in the interval.

E.g. let $\quad f(0)=0, \quad$ and $\quad f(x)=\frac{1}{x}$ when $x>0$.

Then $f(x)$ has a definite value for every x in the interval $0 \leqq x \leqq a$, where a is any given positive number. But $f(x)$ is not bounded in this interval, for we can make $f(x)$ exceed any number we care to name, by letting x approach sufficiently near to zero.

Further, a bounded function need not attain its upper and lower bounds; in other words, M and m need not be members of the aggregate formed by the values of $f(x)$ in the interval.

E.g. let $\quad f(0)=0, \quad$ and $\quad f(x)=1-x$ when $0<x\leqq 1$.

This function, given in the interval $(0, 1)$, attains its lower bound zero, but not its upper bound unity.

25. $\lim\limits_{x\to a} f(x)$. In the previous chapter we have dealt with the limit when $n\to\infty$ of a sequence $u_1, u_2, u_3, \ldots$. In other words, we have been dealing with a function $\phi(n)$, where n is a positive integer, and we have considered the limit of this function as $n\to\infty$.

We pass now to the function of the real variable x and the limit of $f(x)$ when $x\to a$. The idea is familiar enough. The Differential Calculus rests upon it. But for our purpose we must put the matter on a precise arithmetical footing, and a definition of what exactly is meant by the limit of a function of x, as x tends to a definite value, must be given.

$f(x)$ is said to have the limit b as x tends to a, when, any positive number ϵ having been chosen, as small as we please, there is a positive number η such that $|f(x)-b|<\epsilon$, for all values of x for which

$$0<|x-a|\leqq\eta.$$

---

*Hobson, *loc. cit.* 1 (3rd ed., 1927), 280, uses the term *fluctuation*.

In other words $|f(x)-b|$ must be less than ϵ for all points in the interval $(a-\eta,\ a+\eta)$ except the point a.

When this condition is satisfied, we employ the notation $\lim_{x\to a} f(x)=b$, for the phrase *the limit of* $f(x)$, *as* x *tends to* a, *is* b, and we say that $f(x)$ *converges* to b as x tends to a.

One advantage of this notation, as opposed to $\lim_{x=a} f(x)=b$, is that it brings out the fact that we say nothing about what happens when x is *equal* to a. In the definition it will be observed that a statement is made about the behaviour of $f(x)$ for all values of x such that $0<|x-a|\leqq\eta$. The first of these inequalities is inserted expressly to exclude $x=a$.

Sometimes x tends to a from the right-hand only (*i.e.* $x>a$), or from the left-hand only (*i.e.* $x<a$).

In these cases, instead of $0<|x-a|\leqq\eta$, we have $0<(x-a)\leqq\eta$ (right-hand) and $0<(a-x)\leqq\eta$ (left-hand), in the definition.

The notation adopted for these right-hand and left-hand limits is

$$\lim_{x\to a+0} f(x) \quad \text{and} \quad \lim_{x\to a-0} f(x).$$

The assertion that $\lim_{x\to a} f(x)=b$ thus includes

$$\lim_{x\to a+0} f(x)=\lim_{x\to a-0} f(x)=b.$$

It is convenient to use $f(a+0)$ for $\lim_{x\to a+0} f(x)$ when this limit exists, and similarly $f(a-0)$ for $\lim_{x\to a-0} f(x)$ when this limit exists.

When $f(x)$ has not a limit as $x\to a$, it may happen that it diverges to $+\infty$, or to $-\infty$, in the sense in which these terms were used in § 16. Or, more precisely, it may happen that *if any positive number A, however large, is chosen, there corresponds to it a positive number η such that*

$$f(x)>A, \textit{ when } 0<|x-a|\leqq\eta.$$

In this case we say that $\lim_{x\to a} f(x)=+\infty$.

Again, it may happen that *if any negative number $-A$ is chosen, however large A may be, there corresponds to it a positive number η such that $f(x)<-A$, when $0<|x-a|\leqq\eta$.*

In this case we say that $\lim_{x\to a} f(x)=-\infty$.

The modifications when $f(a\pm 0)=\pm\infty$ are obvious.

When $\lim_{x \to a} f(x)$ does not exist, and when $f(x)$ does not *diverge* to $+\infty$, or to $-\infty$, as $x \to a$, it is said to *oscillate* as $x \to a$. It *oscillates finitely* if $f(x)$ is bounded in some neighbourhood of that point.* It *oscillates infinitely* if there is no neighbourhood of a in which $f(x)$ is bounded. (Cf. § 16.)

The modifications to be made in these definitions when $x \to a$ only from the right, or only from the left, are obvious.

26. Some General Theorems on Limits. I. **The Limit of a Sum.**

If $\lim_{x \to a} f(x) = \alpha$ *and* $\lim_{x \to a} g(x) = \beta$, *then* $\lim_{x \to a} [f(x) + g(x)] = \alpha + \beta$.†

Let the positive number ϵ be chosen, as small as we please. Then to $\frac{1}{2}\epsilon$ there correspond the positive numbers η_1, η_2 such that

$$|f(x) - \alpha| < \tfrac{1}{2}\epsilon, \text{ when } 0 < |x - a| \leqq \eta_1,$$

$$|g(x) - \beta| < \tfrac{1}{2}\epsilon, \text{ when } 0 < |x - a| \leqq \eta_2.$$

Thus, if η is not greater than η_1 or η_2,

$$\begin{aligned} |f(x) + g(x) - \alpha - \beta| &\leqq |f(x) - \alpha| + |g(x) - \beta|, \\ &< \tfrac{1}{2}\epsilon + \tfrac{1}{2}\epsilon, \quad \text{when } 0 < |x - a| \leqq \eta, \\ &< \epsilon, \quad \text{when } 0 < |x - a| \leqq \eta. \end{aligned}$$

Therefore $\lim_{x \to a} [f(x) + g(x)] = \alpha + \beta$.

This result can be extended to the sum of any number of functions. The *Limit of a Sum* is equal to the *Sum of the Limits*.

II. **The Limit of a Product.** *If* $\lim_{x \to a} f(x) = \alpha$ *and* $\lim_{x \to a} g(x) = \beta$, *then* $\lim_{x \to a} [f(x)g(x)] = \alpha\beta$.

Let $f(x) = \alpha + \phi(x)$ and $g(x) = \beta + \psi(x)$.

Then $\lim_{x \to a} \phi(x) = 0$ and $\lim_{x \to a} \psi(x) = 0$.

Also $f(x)g(x) = \alpha\beta + \beta\phi(x) + \alpha\psi(x) + \phi(x)\psi(x)$.

From Theorem I our result follows if $\lim_{x \to a} [\phi(x)\psi(x)] = 0$.

---

* $f(x)$ is said to satisfy a certain condition *in the neighbourhood of* $x = a$ when there is a positive number h such that the condition is satisfied when $0 < |x - a| \leqq h$.

Sometimse the *neighbourhood* is meant to include the point $x = a$ itself. In this case it is defined by $|x - a| \leqq h$.

† The corresponding theorem for functions of the positive integer n, as $n \to \infty$, is proved in the same way, and is useful in the argument of certain sections of the previous chapter.

Since $\phi(x)$ tends to zero as $x \to a$ and $\psi(x)$ tends to zero as $x \to a$, a proof of this might appear unnecessary. But if a formal proof is required, it could run as follows :

Given the arbitrary position number ϵ, we have, as in (I),

$$|\phi(x)| < \sqrt{\epsilon}, \text{ when } 0 < |x-a| \leqq \eta_1,$$
$$|\psi(x)| < \sqrt{\epsilon}, \text{ when } 0 < |x-a| \leqq \eta_2.$$

Thus, if η is not greater than η_1 or η_2,

$$|\phi(x)\psi(x)| < \epsilon, \text{ when } 0 < |x-a| \leqq \eta.$$

Therefore $$\lim_{x \to a} [\phi(x)\psi(x)] = 0.$$

This result can be extended to any number of functions. The *Limit of a Product* is equal to the *Product of the Limits.*

III. The Limit of a Quotient.

(i) *If* $\lim_{x \to a} f(x) = a \gtrless 0$, *then* $\lim_{x \to a} \dfrac{1}{f(x)} = \dfrac{1}{a}$.

This follows easily on putting $f(x) = \phi(x) + a$ and examining the expression

$$\frac{1}{a} - \frac{1}{\phi(x) + a}.$$

(ii) *If* $\lim_{x \to a} f(x) = a$, *and* $\lim_{x \to a} g(x) = \beta \gtrless 0$, *then* $\lim_{x \to a} \left[\dfrac{f(x)}{g(x)}\right] = \dfrac{a}{\beta}$.

This follows from (II) and (III (i)).

This result can obviously be generalised as above.

IV. The Limit of a Function of a Function. $\lim_{x \to a} \mathbf{f}[\phi(\mathbf{x})]$.

Let $$\lim_{x \to a} \phi(x) = b \quad \textit{and} \quad \lim_{u \to b} f(u) = f(b).$$

Then $$\lim_{x \to a} f[\phi(x)] = f[\lim_{x \to a} \phi(x)].$$

We are given that $\lim_{u \to b} f(u) = f(b)$.

Therefore to the arbitrary positive number ϵ there corresponds a positive number η_1 such that

$$|f[\phi(x)] - f(b)| < \epsilon, \text{ when } |\phi(x) - b| \leqq \eta_1. \quad \text{..................(1)}$$

Also we are given that $\lim_{x \to a} \phi(x) = b$.

Therefore to this positive number η_1 there corresponds a positive number η such that

$$|\phi(x) - b| < \eta_1, \text{ when } 0 < |x-a| \leqq \eta. \quad \text{......................(2)}$$

Combining (1) and (2), to the arbitrary positive number ϵ there corresponds a positive number η such that

$$|f[\phi(x)] - f(b)| < \epsilon, \text{ when } 0 < |x-a| \leqq \eta.$$

Thus $$\lim_{x \to a} f[\phi(x)] = f(b) = f[\lim_{x \to a} \phi(x)].$$

EXAMPLES.

1. If n is a positive integer, $\lim_{x\to 0} x^n = 0$.

2. If n is a negative integer, $\lim_{x\to +0} x^n = +\infty$; and $\lim_{x\to -0} x^n = -\infty$ or $+\infty$ according as n is odd or even.

[If $n=0$, then $x^n=1$ and $\lim_{x\to 0} x^n = 1$.]

3. $\lim_{x\to 0}(a_0x^n + a_1x^{n-1} + \ldots + a_{n-1}x + a_n) = a_n$.

4. $\lim_{x\to 0}\left(\dfrac{a_0x^m + a_1x^{m-1} + \ldots + a_{m-1}x + a_m}{b_0x^n + b_1x^{n-1} + \ldots + b_{n-1}x + b_n}\right) = \dfrac{a_m}{b_n}$, unless $b_n = 0$.

5. $\lim_{x\to a} x^n = a^n$, if n is any positive or negative integer.

6. If
$$P(x) = a_0x^m + a_1x^{m-1} + \ldots + a_{m-1}x + a_m,$$
then
$$\lim_{x\to a} P(x) = P(a).$$

7. Let
$$P(x) = a_0x^m + a_1x^{m-1} + \ldots + a_{m-1}x + a_m,$$
and
$$Q(x) = b_0x^n + b_1x^{n-1} + \ldots + b_{n-1}x + b_n.$$
Then
$$\lim_{x\to a}\frac{P(x)}{Q(x)} = \frac{P(a)}{Q(a)}, \text{ if } Q(a) \neq 0.$$

8. If $\lim_{x\to a} f(x)$ exists, it is the same as $\lim_{x\to 0} f(x+a)$.

9. If
$$f(x) < g(x) \text{ for } a-h < x < a+h,$$
and
$$\lim_{x\to a} f(x) = \alpha,\quad \lim_{x\to a} g(x) = \beta,$$
then
$$\alpha \leqq \beta.$$

10. If $\lim_{x\to a} f(x) = 0$, then $\lim_{x\to a} |f(x)| = 0$, and conversely.

11. If $\lim_{x\to a} f(x) = l \gtrless 0$, then $\lim_{x\to a} |f(x)| = |l|$.
The converse does not hold.

12. Let $f(x)$ be defined as follows:
$$\left.\begin{aligned} f(x) &= x \sin 1/x, \text{ when } x \gtrless 0 \\ f(0) &= 0 \end{aligned}\right\}.$$
Then $\lim_{x\to a} f(x) = f(a)$ for all values of a.

27. $\lim_{x\to\infty} f(x)$. A precise definition of the meaning of the term "the limit of $f(x)$ when x tends to $+\infty$ (or to $-\infty$)" is also needed.

$f(x)$ is said to have the limit b as x tends to $+\infty$, if, any positive number ϵ having been chosen, as small as we please, there is a positive number X such that
$$|f(x) - b| < \epsilon, \text{ when } x \geqq X.$$

When this condition is satisfied, we write

$$\lim_{x \to +\infty} f(x) = b.$$

A similar notation, $$\lim_{x \to -\infty} f(x) = b,$$

is used when $f(x)$ has the limit b as x tends to $-\infty$, and the precise definition of the term can be obtained by substituting "a negative number $-X$" and "$x \leqq -X$" in the corresponding places in the above.

When it is clear that only positive values of x are in question, the notation $\lim_{x \to \infty} f(x)$ is used instead of $\lim_{x \to +\infty} f(x)$.

From the definition of the limit of $f(x)$ as x tends to $\pm\infty$, it follows that

$$\lim_{x \to +\infty} f(x) = b$$

carries with it $$\lim_{x \to +0} f\left(\frac{1}{x}\right) = b.$$

And, conversely, if $$\lim_{x \to +0} f(x) = b,$$

then $$\lim_{x \to +\infty} f\left(\frac{1}{x}\right) = b.$$

Similarly we have $$\lim_{x \to -\infty} f(x) = \lim_{x \to -0} f\left(\frac{1}{x}\right).$$

The modifications in the above definitions when

(i) $$\lim_{x \to +\infty} f(x) = +\infty \quad \text{or} \quad -\infty,$$

and (ii) $$\lim_{x \to -\infty} f(x) = +\infty \quad \text{or} \quad -\infty$$

will be obvious, on referring to § 25.

And oscillation, finite or infinite, as x tends to $+\infty$ or to $-\infty$, is treated as before.

28. A necessary and sufficient condition for the existence of a limit to f(x) as x tends to a. The general principle of convergence.*

A necessary and sufficient condition for the existence of a limit to $f(x)$ as x tends to a is that, when any positive number ϵ has been chosen, as small as we please, there shall be a positive number η such that $|f(x'') - f(x')| < \epsilon$ for all values of x', x'' for which

$$0 < |x'' - a| < |x' - a| \leqq \eta.$$

(i) *The condition is necessary.*

Let $\lim_{x \to a} f(x) = b$.

---

*See footnote, p. 38. Another treatment of this question is given below in §§ 29. 2 and 29. 3.

Let ϵ be a positive number, as small as we please.

Then to $\frac{1}{2}\epsilon$ there corresponds a positive number η such that

$$|f(x)-b|<\tfrac{1}{2}\epsilon, \text{ when } 0<|x-a|\leqq\eta.$$

Now let x', x'' be any two values of x satisfying

$$0<|x''-a|<|x'-a|\leqq\eta.$$

Then

$$\begin{aligned} |f(x'')-f(x')| &\leqq |f(x'')-b| + |f(x')-b| \\ &< \tfrac{1}{2}\epsilon + \tfrac{1}{2}\epsilon \\ &< \epsilon. \end{aligned}$$

(ii) *The condition is sufficient.*

Let

$$\epsilon_1, \quad \epsilon_2, \quad \epsilon_3, \ldots$$

be a sequence of positive numbers such that

$$\epsilon_{n+1}<\epsilon_n \quad \text{and} \quad \lim_{n\to\infty} \epsilon_n=0.$$

Let

$$\eta_1, \quad \eta_2, \quad \eta_3, \ldots$$

be corresponding positive numbers such that

$$\left.\begin{aligned} |f(x'')-f(x')|<\epsilon_n, \text{ when } 0<|x''-a|<|x'-a|\leqq\eta_n \\ (n=1, 2, 3, \ldots). \end{aligned}\right\} \quad \ldots\ldots(1)$$

Then, since $\epsilon_{n+1}<\epsilon_n$, we can obviously assume that $\eta_n \geqq \eta_{n+1}$.

Now take ϵ_1 and the corresponding η_1.

In the inequalities (1) put $x'=a+\eta_1$ and $x''=x$.

Then we have

$$0<|f(x)-f(a+\eta_1)|<\epsilon_1, \text{ when } 0<|x-a|<\eta_1.$$

Therefore

$$f(a+\eta_1)-\epsilon_1<f(x)<f(a+\eta_1)+\epsilon_1, \text{ when } 0<|x-a|<\eta_1. \ldots(2)$$

In Fig. 5 $f(x)$ lies within the interval A_1 of length $2\epsilon_1$, with centre at $f(a+\eta_1)$, when $0<|x-a|<\eta_1$.

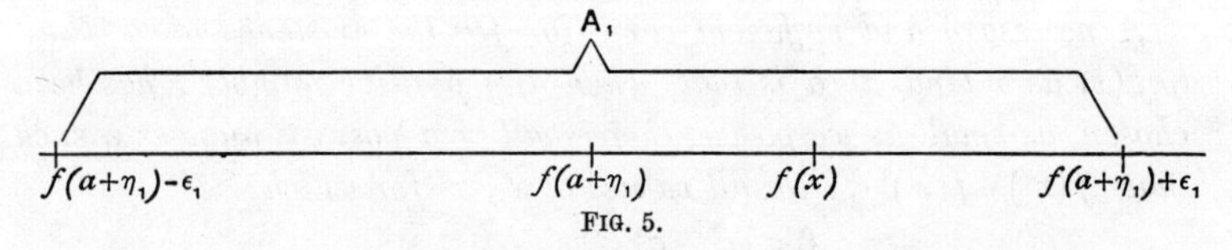

FIG. 5.

Now take ϵ_2 and the corresponding η_2, remembering that $\eta_2 \leqq \eta_1$.

We have, as above,

$$f(a+\eta_2)-\epsilon_2<f(x)<f(a+\eta_2)+\epsilon_2, \text{ when } 0<|x-a|<\eta_2. \ldots(3)$$

Since $\eta_2 \leqq \eta_1$, the interval for x in (3) cannot extend beyond the interval for x in (2), and $f(a+\eta_2)$ is in the open interval A_1.

Therefore, in Fig. 6, $f(x)$ now lies within the interval A_2, which lies entirely within A_1, or lies within it and has with it a common end-point. An overlapping part of

$$\{f(a+\eta_2)-\epsilon_2, \quad f(a+\eta_2)+\epsilon_2\}$$

could be cut off, in virtue of (2).

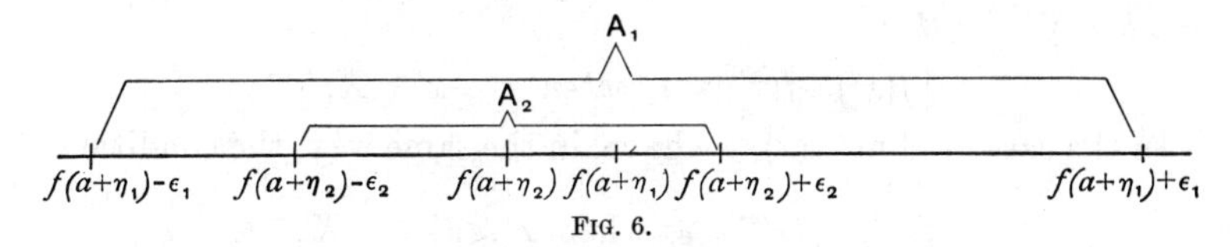

FIG. 6.

In this way we obtain a series of intervals

$$A_1, \quad A_2, \quad A_3, \dots,$$

each lying entirely within the preceding, or lying within it and having with it a common end-point; and, since the length of $A_n \leqq 2\epsilon_n$, we have $\lim_{n\to\infty} A_n = 0$, for we are given that $\lim_{n\to\infty} \epsilon_n = 0$.

If we denote the end-points of these intervals by $\alpha_1, \alpha_2, \alpha_3, \dots$ and $\beta_1, \beta_2, \beta_3, \dots$, where $\beta_n > \alpha_n$, then we know from § 18 that

$$\lim_{n\to\infty} \alpha_n = \lim_{n\to\infty} \beta_n.$$

Denote this common limit by α.

We shall now show that α is the limit of $f(x)$ as $x \to a$.

We can choose ϵ_n in the sequence $\epsilon_1, \epsilon_2, \epsilon_3, \dots$ so that $2\epsilon_n < \epsilon$, where ϵ is any given positive number.

Then we have, as above in (2) and (3),

$$\alpha_n < f(x) < \beta_n, \text{ when } 0 < |x-a| < \eta_n.$$

But $$\alpha_n \leqq \alpha \leqq \beta_n.$$

Therefore
$$\begin{aligned} |f(x)-\alpha| &< \beta_n - \alpha_n \\ &< 2\epsilon_n \\ &< \epsilon, \qquad \text{when } 0<|x-a|<\eta_n. \end{aligned}$$

It follows that $$\lim_{x\to a} f(x) = \alpha.$$

As a matter of fact, we have not obtained

$$|f(x)-\alpha| < \epsilon, \text{ when } 0 < |x-a| \leqq \eta_n$$

in the above, but when $0<|x-a|<\eta_n$.

However, we need only take η smaller than this η_n, and we obtain the inequalities used in our definition of a limit.

29. 1. In the previous section we have supposed that x tends to

a from both sides. The slight modification in the condition for convergence when it tends to a from one side only can easily be made.

Similarly, *a necessary and sufficient condition for the existence of a limit to $f(x)$ as x tends to $+\infty$, is that, if the positive number ϵ has been chosen, as small as we please, there shall be a positive number X such that*

$$|f(x'')-f(x')|<\epsilon, \textit{ when } x''>x'\geqq X.$$

In the case of $\lim_{x\to-\infty} f(x)$, we have, in the same way, the condition

$$|f(x'')-f(x')|<\epsilon, \text{ when } x''<x'\leqq -X.$$

The conditions for the existence of a limit to $f(x)$ as x tends to $+\infty$ or to $-\infty$ can, of course, be deduced from those for the existence of a limit as x tends to $+0$ or to -0.

Actually the argument given in the preceding section is simpler when we deal with $+\infty$ or $-\infty$,* and the case when the variable tends to zero from the right or left can be deduced from these two, by substituting $x=\frac{1}{u}$; when it tends to a, we must substitute $x=a+\frac{1}{u}$.

29. 2. The Upper and Lower Limits of Indetermination of the Bounded Function f(x), when x→a.

As in § 17. 2, there is some advantage to be obtained by using what are called the *upper and lower limits of indetermination* of the function at the point considered. They are defined in much the same way as in that section.

Take a sequence

$$\eta_1, \quad \eta_2, \quad \eta_3, \cdots,$$

where

$$\eta_1>\eta_2>\eta_3 \cdots \quad \text{and} \quad \lim_{n\to\infty} \eta_n=0.$$

Let M_1 be the upper bound of $f(x)$, when $0<|x-a|\leqq\eta_1$, and M_2 its upper bound, when $0<|x-a|\leqq\eta_2$, and so on.

Similarly let $m_1, m_2, \ldots$ be the corresponding lower bounds.

Then $M_1\geqq M_2\geqq M_3 \ldots \geqq m_1$, and this monotonic sequence is bounded below.

Therefore $\lim_{n\to\infty} M_n$ exists, and we denote it by Λ.

It is clear that any other sequence $\eta_1', \eta_2', \ldots$

where

$$\eta_1'>\eta_2'>\eta_3' \cdots \quad \text{and} \quad \lim_{n\to\infty} \eta_n'=0$$

will give the same limit Λ.

---

*Cf. Osgood, *Lehrbuch der Funktionentheorie*, 1 (4 Aufl., Leipzig, 1923), 33. The *general principle of convergence of* § 15 can also be established in this way.

To the arbitrary positive number ϵ, there corresponds a positive integer ν, such that

$$\Lambda \leqq M_n < \Lambda + \epsilon, \text{ when } n \geqq \nu.$$

Thus $$f(x) \leqq M_\nu < \Lambda + \epsilon, \text{ when } 0 < |x-a| \leqq \eta_\nu.$$

Now take a positive number $\alpha < \eta_\nu$, and let $\eta_N \leqq \alpha$.

Then $$\Lambda \leqq M_N \leqq M_\nu.$$

And $f(x) > M_N - \epsilon$ for at least one value x_N in $0 < |x-a| \leqq \eta_N$.

Therefore $f(x) > \Lambda - \epsilon$ for at least one value x_N in $0 < |x-a| \leqq \alpha$.

If we now take $\beta < |x_N - a|$ and proceed in the same way, we see that $f(x) > \Lambda - \epsilon$ for at least one value $x_{N'}$ in $0 < |x-a| \leqq \beta$.

It is thus clear that when $0 < |x-a| \leqq \alpha$, there are an infinite number of points at which $f(x) > \Lambda - \epsilon$.

This number Λ is called the *upper limit of indetermination* of $f(x)$, when $x \to a$, and we write $\Lambda = \overline{\lim}_{x \to a} f(x)$.

If ϵ is an arbitrary positive number, there is a neighbourhood $0 < |x-a| \leqq \eta$, such that for every point of this neighbourhood $f(x) < \Lambda + \epsilon$; and in every neighbourhood of a, however small, there is a point (other than a) at which $f(x) > \Lambda - \epsilon$.

The *lower limit of indetermination* λ of $f(x)$, when $x \to a$, is obtained in a similar way, and we write $\lambda = \underline{\lim}_{x \to a} f(x)$.

If ϵ is an arbitrary positive number, there is a neighbourhood $0 < |x-a| \leqq \eta$, such that for every point of this neighbourhood $f(x) > \lambda - \epsilon$; and in every neighbourhood of a, however small, there is a point (other than a) at which $f(x) < \lambda + \epsilon$.

It is clear that $$\overline{\lim}_{x \to a} f(x) \geqq \underline{\lim}_{x \to a} f(x).$$

These definitions may also be extended to the case when we approach a from the right-hand or the left-hand. In this way we have $\overline{\lim}_{x \to a+0} f(x)$ and $\underline{\lim}_{x \to a+0} f(x)$, which are conveniently written $\overline{f(a+0)}$ and $\overline{f(a-0)}$. Similarly for $\overline{\lim}_{x \to a-0} f(x)$ and $\overline{\lim}_{x \to a-0} f(x)$, which are written $\overline{f(a-0)}$ and $\underline{f(a-0)}$.

29. 3. The following theorems are obtained at once (cf. § 17. 4).

(i) *If* $\lim_{x \to a} f(x)$ *exists, then* $\overline{\lim}_{x \to a} f(x) = \underline{\lim}_{x \to a} f(x) = \lim_{x \to a} f(x)$.

(ii) *Conversely, if* $\overline{\lim}_{x \to a} f(x) = \underline{\lim}_{x \to a} f(x)$, *then* $\lim_{x \to a} f(x)$ *exists and is equal to their common value.*

(iii) The general principle of convergence for $\lim f(x)$, when $x \to a$:

A necessary and sufficient condition for the existence of a limit to $f(x)$ as x tends to a is that, when any positive number ϵ has been chosen, as small as we please, there shall be a positive number η such that $|f(x'') - f(x')| < \epsilon$ for all values of x', x'' for which $0 < |x''-a| < |x'-a| \leqq \eta$.

When the upper and lower limits of indetermination are used the proof that this is a sufficient condition for the convergence of $f(x)$ as $x \to a$ is much shorter and simpler than that given in § 28. See also § 17. 4 (iii).

29. 4. The Oscillation of a Bounded Function at a Point. In § 24 we have defined the *oscillation of a function in an interval.* We now define the *oscillation at a point.*

Let a be a point in the interval in which $f(x)$ is given, and $\eta_1, \eta_2, \eta_3, \ldots$ a sequence, where $\eta_1 > \eta_2 > \eta_3 \ldots$ and $\lim_{n \to \infty} \eta_n = 0$. For any positive integer n let M_n and m_n be the upper and lower bounds of $f(x)$ in the neighbourhood of $x = a$, defined by $|a - x| \leqq \eta_n$, the point a itself now being a point of the neighbourhood.

Then as in § 29. 2, we have

$$M_1 \geqq M_2 \geqq M_3 \ldots \geqq m_1,$$
$$m_1 \leqq m_2 \leqq m_3 \ldots \leqq M_1.$$

Also $\lim_{n \to \infty} M_n$ and $\lim_{n \to \infty} m_n$ exist and are independent of the particular sequence $\eta_1, \eta_2, \eta_3 \ldots$ chosen.

*If these limits are M and m respectively, then $(M - m)$ is called the oscillation of $f(x)$ at the point a.**

It is clear that the oscillation at a is the limit of the oscillation of the function in the interval $a - \eta \leqq x \leqq a + \eta$, as $\eta \to 0$.

Also the oscillation at a is the difference between (i) the greater of $f(a)$ and $\overline{\lim_{x \to a}} f(x)$ and (ii) the smaller of $f(a)$ and $\underline{\lim_{x \to a}} f(x)$.

At a point where $f(x)$ is continuous, the oscillation is zero, and at any other point it is different from zero.

If the oscillation at $x = a$ is k, then in every neighbourhood $a - \eta \leqq x \leqq a + \eta$, the oscillation of $f(x)$ is greater than or equal to k.

30. Continuous Functions. The function $f(x)$ is said to be continuous when $x = x_0$, if $f(x)$ has a limit as x tends to x_0 from either side, and each of these limits is equal to $f(x_0)$.

Thus $f(x)$ *is continuous when $x = x_0$, if, to the arbitrary positive number ϵ, there corresponds a positive number η such that*

$$|f(x) - f(x_0)| < \epsilon, \text{ when } |x - x_0| \leqq \eta.$$

When $f(x)$ is defined in an interval (a, b), we shall say that *it is continuous in the interval (a, b), if it is continuous for every value of x between a and b $(a < x < b)$, and if $f(a+0)$ exists and is equal to $f(a)$, and $f(b-0)$ exists and is equal to $f(b)$.*

In such cases it is convenient to make a slight change in our definition of continuity at a point, and to say that $f(x)$ is continuous at the end-points a and b when these conditions are satisfied.

It follows from the definition of continuity that the sum or

*It may be noted that Hobson (*loc. cit.* **1** (3rd ed., 1927), 300) uses the term *saltus*.

product of any number of functions, which are continuous at a point, is also continuous at that point. The same holds for the quotient of two functions, continuous at a point, unless the denominator vanishes at that point (cf. § 26). A continuous function of a continuous function is also a continuous function (cf. § 26 (IV)).

The polynomial

$$P(x) = a_0x^n + a_1x^{n-1} + \ldots + a_{n-1}x + a_n$$

is continuous for all values of x.

The rational function

$$R(x) = P(x)/Q(x)$$

is continuous in any interval which does not include values of x making the polynomial $Q(x)$ zero.

The functions $\sin x$, $\cos x$, $\tan x$, *etc.* and the corresponding functions $\sin^{-1}x$, $\cos^{-1}x$, $\tan^{-1}x$, *etc.* are continuous except, in certain cases, at particular points.

e^x is continuous everywhere; $\log x$ is continuous for the interval $x > 0$.

31. 1. Properties of Continuous Functions.* We shall now prove several important theorems on continuous functions, to which reference will frequently be made later. It will be seen that in these proofs we rely only on the definition of continuity and the results obtained in the previous pages.

THEOREM I. *Let $f(x)$ be continuous in the interval (a, b)†, and let the positive number ϵ be chosen, as small as we please. Then the interval (a, b) can always be broken up into a finite number of partial intervals, such that $|f(x') - f(x'')| < \epsilon$, when x' and x'' are any two points in‡ the same partial interval.*

Let us suppose that this is not true. Then let $c = \frac{1}{2}(a+b)$. At least one of the intervals (a, c), (c, b) must be such that it is impossible to break it up into a finite number of partial intervals which satisfy the condition named in the theorem. Denote by (a_1, b_1) this new interval, which is half of (a, b). Operating on

*This section follows closely the treatment given by Goursat, *loc. cit.*, 1 (4e éd., 1923), § 8.

†In these theorems the continuity of $f(x)$ is supposed given in the closed interval $(a \leq x \leq b)$, as explained in § 30.

‡ *i.e.* in or at the ends of the partial interval.

(a_1, b_1) in the same way as we have done with (a, b), and then proceeding as before, we obtain an infinite set of intervals such as we have met in the theorem of § 18. The sequence of end-points $a, a_1, a_2, \ldots$ converges, and the sequence of end-points $b, b_1, b_2, \ldots$ also converges, the limit of each being the same, say α. Also each of the intervals (a_n, b_n) has the property we have ascribed to the original interval (a, b). It is impossible to break it up into a finite number of partial intervals which satisfy the condition named in the theorem.

Let us suppose that α does not coincide with a or b. Since the function $f(x)$ is continuous when $x=\alpha$, we know that there is a positive number η such that $|f(x)-f(\alpha)|<\frac{1}{2}\epsilon$ when $|x-\alpha| \leqq \eta$. Let us choose n so large that $(b_n - a_n)$ is less than η. Then the interval (a_n, b_n) is contained entirely within $(\alpha-\eta, \alpha+\eta)$, for we know that $a_n \leqq \alpha \leqq b_n$. Therefore, if x' and x'' are any two points in the interval (a_n, b_n), it follows from the above that

$$|f(x')-f(\alpha)|<\tfrac{1}{2}\epsilon \quad \text{and} \quad |f(x'')-f(\alpha)|<\tfrac{1}{2}\epsilon.$$

But $$|f(x')-f(x'')| \leqq |f(x')-f(\alpha)| + |f(x'')-f(\alpha)|.$$

Thus we have $$|f(x')-f(x'')|<\epsilon,$$

and our hypothesis leads to a contradiction.

There remains the possibility that α might coincide with either a or b. The slight modification required in the above argument is obvious.

Hence the assumption that the theorem is untrue leads in every case to a contradiction, and its truth is established.

COROLLARY I. Let $a, x_1, x_2, \ldots x_{n-1}, b$ be a mode of subdivision of (a, b) into partial intervals satisfying the conditions of Theorem I.

Then

$$\begin{aligned} |f(x)| &\leqq |f(a)| + |f(x)-f(a)| \\ &< |f(a)| + \epsilon, \qquad \text{when } 0<(x-a) \leqq (x_1-a). \end{aligned}$$

Therefore

$$|f(x_1)| < |f(a)| + \epsilon.$$

In the same way

$$\begin{aligned} |f(x)| &\leqq |f(x_1)| + |f(x)-f(x_1)| \\ &< |f(x_1)| + \epsilon, \qquad \text{when } 0<(x-x_1) \leqq (x_2-x_1) \\ &< |f(a)| + 2\epsilon, \qquad \text{when } 0<(x-x_1) \leqq (x_2-x_1). \end{aligned}$$

Therefore

$$|f(x_2)| < |f(a)| + 2\epsilon.$$

Proceeding in the same way for each successive partial interval we obtain from the nth interval

$$|f(x)| < |f(a)| + n\epsilon, \text{ when } 0 < (x - x_{n-1}) \leqq (b - x_{n-1}).$$

Thus we see that in the whole interval (a, b)

$$|f(x)| < |f(a)| + n\epsilon.$$

It follows that *a function which is continuous in a given interval is bounded in that interval.*

COROLLARY II. Let us suppose the interval (a, b) divided up into n partial intervals (a, x_1), (x_1, x_2), ... (x_{n-1}, b), such that $|f(x') - f(x'')| < \frac{1}{2}\epsilon$ for any two points in the same partial interval. Let η be a positive number smaller than the least of the numbers $(x_1 - a)$, $(x_2 - x_1)$, ... $(b - x_{n-1})$. Now take any two points x' and x'' in the interval (a, b), such that $|x' - x''| \leqq \eta$. If these two points belong to the same partial interval, we have

$$|f(x') - f(x'')| < \tfrac{1}{2}\epsilon.$$

On the other hand, if they do not belong to the same partial interval, they must lie in two consecutive partial intervals. In this case it is clear that $|f(x') - f(x'')| < \frac{1}{2}\epsilon + \frac{1}{2}\epsilon = \epsilon$.

Hence, *the positive number ϵ having been chosen, as small as we please, there is a positive number η such that $|f(x') - f(x'')| < \epsilon$, when x', x'' are any two values of x in the interval (a, b) for which $|x' - x''| \leqq \eta$.*

We started with the assumption that $f(x)$ was continuous in (a, b). It follows from this assumption that if x is any point in this interval, and ϵ any arbitrary positive number, then there is a positive number η such that

$$|f(x') - f(x)| < \epsilon, \text{ when } |x' - x| \leqq \eta.$$

To begin with, we have no justification for supposing that the same η could do for all values of x in the interval. But the theorem proved in this corollary establishes that this is the case. This result is usually expressed by saying that $f(x)$ is *uniformly continuous* in the interval (a, b).

We have thus shown that *a function which is continuous in an interval is also uniformly continuous in the interval.*

THEOREM II. *If $f(a)$ and $f(b)$ are unequal and $f(x)$ is continuous in the interval (a, b), as x passes from a to b, $f(x)$ takes at least once every value between $f(a)$ and $f(b)$.*

First, let us suppose that $f(a)$ and $f(b)$ have different signs, *e.g.* $f(a)<0$ and $f(b)>0$. We shall show that for at least one value of x between a and b, $f(x)=0$.

From the continuity of $f(x)$, we see that it is negative in the neighbourhood of a and positive in the neighbourhood of b. Consider the set of values of x between a and b which makes $f(x)$ positive. Let λ be the lower bound of this aggregate. Then $a<\lambda<b$. From the definition of the lower bound $f(x)$ is negative or zero in $a \leqq x<\lambda$. But $\lim_{x\to\lambda-0} f(x)$ exists and is equal to $f(\lambda)$. Therefore $f(\lambda)$ is also negative or zero. But $f(\lambda)$ cannot be negative. For if $f(\lambda)=-m$, m being a positive number, then there is a positive number η such that

$$|f(x)-f(\lambda)|<m, \text{ when } |x-\lambda| \leqq \eta,$$

since $f(x)$ is continuous when $x=\lambda$. The function $f(x)$ would then be negative for the values of x in (a, b) between λ and $\lambda+\eta$, and λ would not be the lower bound of the above aggregate. We must therefore have $f(\lambda)=0$.

Now let N be any number between $f(a)$ and $f(b)$, which may be of the same or different signs. The continuous function $\phi(x)=f(x)-N$ has opposite signs when $x=a$ and $x=b$. By the case we have just discussed, $\phi(x)$ vanishes for at least one value of x between a and b, *i.e.* in the *open* interval (a, b).

Thus our theorem is established.

Again, if $f(x)$ is continuous in (a, b), we know from Corollary I above that it is bounded in that interval. In the next theorem we show that it attains these bounds.

THEOREM III. *If $f(x)$ is continuous in the interval (a, b), and M, m are its upper and lower bounds, then $f(x)$ takes the value M and the value m at least once in the interval.*

We shall show first that $f(x)=M$ at least once in the interval.

Let $c=\frac{1}{2}(a+b)$; the upper bound of $f(x)$ is equal to M, for at least one of the intervals (a, c), (c, b). Replacing (a, b) by this interval we bisect it, and proceed as before. In this way, as in Theorem I, we obtain an infinite set of intervals (a, b), (a_1, b_1),

(a_2, b_2), ... tending to zero in the limit, each lying entirely within the preceding, or lying within it and having with it a common end-point, the upper bound of $f(x)$ in each being M.

Let λ be the common limit of the sequences $a, a_1, a_2, \ldots$ and $b, b_1, b_2, \ldots$. We shall show that $f(\lambda)=M$.

For suppose $f(\lambda)=M-h$, where $h>0$. Since $f(x)$ is continuous at $x=\lambda$, there is a positive number η such that

$$|f(x)-f(\lambda)|<\tfrac{1}{2}h, \text{ when } |x-\lambda| \leqq \eta.$$

Thus $f(x)<M-\tfrac{1}{2}h$, when $|x-\lambda| \leqq \eta$.

Now take n so large that (b_n-a_n) will be less than η. The interval (a_n, b_n) will be contained wholly within $(\lambda-\eta, \lambda+\eta)$. The upper bound of $f(x)$ in the interval (a_n, b_n) would then be different from M, contrary to our hypothesis.

Combining this theorem with the preceding we obtain the following additional result:

THEOREM IV. *If $f(x)$ is continuous in the interval (a, b), and M, m are its upper and lower bounds, then it takes at least once in this interval the values M, m, and every value between M and m.*

Also, since the *oscillation* of a function in an interval was defined as the difference between its upper and lower bounds (cf. § 24), and since the function attains its bounds at least once in the interval, we can state Theorem I afresh as follows:

If $f(x)$ is continuous in the interval (a, b), then we can divide (a, b) into a finite number of partial intervals

$$(a, x_1), \quad (x_1, x_2), \ldots (x_{n-1}, b),$$

*in each of which the oscillation of $f(x)$ is less than any given positive number.**

And a similar change can be made in the statement of the property known as uniform continuity.

31. 2. The Heine-Borel Theorem. *Let an interval (a, b) and an infinite set Δ of intervals, all in (a, b), be given such that every point x of $a \leqq x \leqq b$ is an interior point of at least one of the intervals of Δ. (The ends a and b being regarded as interior to an interval of the set, when a is the left-hand end of an interval, and b the right-hand end of another interval.)*

Then a set consisting of a finite number of the intervals of Δ has the same

*The argument of Theorem I, adapted to this case, leads to the theorem : *If the oscillation at every point of $a \leqq x \leqq b$ is less than a given number k, then the interval can be divided up into a finite number of partial intervals in each of which the oscillation is less than k.*

property; namely, every point of the closed interval (a, b) *is an interior point of at least one of the intervals of this finite set (with the same convention as to the ends* a *and* b*).*

Let us suppose that this is not true. Then let $c=\frac{1}{2}(a+b)$.

At least one of the closed intervals (a, c) and (c, b) must be such that all its points are not interior points of at least one interval of a finite set of intervals of the set Δ.

Denote by (a_1, b_1) this new interval, which is half of (a, b).

Operating on (a_1, b_1) as we have done on (a, b), and then proceeding as before, we obtain an infinite set of closed intervals (a, b), (a_1, b_1), (a_2, b_2) ... such that their ends $a, a_1, a_2, \ldots$ form a bounded monotonic ascending sequence (or from and after some value of n are all identical), and their ends $b, b_1, b_2, \ldots$ form a bounded monotonic descending sequence (or from and after some value of n are all identical).

Thus $\lim_{n\to\infty} a_n$ and $\lim_{n\to\infty} b_n$ exist, and they are equal, since $b_n - a_n = \dfrac{1}{2^n}(b-a)$.

Suppose that $\lim_{n\to\infty} a_n = \alpha$, different from a and b.

Then α is an interior point of one of the intervals of Δ, say (a', b'). And by taking n large enough, we can bring a_n and b_n inside (a', b'). Then all the points of this interval (a_n, b_n) are interior points of one of the intervals of Δ, contrary to our hypothesis.

A similar argument applies to the case when α coincides with a or b.

Thus our theorem is proved.

Special cases of the Heine-Borel Theorem are the first theorem of § 31. 1 and the theorem stated in the footnote on p. 71.

It can be at once extended to the case of a rectangular domain instead of the linear interval; and is equally useful in dealing with the properties of functions of two variables (Cf. § 37). Indeed the proof is independent of the number of dimensions. And it finds a place also in the general theory of sets of points.*

The title Heine-Borel Theorem is so generally used by English writers that it has been adopted in the text. But German and French mathematicians now refer to it as Borel's Theorem, and there is no doubt that this is the better name. The theorem (for the case of a countably infinite set of intervals) was first enunciated and proved by Borel. [Thèse, Paris, 1894; *Annales Sci. de l'École Normale* (3), 12 (1895), 51: *Leçons sur la théorie des fonctions* (Paris, 1898).] Because of the similarity of Borel's proof and that by means of which Heine † established the uniformity of the continuity of a function, given as continuous in a closed interval, it became customary to call it the Heine-Borel Theorem. But it may well be the case that the theorem is contained implicitly in similar demonstrations by authors previous to Heine. And, as Lebesgue remarks, the theorem is not one of those of which the demonstration

---

* Cf. Hobson, *loc. cit.* 1 (3rd ed., 1927), § 73.

† *Journal für Math.*, 74 (1872), 188.

offers great difficulties.* The merit lay in perceiving it, enunciating it, and divining its interest, not in demonstrating it. He refers to it always as Borel's Theorem, and regards the other title as unsuitable.

32. Continuity in an Infinite Interval. Some of the results of the last section can be extended to the case when $f(x)$ is continuous in $x \geqq a$, where a is some definite positive number, and $\lim_{x\to\infty} f(x)$ exists.

Let $u = a/x$. When $x \geqq a$, we have $0 < u \leqq 1$.

With the values of u in $0 < u \leqq 1$, associate the values of $f(x)$ at the corresponding points in $x \geqq a$, and to $u=0$ assign $\lim_{x\to\infty} f(x)$.

We thus obtain a function of u, which is continuous in the closed interval $(0, 1)$.

Therefore it is bounded in this interval, and attains its bounds M, m. Also it takes at least once every value between M and m, as u passes over the interval $(0, 1)$.

Thus we may say that $f(x)$ is bounded in the range† given by $x \geqq a$ and the new "point" $x = \infty$, at which $f(x)$ is given the value $\lim_{x\to\infty} f(x)$.

Also $f(x)$ takes at least once in this range its upper and lower bounds, and every value between these bounds.

For example, the function $\dfrac{x^2}{a^2+x^2}$ is continuous in $(0, \infty)$. It does not attain its upper bound—unity—when $x \geqq 0$, but it takes this value when $x = \infty$, as defined above.

33. Discontinuous Functions. When $f(x)$ is defined for x_0 and the neighbourhood of x_0 (*e.g.* $0 < |x - x_0| \leqq h$), and $f(x_0+0) = f(x_0-0) = f(x_0)$, then $f(x)$ is continuous at x_0.

On the other hand, when $f(x)$ is defined for the neighbourhood of x_0, and it may be also for x_0, while $f(x)$ is not continuous at x_0, it is natural to say that $f(x)$ is *discontinuous* at x_0, and to call x_0 a *point of discontinuity* of $f(x)$.

Points of discontinuity may be classified as follows:

I. $f(x_0+0)$ and $f(x_0-0)$ may exist and be equal. If their common value is different from $f(x_0)$, or if $f(x)$ is not defined for x_0, then we have a point of discontinuity there.

Ex. $\qquad f(x) = (x - x_0) \sin 1/(x - x_0)$, when $x \lessgtr x_0$.

Here $f(x_0+0) = f(x_0-0) = 0$, and if we give $f(x)_0$ any value other than zero, or if we leave $f(x_0)$ undefined, x_0 is a point of discontinuity of $f(x)$.

---

* In a review of W. H. and G. C. Young's "Theory of Sets of Points" in the *Bull. des sciences math.* (2) 31 (1907), 134.

†It is convenient to speak of this range as the interval (a, ∞), and to write $f(\infty)$ for $\lim_{x\to\infty} f(x)$.

II. $f(x_0+0)$ and $f(x_0-0)$ may exist and be unequal. Then x_0 is a point of discontinuity of $f(x)$, whether $f(x_0)$ is defined or not.

Ex. $$f(x)=\frac{1}{1-e^{1/(x-x_0)}}, \text{ when } x \gtrless x_0.$$

Here $$f(x_0+0)=0 \text{ and } f(x_0-0)=1.$$

In both these cases $f(x)$ is said to have an *ordinary* or *simple* discontinuity at x_0. And the same term is applied when the point x_0 is an end-point of the interval in which $f(x)$ is given, and $f(x_0+0)$, or $f(x_0-0)$, exists and is different from $f(x_0)$, if $f(x)$ is defined for x_0.

III. $f(x)$ may have the limit $+\infty$, or $-\infty$, as $x \to x_0$ on either side, and it may oscillate on one side or the other. Take in this section the cases in which there is no oscillation. These may be arranged as follows:

(i) $f(x_0+0)=f(x_0-0)=+\infty$ (or $-\infty$).

Ex. $$f(x)=1/(x-x_0)^2, \text{ when } x \gtrless x_0.$$

(ii) $f(x_0+0)=+\infty$ (or $-\infty$) and $f(x_0-0)=-\infty$ (or $+\infty$).

Ex. $$f(x)=1/(x-x_0), \text{ when } x \lesseqgtr x_0.$$

(iii) $\left.\begin{matrix} f(x_0+0)=+\infty \text{ (or } -\infty) \\ f(x_0-0) \text{ exists} \end{matrix}\right\}$ or $\left.\begin{matrix} f(x_0-0)=+\infty \text{ (or } -\infty) \\ f(x_0+0) \text{ exists} \end{matrix}\right\}.$

Ex. $$\left.\begin{matrix} f(x)=1/(x-x_0), & \text{when } x > x_0 \\ f(x)=x-x_0, & \text{when } x \leqq x_0 \end{matrix}\right\}.$$

In these cases we say that the point x_0 is *an infinity* of $f(x)$, and the same term is used when x_0 is an end-point of the interval in which $f(x)$ is given, and $f(x_0+0)$, or $f(x_0-0)$, is $+\infty$ or $-\infty$.

It is usual to say that $f(x)$ *becomes infinite* at a point x_0 of the kind given in (i), and that $f(x_0)=+\infty$ (or $-\infty$). But this must be regarded as simply a short way of expressing the fact that $f(x)$ diverges to $+\infty$ (or to $-\infty$) as $x \to x_0$.

It will be noticed that $\tan x$ has an infinity at $\frac{1}{2}\pi$, but that $\tan \frac{1}{2}\pi$ is not defined. On the other hand,

$$\tan(\tfrac{1}{2}\pi-0)=+\infty \quad \text{and} \quad \tan(\tfrac{1}{2}\pi+0)=-\infty.$$

IV. When $f(x)$ oscillates at x_0 on one side or the other, x_0 is said to be a *point of oscillatory discontinuity*. The oscillation is *finite* when $f(x)$ is bounded in some neighbourhood of x_0; it is *infinite* when there is no neighbourhood of x_0 in which $f(x)$ is bounded (cf. § 25).

Ex. (i) $f(x) = \sin 1/(x - x_0)$, when $x \gtrless x_0$.

(ii) $f(x) = 1/(x - x_0) \sin 1/(x - x_0)$, when $x \gtrless x_0$.

In both these examples x_0 is a point of oscillatory discontinuity. The first oscillates finitely at x_0, the second oscillates infinitely. The same remark would apply if the function had been given only for one side of x_0.

The *infinities* defined in (III) and the points at which $f(x)$ *oscillates infinitely* are said to be *points of infinite discontinuity.*

34. Monotonic Functions. *The function $f(x)$, given in the interval (a, b), is said to be monotonic in that interval if*

$$\textit{either (i) } f(x') \leqq f(x''), \textit{ when } a \leqq x' < x'' \leqq b;$$

$$\textit{or (ii) } f(x') \geqq f(x''), \textit{ when } a \leqq x' < x'' \leqq b.$$

In the first case, the function never decreases as x increases and it is said to be *monotonic increasing*; in the second case, it never increases as x increases, and it is said to be *monotonic decreasing.**

The monotonic character of the function may fail at the end-points of the interval, and in this case it is said to be *monotonic in the open interval.*

The properties of monotonic functions are very similar to those of monotonic sequences, treated in § 17.1, and they may be established in precisely the same way:

(i) *If $f(x)$ is monotonic increasing when $x \geqq X$, and $f(x)$ is less than some fixed number A when $x \geqq X$, then $\lim_{x \to +\infty} f(x)$ exists and is less than or equal to A.*

(ii) *If $f(x)$ is monotonic increasing when $x \leqq X$, and $f(x)$ is greater than some fixed number A when $x \leqq X$, then $\lim_{x \to -\infty} f(x)$ exists and is greater than or equal to A.*

(iii) *If $f(x)$ is monotonic increasing in an open interval (a, b), and $f(x)$ is greater than some fixed number A in that open interval, then $f(a + 0)$ exists and is greater than or equal to A.*

(iv) *If $f(x)$ is monotonic increasing in an open interval (a, b), and $f(x)$ is less than some fixed number A in that open interval, then $f(b - 0)$ exists and is less than or equal to A.*

---

*The footnote, p. 43, also applies here.

These results can be readily adapted to the case of monotonic decreasing functions, and it follows at once from (iii) and (iv) that *if $f(x)$ is bounded and monotonic in an open interval, it can only have ordinary discontinuities in that interval, or at its ends.*

It may be worth observing that if $f(x)$ is monotonic in a closed interval, the same result follows, but that if we are only given that it is monotonic in an open interval, and not told that it is bounded, the function may have an infinity at either end.

E.g. $f(x)=1/x$ is monotonic in the open interval (0, 1), but not bounded.

At first one might be inclined to think that a function which is bounded and monotonic in an interval can have only a finite number of points of discontinuity in that interval.

The following example shows that this is not the case:

$$\text{Let } f(x)=1, \quad \text{when } \tfrac{1}{2}<x\leqq 1;$$

$$\text{let } f(x)=\tfrac{1}{2}, \quad \text{when } \frac{1}{2^2}<x\leqq \tfrac{1}{2};$$

and, in general,

$$\text{let } f(x)=\frac{1}{2^n}, \quad \text{when } \frac{1}{2^{n+1}}<x\leqq\frac{1}{2^n},$$

(n being any positive integer).

Also let $f(0)=0$.

Then $f(x)$ is monotonic in the interval (0, 1).

This function has an infinite number of points of discontinuity, namely at $x=\dfrac{1}{2^n}$ (n being any positive integer).

Obviously there can only be a finite number of points of discontinuity at which the jump would be greater than or equal to k, where k is any fixed positive number, if the function is monotonic (and bounded) in an interval.*

35. Inverse Functions. Let the function $f(x)$, defined in the interval (a, b), be continuous and monotonic in the *stricter sense*† in (a, b).

For example, let $y=f(x)$ be continuous and continually increase from A to B as x passes from a to b.

Then to every value of y in (A, B) there corresponds one and only one value of x in (a, b). [§ 31. 1, Theorem II.]

*It follows that if there are an infinite number of points η discontinuity, this set of points is countably infinite.

†Cf. footnote, p. 75.

This value of x is a function of y—say $\phi(y)$—which is itself continually increasing in the interval (A, B).

The function $\phi(y)$ is called the *inverse* of the function $f(x)$.

We shall now show that *$\phi(y)$ is a continuous function of y* in the interval in which it is defined.

For let y_0 be any number between A and B, and x_0 the corresponding value of x. Also let ϵ be an arbitrary positive number such that $x_0 - \epsilon$ and $x_0 + \epsilon$ lie in (a, b) (Fig. 7).

Let $y_0 - \eta_1$ and $y_0 + \eta_2$ be the corresponding values of y.

Then, if the positive number η is less than the smaller of η_1 and η_2, it is clear that

$$|x - x_0| < \epsilon, \text{ when } |y - y_0| \leqq \eta.$$

Therefore $$|\phi(y) - \phi(y_0)| < \epsilon, \text{ when } |y - y_0| \leqq \eta.$$

Thus $\phi(y)$ is continuous at y_0.

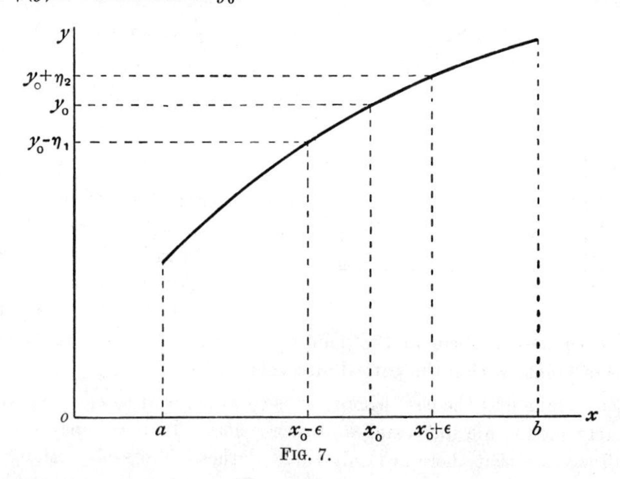

FIG. 7.

A similar proof applies to the end-points A and B, and it is obvious that the same argument applies to a function which is monotonic in the *stricter sense* and *decreasing*.

The functions $\sin^{-1}x$, $\cos^{-1}x$, etc., thus arise as the inverses of the functions $\sin x$ and $\cos x$, where $0 \leqq x \leqq \frac{1}{2}\pi$, and so on.

In the first place these appear as functions of y, namely $\sin^{-1}y$, where $0 \leqq y \leqq 1$, $\cos^{-1}y$, where $0 \leqq y \leqq 1$, etc. The symbol y is then replaced by the usual symbol x for the independent variable.

In the same way $\log x$ appears as the inverse of the function e^x.

There is a simple rule for obtaining the graph of the inverse function $f^{-1}(x)$ when the graph of $f(x)$ is known. *$f^{-1}(x)$ is the image of $f(x)$ in the line $y = x$.* The proof of this may be left to the reader.

The following theorem may be compared with that of § 26 (IV):

Let $f(u)$ be a continuous function, monotonic in the stricter sense, and let

$$\lim_{x \to a} f[\phi(x)] = f(b)$$

Then $\lim_{x \to a} \phi(x)$ *exists and is equal to* b.

A strict proof of this result may be obtained, relying on the property proved above, that the inverse of the function $f(u)$ is a continuous function.

The theorem is almost intuitive, if we are permitted to use the graph of $f(u)$. The reader is familiar with its application in finding certain limiting values, where logarithms are taken. In these cases it is shown that $\lim \log u = \log b$, and it is inferred that $\lim u = b$.*

36. 1.† Let the bounded function $f(x)$, given in the interval (a, b), be such that this interval can be divided up into a finite number of open partial intervals, in each of which $f(x)$ is monotonic; or, in accordance with the more usual but less exact expression, let the function have only a finite number of maxima and minima in the interval.

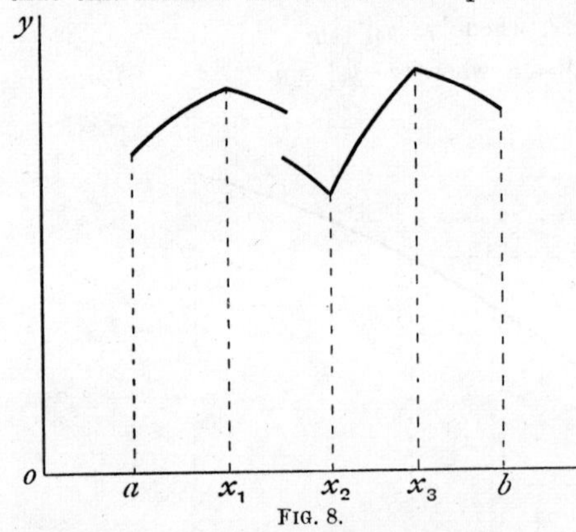

FIG. 8.

Suppose that the points $x_1, x_2, \ldots x_{n-1}$ divide this interval into the open intervals $(a, x_1), (x_1, x_2), \ldots (x_{n-1}, b)$, in which $f(x)$ is monotonic. Then we know that $f(x)$ can only have ordinary discontinuities, which can occur at the points $a, x_1, x_2, \ldots x_{n-1}, b$, and also at any number of points within the partial intervals. (§ 34.)

I. Let us take first the case where $f(x)$ is continuous at $a, x_1, \ldots x_{n-1}, b$, and alternately monotonic increasing and decreasing. To make matters clearer, we shall assume that there are only three of these points of section, namely x_1, x_2, x_3, $f(x)$ being monotonic *increasing* in the first interval (a, x_1), *decreasing* in the second (x_1, x_2), and so on (Fig. 8).

It is obvious that the intervals may in this case be regarded as closed, the monotonic character of $f(x)$ extending to the ends of each.

Consider the functions $F(x)$, $G(x)$ given by the following scheme:

$F(x)$	$G(x)$	
$f(x)$	0	$a \leqq x \leqq x_1$
$f(x_1)$	$f(x_1) - f(x)$	$x_1 \leqq x \leqq x_2$
$f(x_1) - f(x_2) + f(x)$	$f(x_1) - f(x_2)$	$x_2 \leqq x \leqq x_3$
$f(x_1) - f(x_2) + f(x_3)$	$f(x_1) - f(x_2) + f(x_3) - f(x)$	$x_3 \leqq x \leqq b$

*Cf. Hobson, *Plane Trigonometry* (7th ed., 1928), p. 130.

†For an alternative treatment of the subject matter of this section, see **§ 36. 2.**

It is clear that $F(x)$ and $G(x)$ are *monotonic increasing* in the closed interval (a, b), and that $f(x) = F(x) - G(x)$ in (a, b).
If $f(x)$ is decreasing in the first partial interval, we start with

$$F(x) = f(a), \quad G(x) = f(a) - f(x), \quad \text{when } a \leqq x \leqq x_1,$$

and proceed as before, *i.e.* we begin with the second line of the above diagram, and substitute a for x_1, etc.

Also, since the function $f(x)$ is bounded in (a, b), by adding some number to both $F(x)$ and $G(x)$, we can make both these functions positive if, in the original discussion, one or both were negative.

It is clear that the process outlined above applies equally well to n partial intervals.

We have thus shown that *when the bounded function $f(x)$, given in the interval (a, b), is such that this interval can be divided up into a finite number of partial intervals, $f(x)$ being alternately monotonic increasing and monotonic decreasing in these intervals, and continuous at their ends, then we can express $f(x)$ as the difference of two (bounded) functions, which are positive and monotonic increasing in the interval (a, b).*

II. There remains the case when some or all of the points $a, x_1, x_2, \ldots x_{n-1}, b$ are points of discontinuity of $f(x)$, and the proviso that the function is alternately monotonic increasing and decreasing is dropped.

We can obtain from $f(x)$ a new function $\phi(x)$, with the same monotonic properties as $f(x)$ in the open partial intervals $(a, x_1), (x_1, x_2), \ldots (x_{n-1}, b)$, but continuous at their ends.

The process is obvious from Fig. 9. We need only keep the first part of the curve fixed, move the second up or down till its end-point $(x_1, f(x_1+0))$

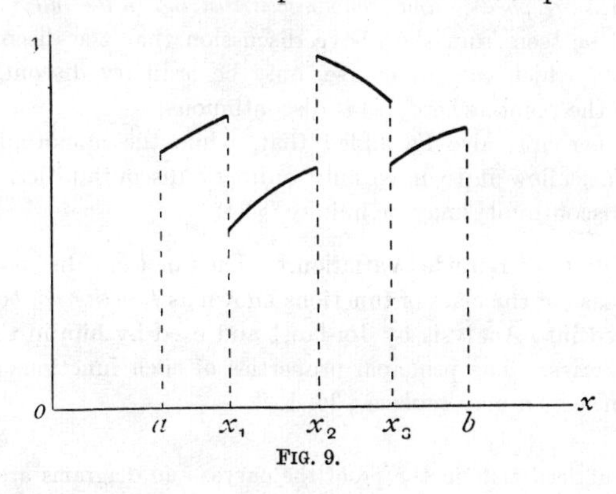

FIG. 9.

coincides with $(x_1, f(x_1-0))$, then proceed to the third curve and move it up or down to the new position of the second, and so on.

If the values of $f(x)$ at $a, x_1, x_2, \ldots x_{n-1}, b$ are not the same as

$$f(a+0), \quad f(x_1+0) \quad \text{or} \quad f(x_1-0), \text{ etc.},$$

we must treat these points separately.

In this way, but arithmetically,* we obtain the function $\phi(x)$ defined as follows:

In $a \leqq x < x_1$, $\phi(x) = f(x)$, supposing for clearness $f(a) = f(a+0)$.
At x_1, $\phi(x) = f(x) + a_1$.
In $x_1 < x < x_2$, $\phi(x) = f(x) + a_1 + a_2$.
At x_2, $\phi(x) = f(x) + a_1 + a_2 + a_3$.

And so on,
$a_1, a_2, a_3, \ldots$ being definite numbers depending on $f(x_1 \pm 0)$, $f(x_1)$, etc.

We can now apply the theorem proved above to the function $\phi(x)$ and write

$$\phi(x) = \Phi(x) - \Psi(x) \text{ in } (a, b),$$

$\Phi(x)$ and $\Psi(x)$ being positive and monotonic increasing in this interval.

It follows that:

In $a \leqq x < x_1$, $f(x) = \Phi(x) - \Psi(x)$.
At x_1, $f(x) = \Phi(x) - \Psi(x) - a_1$.
In $x_1 < x < x_2$, $f(x) = \Phi(x) - \Psi(x) - a_1 - a_2$.

And so on.

If any of the terms $a_1, a_2, \ldots$ are negative we put them with $\Phi(x)$: the positive terms we leave with $\Psi(x)$. Thus finally we obtain, as before, that

$$f(x) = F(x) - G(x) \text{ in } (a, b),$$

where $F(x)$ and $G(x)$ are positive and monotonic increasing in this interval.

We have thus established the important theorem:

If the bounded function $f(x)$, given in the interval (a, b), is such that this interval can be divided up into a finite number of open partial intervals, in each of which $f(x)$ is monotonic, then we can express $f(x)$ as the difference of two (bounded) functions, which are positive and monotonic increasing in the interval (a, b).

Also it will be seen from the above discussion that the discontinuities of $F(x)$ and $G(x)$, which can, of course, only be ordinary discontinuities, can occur only at the points where $f(x)$ is discontinuous.

It should, perhaps, also be added that, while the monotonic properties ascribed to $f(x)$ allow it to have only ordinary discontinuities, the number of points of discontinuity may be infinite (§ 34).

36. 2. Functions of Bounded Variation.† The functions discussed in § 36. 1 are a special case of the class of functions known as *functions of bounded variation* introduced into Analysis by Jordan,‡ and used by him in the treatment of Fourier's Series. The principal properties of such functions are obtained in this section, which may replace § 36. 1.

---

*It will be noticed that in the proof the curves and diagrams are used simply as illustrations.

†There is a very complete treatment of functions of bounded variation (fonctions à variation bornée) in Lebesgue's *Leçons sur l'Integration*, (2e éd., Paris, 1928), Ch. IV. See also de la Vallée Poussin's *Cours d'Analyse* 2 (4e éd., 1922), Ch. II, § 2.

‡Cf. Jordan, *Cours d'Analyse*, 1 (2e éd., Paris, 1893), 54.

Definition. *Let $f(x)$ be bounded in (a, b) and let $a=x_0, x_1, x_2, \dots x_{n-1}, x_n=b$, be a mode of division of this interval. Let $y_0, y_1, \dots y_{n-1}, y_n$ be the values of $f(x)$ at these points.*

Then
$$\sum_0^{n-1}(y_{r+1}-y_r)=f(b)-f(a)=p-n,$$
where p is the sum of the positive differences, and $-n$ the sum of the negative differences.

The sum $\sum\limits_0^{n-1}|y_{r+1}-y_r|$ is denoted by t, and we have
$$t=\sum_0^{n-1}|y_{r+1}-y_r|=p+n.$$

To every mode of division of (a, b) into such partial intervals, there correspond sums t, p and n.

When the sums t, corresponding to all possible modes of division of (a, b), are bounded and their upper bound is T, we say that T is the total variation of $f(x)$ in (a, b), and that $f(x)$ is of **bounded variation** *in this interval.*

Since
$$2p=t+f(b)-f(a),$$
and
$$2n=t-f(b)+f(a),$$
it is clear that, if $f(x)$ is of bounded variation, the sums p and n are also bounded.

If their upper bounds are P and N, we have
$$2P=T+f(b)-f(a),$$
and
$$2N=T-f(b)+f(a),$$
P and $-N$ are sometimes called the *positive variation* and the *negative variation* in the given interval.

Again it is clear that a bounded monotonic function is of bounded variation. And that, if $f(x)$ is of bounded variation in (a, b), it is also of bounded variation in (α, β) where $a \leqq \alpha < \beta \leqq b$. Further $f(x)$ does not cease to be of bounded variation, if its value is altered at a finite number of points.

These facts follow at once from the definition just given, and we proceed now to establish further properties of such functions.

I. *If $f(x)$ is of bounded variation in (a, c) and (c, b), where $a < c < b$, it is of bounded variation in (a, b).*

Take any mode of division of (a, b), say, $a=x_0, x_1, x_2, \dots x_{n-1}, x_n=b$.

If one of these points coincides with c, then its sum t satisfies $t=t_1+t_2$, where t_1 and t_2 are the sums for this mode of division of (a, c) and (c, b) respectively.

If c lies between two points x_r and x_{r+1}, since
$$|y_{r+1}-y_r| \leqq |y_{r+1}-f(c)|+|f(c)-y_r|,$$
it is clear that $t \leqq t_1+t_2$, with the same notation as before.

But if T_1 and T_2 are the total variations in (a, c) and (c, b) respectively, $t_1 \leqq T_1$ and $t_2 \leqq T_2$.

Therefore
$$t \leqq T_1+T_2.$$

Thus we have shown that $f(x)$ is of bounded variation in (a, b).

II. *Let c be a point between a, b and $f(x)$ be of bounded variation in (a, b), T, T_1 and T_2 being its total variations in (a, b), (a, c) and (c, b).*

Then $$T = T_1 + T_2.$$

We have seen that if $f(x)$ is of bounded variation in (a, b), it is also of bounded variation in (a, c) and (c, b), and the concluding line of the argument of Theorem I shows that

$$T \leqq T_1 + T_2. \quad \text{.....(i)}$$

Now take the usual arbitrary positive number ϵ.

There is a mode of division of (a, c) and also of (c, b), for which the sums t_1 and t_2 satisfy

$$T_1 - \tfrac{1}{2}\epsilon < t_1, \quad T_2 - \tfrac{1}{2}\epsilon < t_2.$$

Thus $$T_1 + T_2 - \epsilon < t_1 + t_2.$$

But these modes of division of (a, c) and (c, b) form a mode of division of (a, b).

Therefore $$t_1 + t_2 \leqq T.$$

Thus $$T_1 + T_2 - \epsilon \leqq T. \quad \text{.....(ii)}$$

It follows from the inequalities (i) and (ii) that

$$T = T_1 + T_2.$$

III. *If $f(x)$ is bounded in (a, b), and this interval can be broken up into a finite number of partial intervals (open or closed) in which the function is monotonic, then $f(x)$ is of bounded variation in (a, b).*

This follows from Theorem I, since in all these intervals $f(x)$ is of bounded variation.*

---

*It might be thought that a function with an infinite number of turning points in a finite interval could not be of bounded variation in that interval. But the following example shows that this is not the case.

Let $f(x) = x^2 \sin \dfrac{1}{x^{\frac{4}{3}}}$, when $x > 0$, and $f(0) = 0$.

It is easy to show that there is one turning point and only one in

$$\left(\frac{1}{((n+1)\pi)^{\frac{3}{4}}}, \frac{1}{(n\pi)^{\frac{3}{4}}}\right),$$

where n is any positive integer.

If the interval extends to the origin, the number of turning points is infinite.

But the absolute value of the maximum or minimum in the interval

$$\left(\frac{1}{((n+1)\pi)^{\frac{3}{4}}}, \frac{1}{(n\pi)^{\frac{3}{4}}}\right) \text{ is less than } \frac{1}{(n\pi)^{\frac{3}{2}}}.$$

It follows that the total variation in any interval, $(0, a)$, where $a > 0$, is less than

$$\frac{2}{\pi^{\frac{3}{2}}} \sum_1^\infty \frac{1}{n^{\frac{3}{2}}}.$$

On the other hand the functions

$$\left.\begin{aligned} &f(x) = \sin \frac{1}{x} \text{ when } x > 0 \\ &f(0) = 0 \end{aligned}\right\} \text{ and } \left.\begin{aligned} &f(x) = x \sin \frac{1}{x} \text{ when } x > 0 \\ &f(0) = 0 \end{aligned}\right\}$$

are not of bounded variation in such an interval.

IV. *If $f(x)$ is of bounded variation in (a, b), it is the difference of two positive, monotonic increasing functions; and the difference of two bounded monotonic increasing functions is a function of bounded variation.*

(i) First, let $P(x)$ and $-N(x)$ be the positive and negative variations for the interval (a, x), where $a < x < b$.

Then, by the definition of P and N, we have

$$f(x) - f(a) = P(x) - N(x).$$

Also $P(x)$ and $N(x)$ are positive and monotonic increasing functions of x.

Thus $$f(x) = [P(x) + f(a) + |f(a)|] - [N(x) + |f(a)|],$$
which establishes the first part of the theorem.

(ii) Next let $$f(x) = F(x) - G(x),$$
where $F(x)$ and $G(x)$ are bounded, monotonic increasing functions in (a, b).

Take any mode of division of the interval

$$a = x_0,\ x_1,\ x_2,\ \dots\ x_{n-1},\ x_n = b.$$

Then $$f(x_{r+1}) - f(x_r) = [F(x_{r+1}) - F(x_r)] - [G(x_{r+1}) - G(x_r)],$$

and $$\sum_0^{n-1} |f(x_{r+1}) - f(x_r)| \leqq \sum_0^{n-1} [F(x_{r+1}) - F(x_r)] + \sum_0^{n-1} [G(x_{r+1}) - G(x_r)]$$
$$\leqq [F(b) - F(a)] + [G(b) - G(a)].$$

Therefore $f(x)$ is of bounded variation in (a, b).

V. *If $f_1(x)$ and $f_2(x)$ are two functions of bounded variation, so are $f_1(x) \pm f_2(x)$ and $f_1(x)\, f_2(x)$.*

Also, if $f(x)$ is of bounded variation, and $|f(x)| <$ some definite positive number in the interval, then $1/f(x)$ is of bounded variation.

(i) Let $$f_1(x) = F_1(x) - G_1(x) \quad \text{and} \quad f_2(x) = F_2(x) - G_2(x),$$
where $F_1(x)$, $F_2(x)$, $G_1(x)$ and $G_2(x)$ are positive monotonic increasing functions.

Then $$f_1(x) + f_2(x) = [F_1(x) + F_2(x)] - [G_1(x) + G_2(x)],$$
and the sum of the given functions is of bounded variation by Theorem IV.

Similarly for the difference and product.

(ii) With the usual notation,

$$\left| \frac{1}{y_{r+1}} - \frac{1}{y_r} \right| = \frac{|y_{r+1} - y_r|}{|y_{r+1}|\,|y_r|} < \frac{1}{\mu^2} |y_{r+1} - y_r|,$$
if $|f(x)| > \mu > 0$.

Thus the sum t for $1/f(x)$ is less than T/μ^2, where T is the total variation of $f(x)$ in the interval.

VI. *A function of bounded variation has only ordinary discontinuities.*

This follows at once from Theorem IV, for the discontinuities of $f(x)$ must also be discontinuities of $F(x)$ and $G(x)$, and these are ordinary discontinuities. (§ 34.)

If the discontinuities of $f(x)$ are infinite in number, they are countably infinite.

For, if we have a sequence $\eta_1, \eta_2, \eta_3, \ldots$ when

$$\eta_1 > \eta_2 > \eta_3 \ldots \quad \text{and} \quad \lim_{n\to\infty} \eta_n = 0,$$

we know that there is only a finite number of points at which

$$F(x+0) - F(x-0)| > \eta_n,$$

where n is any positive integer.

VII. *If $f(x)$ is of bounded variation in (a, b) and continuous at a point c of that interval, then $T(x)$, $P(x)$ and $N(x)$ are also continuous at c.*

Take the arbitrary positive number ϵ. There is a mode of division of (a, c), say $a, x_1, x_2, \ldots x_{n-1}, c$, such that the sum t for it satisfies

$$T(c) - \tfrac{1}{2}\epsilon < t \leqq T(c).$$

Also there is a neighbourhood of c such that

$$|f(x) - f(c)| < \tfrac{1}{2}\epsilon, \quad \text{when} \quad 0 < c - x \leqq \eta.$$

It is clear that we can take x_{n-1} within this neighbourhood, for if it were outside it, by adding a point of section within it, we do not diminish the sum t.

Also
$$t = \sum_{0}^{n-2} |y_{r+1} - y_r| + |f(c) - f(x_{n-1})|$$
$$\leqq T(x_{n-1}) + \tfrac{1}{2}\epsilon$$
$$\leqq T(c-0) + \tfrac{1}{2}\epsilon,$$

since $T(x)$ is a monotonic increasing function.

Thus
$$T(c) - \tfrac{1}{2}\epsilon < T(c-0) + \tfrac{1}{2}\epsilon,$$

and
$$T(c) < T(c-0) + \epsilon.$$

It follows that
$$T(c) \leqq T(c-0).$$

But, since $T(x)$ is monotonic increasing, $T(c) \nless T(c-0)$.

Therefore we must have $T(c) = T(c-0)$.

Taking a neighbourhood to the right of a, we find in the same way that

$$T(c) = T(c+0).$$

Thus $T(x)$ is continuous at $x=c$, and the same holds for $P(x)$ and $N(x)$.

37. Functions of Several Variables. So far we have dealt only with functions of a single variable. If to every value of x in the interval $a \leqq x \leqq b$ there corresponds a number y, then we have said that y is a function of x in the interval (a, b), and we have written $y = f(x)$.

The extension to functions of two variables is immediate:

To every pair of values of x and y, such that

$$a \leqq x \leqq a', \quad b \leqq y \leqq b',$$

let there correspond a number z. Then z is said to be a function of x and y in this domain, and we write $z = f(x, y)$.

If we consider x and y as the coordinates of a point in a plane, to every pair of values of x and y there corresponds a point in the plane, and the region defined by $a \leqq x \leqq a'$, $b \leqq y \leqq b'$ will be a rectangle.

In the case of the single variable, it is necessary to distinguish between the *open* interval $(a < x < b)$ and the *closed* interval $(a \leqq x \leqq b)$. So, in the case of two dimensions, it is well to distinguish between *open* and *closed* domains. In the former the boundary of the region is not included in the domain; in the latter it is included.

In the above definition we have taken a rectangle for the domain of the variables. A function of two variables may be defined in the same way for a domain of which the boundary is a curve C: or again, the domain may have a curve C for its external boundary, and other curves, C', C'', etc., for its internal boundary.

A function of three variables, or any number of variables, will be defined as above. For three variables, we can still draw upon the language of geometry, and refer to the domain as contained within a surface S, etc.

We shall now refer briefly to some properties of functions of two variables.

A function is said to be *bounded* in the domain in which it is defined, if the set of values of z, for all the points of this domain, forms a bounded aggregate. The *upper* and *lower bounds*, M and m, and the *oscillation*, are defined as in § 24.

$f(x, y)$ is said to have the limit l as (x, y) tends to (x_0, y_0), when, any positive number ϵ having been chosen, as small as we please, there is a positive number η such that $|f(x, y) - l| < \epsilon$ for all values of (x, y) for which

$$|x - x_0| \leqq \eta, \quad |y - y_0| \leqq \eta \quad \textit{and} \quad 0 < |x - x_0| + |y - y_0|.$$

In other words, $|f(x, y) - l|$ must be less than ϵ for all points in the square, centre at (x_0, y_0), whose sides are parallel to the coordinate axes and of length 2η, the centre itself being excluded from the domain.

A necessary and sufficient condition for the existence of a limit to $f(x, y)$ as (x, y) tends to (x_0, y_0) is that, to the arbitrary positive number ϵ, there shall correspond a positive number η such that $|f(x', y') - f(x'', y'')| < \epsilon$, where (x', y') and (x'', y'') are any two *points other than* (x_0, y_0) *in the square, centre at* (x_0, y_0)*, whose sides are parallel to the coordinate axes and of length* 2η.

The proof of this theorem can be obtained in exactly the same way as in the one-dimensional case, squares taking the place of the intervals in the preceding proof.

A function $f(x, y)$ is said to be continuous when $x=x_0$ and $y=y_0$, if $f(x, y)$ has the limit $f(x_0, y_0)$ as (x, y) tends to (x_0, y_0).

Thus, $f(x, y)$ *is continuous when* $x=x_0$ *and* $y=y_0$*, if, to the arbitrary positive number* ϵ*, there corresponds a positive number* η *such that* $|f(x, y)-f(x_0, y_0)|<\epsilon$ *for all values of* (x, y) *for which*

$$|x-x_0|\leqq\eta \quad \text{and} \quad |y-y_0|\leqq\eta.$$

In other words, $|f(x, y)-f(x_0, y_0)|$ must be less than ϵ for all points in the square, centre at (x_0, y_0), whose sides are parallel to the coordinate axes and of length 2η.*

It is convenient to speak of a function as continuous *at a point* (x_0, y_0) instead of *when* $x=x_0$ *and* $y=y_0$. Also when a function of two variables is continuous at (x, y), as defined above, for every point of a domain, we shall say that it is a continuous function of (x, y) in the domain.

It is easy to see that we can substitute for the square, with centre at (x_0, y_0), referred to above, a circle with the same centre.† The definition of a limit would then read as follows:

$f(x, y)$ *is said to have the limit* l *as* (x, y) *tend to* (x_0, y_0)*, if, to the arbitrary positive number* ϵ*, there corresponds a positive number* η *such that* $|f(x, y)-l|<\epsilon$ *for all values of* (x, y) *for which*

$$0<\sqrt{\{(x-x_0)^2+(y-y_0)^2\}}\leqq\eta.$$

If a function converges at (x_0, y_0) according to this definition (based on the circle), it converges according to the former definition (based on the square); and conversely. And the limits in both cases are the same.

Also continuity at (x_0, y_0) would now be defined as follows:

$f(x, y)$ *is continuous at* (x_0, y_0)*, if, to the arbitrary positive number* ϵ*,*

---

*There are obvious changes to be made in these statements when we are dealing with a point (x_0, y_0) on one of the boundaries of the domain in which the function is defined.

†We may also use a rectangle, centre at (x_0, y_0), and sides 2η, $2\eta'$ say.

there corresponds a positive number η *such that* $|f(x, y)-f(x_0, y_0)|<\epsilon$ *for all values of* (x, y) *for which*

$$\sqrt{\{(x-x_0)^2+(y-y_0)^2\}} \leqq \eta.$$

Every function, which is continuous at (x_0, y_0) under this definition, is continuous at (x_0, y_0) under the former definition, and conversely.

It is important to notice that if a function of x and y is continuous with respect to the two variables, as defined above, it is also continuous when considered as a function of x alone, or of y alone.

For example, let $f(x, y)$ be defined as follows:

$$\begin{cases} f(x, y)=\dfrac{2xy}{x^2+y^2}, & \text{when at least one of the variables is not zero,} \\ f(0, 0)=0. \end{cases}$$

Then $f(x, y)$ is a continuous function of x, for all values of x, when y has any fixed value, zero or not; and it is a continuous function of y, for all values of y, when x has any fixed value, zero or not.

But it is not a continuous function of (x, y) in any domain which includes the origin, since $f(x, y)$ is not continuous when $x=0$ and $y=0$.

For, if we put $x=r \cos \theta$, $y=r \sin \theta$, we have $f(x, y)=\sin 2\theta$, which is independent of r, and varies from -1 to $+1$.

However, it is a continuous function of (x, y) in any domain which does not include the origin.

On the other hand, the function defined by

$$\begin{cases} f(x, y)=\dfrac{2xy}{\sqrt{(x^2+y^2)}}, & \text{when at least one of the variables is not zero,} \\ f(0, 0)=0, \end{cases}$$

is a continuous function of (x, y) in any domain which includes the origin.

The theorems as to the continuity of the sum, product and, in certain cases, quotient of two or more continuous functions, can be readily extended to the case of functions of two or more variables. A continuous function of one or more continuous functions is also continuous.

In particular we have the theorem:

Let $u=\phi(x, y)$, $v=\psi(x, y)$ *be continuous at* (x_0, y_0), *and let* $u_0=\phi(x_0, y_0)$, $v_0=\psi(x_0, y_0)$.

Let $z=f(u, v)$ *be continuous in* (u, v) *at* (u_0, v_0).

Then $z=f[\phi(x, y), \psi(x, y)]$ *is continuous in* (x, y) *at* (x_0, y_0).

Further, the general theorems on continuous functions, proved in § 31, hold, with only verbal changes, for functions of two or more variables.

For example:

If a function of two variables is continuous at every point of a closed domain, it is uniformly continuous in the domain.

In other words, when the positive number ϵ *has been chosen, as small as we please, there is a positive number* η *such that* $|f(x', y')-f(x'', y'')|<\epsilon$, *when* (x', y') *and* (x'', y'') *are any two points in the domain for which*

$$\sqrt{\{(x'-x'')^2+(y'-y'')^2\}} \leqq \eta.$$

REFERENCES.

De la Vallée Poussin, *loc. cit.*, 1 (5e éd., 1923), Introduction, § 3.

Goursat, *loc. cit.*, 1 (4e éd., 1923), Ch. I.

Hardy, *loc. cit.* (5th ed., 1928), Ch. V.

Osgood, *Lehrbuch der Funktionentheorie,* 1 (4 Aufl., Leipzig, 1923), Kap. I.

Pierpont, *Theory of Functions of Real Variables,* 1 (Boston, 1905), Ch. VI and VII.

And

Pringsheim, "Grundlagen der allgemeinen Funktionenlehre," *Enc. d. math. Wiss.*, Bd. II, Tl. I (Leipzig, 1899).

CHAPTER IV

THE DEFINITE INTEGRAL

38. In the usual elementary treatment of the Definite Integral, defined as the limit of a sum, it is assumed that the function of x considered may be represented by a curve. The limit is the area between the curve, the axis of x and the two bounding ordinates.

For long this demonstration was accepted as sufficient. To-day, however, analysis is founded on a more solid basis. No appeal is made to the intuitions of geometry. Further, even among the continuous functions of analysis, there are many which cannot be represented graphically.

E.g. let $\quad f(x) = x \sin \frac{1}{x}$, when $x \gtrless 0$,

and $\quad f(0) = 0$.

Then $f(x)$ is continuous for every value of x, but it has not a differential coefficient when $x = 0$.

It is continuous at $x = 0$, because

$$|f(x) - f(0)| = |f(x)| \leqq |x|;$$

and $\quad |f(x) - f(0)| < \epsilon$, when $0 < |x| \leqq \eta$, if $\eta < \epsilon$.

Also it is continuous when $x \gtrless 0$, since it is the product of two continuous functions [cf. § 30].

It has not a differential coefficient at $x = 0$, because

$$\frac{f(h) - f(0)}{h} = \sin \frac{1}{h},$$

and $\sin 1/h$ has not a limit as $h \to 0$.

It has a differential coefficient at every point where $x \gtrless 0$, and at such points

$$f'(x) = \sin \frac{1}{x} - \frac{1}{x} \cos \frac{1}{x}.$$

More curious still, Weierstrass discovered a function, which is continuous for every value of x, while it has not a differential coefficient anywhere.* This function is defined by the sum of the infinite series

$$\sum_{0}^{\infty} a^n \cos b^n \pi x,$$

a being a positive odd integer and b a positive number less than unity, connected with a by the inequality $ab > 1 + \frac{3}{2}\pi$.†

Other examples of such extraordinary functions have been given since Weierstrass's time. And for this reason alone it would have been necessary to substitute an exact arithmetical treatment for the traditional discussion of the Definite Integral.

Riemann‡ was the first to give such a rigorous arithmetical treatment. The definition adopted in this chapter is due to him. The limitations imposed upon the integrand $f(x)$ will be indicated as we proceed.

In Higher Analysis the Riemann Integral has now been superseded by the Lebesgue Integral, or by one of the others allied to it. This advance dates from Lebesgue's first memoir, which appeared in 1902.§ Much has since then been done in this field, but the ideas involved are far from elementary; and, though it is especially in the rigorous treatment of the Theory of Fourier's Series that the advantages of the new definition of the integral are to be found, it does not seem proper to introduce it into this work. In an Appendix|| the Lebesgue Integral is defined and some of its distinctive properties are obtained; also it is shown in what way its introduction simplifies and completes the more elementary treatment of the text.

---

*It seems impossible to assign an exact date to this discovery. Weierstrass himself did not at once publish it, but communicated it privately, as was his habit, to his pupils and friends. Du Bois-Reymond quotes it in a paper published in 1874.

†Hardy has shown that this relation can be replaced by $0 < a < 1$, $b > 1$ and $ab \geqq 1$ [cf. *Trans. Amer. Math. Soc.*, **17** (1916)]. An interesting discussion of Weierstrass's function is to be found in a paper, "Infinite Derivates," *Quart. J. of Math.*, **47** (1916), 127, by Grace Chisholm Young.

‡In his classical paper, *Über die Darstellbarkeit einer Function durch eine trigonometrische Reihe.* See above, p. 6.

But the earlier work of Cauchy and Dirichlet must not be forgotten.

§"Intégrale, Longueur, Aire," *Annali di Mat.* (3), **7** (1902), 230.

||See Appendix II, where references to books and memoirs on this subject will be found.

39. The Sums S and s.* Let $f(x)$ be a bounded function, given in the interval (a, b).

Suppose this interval broken up into n partial intervals

$$(a, x_1), \quad (x_1, x_2), \dots (x_{n-1}, b),$$

where $$a < x_1 < x_2 \dots < x_{n-1} < b.$$

Let M, m be the upper and lower bounds of $f(x)$ in the whole interval, and M_r, m_r those in the closed interval (x_{r-1}, x_r), writing $a = x_0$ and $b = x_n$.

Let $$\left.\begin{aligned} S &= M_1(x_1 - a) + M_2(x_2 - x_1) + \dots + M_n(b - x_{n-1}), \\ s &= m_1(x_1 - a) + m_2(x_2 - x_1) + \dots + m_n(b - x_{n-1}). \end{aligned}\right\}$$
and

To every mode of subdivision of (a, b) into such partial intervals, there corresponds a sum S and a sum s such that $s \leqq S$.

The sums S have a lower bound, since they are all greater than $m(b - a)$, and the sums s have an upper bound, since they are all less than $M(b - a)$.

Let the lower bound of the sums S be J, and the upper bound of the sums s be I.

We shall now show that $I \leqq J$.

Let $$a, \quad x_1, \quad x_2, \dots x_{n-1}, \quad b$$

be the set of points to which a certain S and s correspond.

Suppose some or all of the intervals $(a, x_1), (x_1, x_2), \dots (x_{n-1}, b)$ to be divided into smaller intervals, and let

$$a, \quad y_1, \quad y_2, \dots y_{k-1}, \quad x_1, \quad y_k, \quad y_{k+1}, \dots y_{l-1}, \quad x_2, \quad y_l, \dots b$$

be the set of points thus obtained.

The second mode of division will be called *consecutive* to the first, when it is obtained from it in this way.

Let Σ, σ be the sums for the new division.

Compare, for example, the parts of S and Σ which come from the interval (a, x_1).

Let M'_1, m'_1 be the upper and lower bounds of $f(x)$ in (a, y_1), M'_2, m'_2 in (y_1, y_2), and so on.

The part of Σ which comes from (a, x_1) is then

$$M'_1(y_1 - a_1) + M'_2(y_2 - y_1) \dots + M'_k(x_1 - y_{k-1}).$$

But the numbers M'_1, M'_2, ... cannot exceed M_1.

Thus the part of Σ which we are considering is at most equal to $M_1(x_1 - a)$.

---

*The argument which follows is taken, with slight modifications, from Goursat's *Cours d'Analyse*, 1 (4e éd., 1923), pp. 171 *et seq.*

Similarly the part of Σ which comes from (x_1, x_2) is at most equal to $M_2(x_2 - x_1)$, and so on.

Adding these results we have $\Sigma \leqq S$.

Similarly we obtain $\sigma \geqq s$.

Consider now any two modes of division of (a, b).

Denote them by

$$a, \quad x_1, \quad x_2, \dots x_{m-1}, \quad b, \quad \text{with sums } S \text{ and } s, \quad \dots\dots\dots(1)$$

and

$$a, \quad y_1, \quad y_2, \dots y_{n-1}, \quad b, \quad \text{with sums } S' \text{ and } s'. \quad \dots\dots\dots(2)$$

On superposing these two, we obtain a third mode of division (3), *consecutive* to both (1) and (2).

Let the sums for (3) be Σ and σ.

Then, since (3) is consecutive to (1),

$$S \geqq \Sigma \quad \text{and} \quad \sigma \geqq s.$$

Also, since (3) is consecutive to (2),

$$S' \geqq \Sigma \quad \text{and} \quad \sigma \geqq s'.$$

But $\qquad \Sigma \geqq \sigma.$

Therefore $\qquad S \geqq s' \quad \text{and} \quad S' \geqq s.$

Thus the sum S arising from any mode of division of (a, b) is not less than the sum s arising from the same, or any other, mode of division.

It follows at once that $I \leqq J$.

For we can find a sum s as near I as we please, and a sum S (not necessarily from the same mode of division) as near J as we please. If $I > J$, this would involve the existence of an s and an S for which $s > S$.

The argument of this section will offer less difficulty, if the reader follow it for an ordinary function represented by a curve, when the sums S and s will refer to certain rectangles associated with the curve.

40. Darboux's Theorem. *The sums S and s tend respectively to J and I, when the points of division are multiplied indefinitely, in such a way that all the partial intervals tend to zero.*

Stated more precisely, the theorem reads as follows:

If the positive number ϵ is chosen, as small as we please, there is a positive number η such that, for all modes of division in which all the partial intervals are less than or equal to η, the sum S is greater than J by less than ϵ, and the sum s is smaller than I by less than ϵ.

Let ϵ be any positive number as small as we please.

Since the sums S and s have J and I for lower and upper bounds respectively, there is a mode of division such that the sum S for it exceeds J by less than $\frac{1}{2}\epsilon$.

Let this mode of division be

$$a, \quad a_1, \quad a_2, \dots a_{p-1}, \quad b, \quad \text{with sums } S_1 \text{ and } s_1. \quad \dots\dots\text{(1)}$$

Then $$S_1 < J + \tfrac{1}{2}\epsilon.$$

Let η be a positive number such that all the partial intervals of (1) are greater than η.

Let

$$a = x_0, \quad x_1, \quad x_2, \dots x_{n-1}, \quad b = x_n, \quad \text{with sums } S_2 \text{ and } s_2, \dots\text{(2)}$$

be any mode of division such that

$$(x_r - x_{r-1}) \leqq \eta, \quad \text{when } r = 1, 2, \dots n.$$

The mode of division obtained by superposing (1) and (2), *e.g.* $a, x_1, x_2, a_1, x_3, a_2, x_4, \dots x_{n-1}, b,$ with sums S_3 and s_3, ...(3) is *consecutive* to (1) and (2).

Then, by § 39, we have $S_1 \geqq S_3$.

But $$S_1 < J + \tfrac{1}{2}\epsilon.$$

Therefore $$S_3 < J + \tfrac{1}{2}\epsilon.$$

Further,

$$S_2 - S_3 = \Sigma\,[M(x_{r-1}, x_r)(x_r - x_{r-1}) - M(x_{r-1}, a_k)(a_k - x_{r-1}) - M(a_k, x_r)(x_r - a_k)],$$

$M(x', x'')$ denoting the upper bound of $f(x)$ in the interval (x', x''), and the symbol Σ standing as usual for a summation, extending in this case to all the intervals (x_{r-1}, x_r) of (2) which have one of the points $a_1, a_2, \dots a_{p-1}$ as an internal point, and not an end-point. From the fact that each of the partial intervals of (1) is greater than η, and that each of those of (2) does not exceed η, we see that no two of the a's can lie between two consecutive x's of (2).

There are at most $(p-1)$ terms in the summation denoted by Σ. Let $|f(x)|$ have A for its upper bound in (a, b).

We can rewrite $S_2 - S_3$ above in the form

$$S_2 - S_3 = \Sigma\,[\{M(x_{r-1}, x_r) - M(x_{r-1}, a_k)\}(a_k - x_{r-1}) + \{M(x_{r-1}, x_r) - M(a_k, x_r)\}(x_r - a_k)].$$

But $\{M(x_{r-1}, x_r) - M(x_{r-1}, a_k)\}$ and $\{M(x_{r-1}, x_r) - M(a_k, x_r)\}$ are both positive or zero, and they cannot exceed $2A$.

Therefore $$S_2 - S_3 \leqq 2A\Sigma(x_r - x_{r-1}),$$
the summation having at most $(p-1)$ terms, and $(x_r - x_{r-1})$ being at most equal to η.

Thus $$S_2 - S_3 \leqq 2(p-1)A\eta.$$

Therefore $$S_2 < J + \tfrac{1}{2}\epsilon + 2(p-1)A\eta,$$
since we have seen that $S_3 < J + \tfrac{1}{2}\epsilon$.

So far the only restriction placed upon the positive number η has been that the partial intervals of (1) are each greater than η.

We can thus choose η so that
$$\eta < \frac{\epsilon}{4(p-1)A}.$$

With such a choice of η, $S > J + \epsilon$.

Thus *we have shown that for any mode of division such that the greatest of the partial intervals is less than or equal to a certain positive number η, dependent on ϵ, the sum S exceeds J by less than ϵ.*

Similarly for s and I; and it is obvious that we can make the same η satisfy both S and s, by taking the smaller of the two to which we are led in this argument.

41. The Definite Integral of a Bounded Function. We now come to the definition of the definite integral of a bounded function $f(x)$, given in an interval (a, b).

A bounded function $f(x)$, given in the interval (a, b), is said to be integrable in that interval, when the lower bound J of the sums S and the upper bound I of the sums s of § 39 are equal.

The common value of these bounds I and J is called the definite integral of $f(x)$ between the limits a and b, and is written
$$\int_a^b f(x)\,dx.^*$$

It follows from the definition that $\int_a^b f(x)\,dx$ cannot be greater than the sum S or less than the sum s corresponding to any mode of division of (a, b). These form approximations by excess and defect to the integral.

---

*The bound J of the sums S is usually called the *upper integral* of $f(x)$ and the bound I of the sums s the *lower integral*.

We can replace the sums

$$\left.\begin{aligned} S &= M_1(x_1-a)+M_2(x_2-x_1)+\ldots+M_n(b-x_{n-1}), \\ s &= m_1(x_1-a)+m_2(x_2-x_1)+\ldots+m_n(b-x_{n-1}), \end{aligned}\right\}$$

by more general expressions, as follows:

Let $\xi_1, \xi_2, \ldots \xi_r, \ldots \xi_n$ be any values of x in the partial intervals $(a, x_1), (x_1, x_2), \ldots (x_{r-1}, x_r), \ldots (x_{n-1}, b)$ respectively.

The sum

$$f(\xi_1)(x_1-a)+f(\xi_2)(x_2-x_1)+\ldots+f(\xi_n)(b-x_{n-1}) \quad \ldots\ldots\ldots(1)$$

obviously lies between the sums S and s for this mode of division, since we have $m_r \leqq f(\xi_r) \leqq M_r$ for each of the partial intervals.

But, when the number of points of division (x_r) increases indefinitely in such a way that all the partial intervals tend to zero, the sums S and s have a common limit, namely $\int_a^b f(x)\,dx$.

Therefore the sum (1) has the same limit.

Thus we have shown that, for an integrable function $f(x)$, *the sum* $\quad f(\xi_1)(x_1-a)+f(\xi_2)(x_2-x_1)+\ldots+f(\xi_n)(b-x_{n-1})$

has the definite integral $\int_a^b f(x)\,dx$ *for its limit, when the number of points of division* (x_r) *increases indefinitely in such a way that all the partial intervals tend to zero,* $\xi_1, \xi_2, \ldots \xi_n$ *being any values of* x *in these partial intervals.**

In particular, we may take $a, x_1, x_2, \ldots x_{n-1}$, or $x_1, x_2, \ldots x_{n-1}, b$, for the values of $\xi_1, \xi_2, \ldots \xi_n$.

42. Necessary and Sufficient Conditions for Integrability. Any one of the following is a necessary and sufficient condition for the integrability of the bounded function $f(x)$ given in the interval (a, b):

I. *When any positive number* ϵ *has been chosen, as small as we please, there shall be a positive number* η *such that* $S-s<\epsilon$ *for every mode of division of* (a, b) *in which all the partial intervals are less than or equal to* η.

We have $\qquad S-s<\epsilon$, as stated above.

But $\qquad S \geqq J \quad \text{and} \quad s \leqq I.$

---

*We may substitute in the above, for $f(\xi_1), f(\xi_2), \ldots f(\xi_n)$, any values $\mu_1, \mu_2, \ldots \mu_n$ intermediate between $(M_1, m_1), (M_2, m_2)$, etc., the upper and lower bounds of $f(x)$ in the partial intervals.

Therefore. $J - I < \epsilon$.

And J must be equal to I.

Thus the condition is *sufficient.*

Further, if $I = J$, the condition is satisfied.

For, given ϵ, by Darboux's Theorem, there is a positive number η such that $S - J < \frac{1}{2}\epsilon$ and $I - s < \frac{1}{2}\epsilon$ for every mode of division in which all the partial intervals are less than or equal to η.

But $S - s = (S - J) + (I - s)$, since $I = J$.

Therefore $S - s < \epsilon$.

II. *When any positive number ϵ has been chosen, as small as we please, there shall be a mode of division of (a, b) such that $S - s < \epsilon$.*

It has been proved in (I) that this condition is *sufficient.* Also it is *necessary.* For we are given $I = J$, as $f(x)$ is integrable, and we have shown that in this case there are any number of modes of division, such that $S - s < \epsilon$.

III. *Let ω, σ be any pair of positive numbers. There shall be a mode of division of (a, b) such that the sum of the lengths of the partial intervals in which the oscillation is greater than or equal to ω shall be less than σ.**

This condition is *sufficient.* For, having chosen the arbitrary positive number ϵ, take

$$\sigma = \frac{\epsilon}{2(M - m)} \quad \text{and} \quad \omega = \frac{\epsilon}{2(b - a)},$$

where M, m are the upper and lower bounds respectively of $f(x)$ in (a, b).

Then there is a mode of division such that the sum of the lengths of the partial intervals in which the oscillation is greater than or equal to ω shall be less than σ. Let the intervals (x_{r-1}, x_r) in which the oscillation is greater than or equal to ω be denoted by D_r, and those in which it is less than ω by d_r, and let the oscillation $(M_r - m_r)$ in (x_{r-1}, x_r) be denoted by ω_r.

Then we have, for this mode of division,

$$\begin{aligned} S - s &= \Sigma\omega_r D_r + \Sigma\omega_r d_r \\ &< (M - m)\frac{\epsilon}{2(M - m)} + \frac{\epsilon}{2(b - a)}(b - a) \\ &< \frac{\epsilon}{2} + \frac{\epsilon}{2} \\ &< \epsilon, \end{aligned}$$

and, by (II), $f(x)$ is integrable in (a, b).

Also the condition is *necessary.* For, by (II), if $f(x)$ is integrable in (a, b), there is a mode of division such that $S - s < \omega\sigma$. Using D_r, d_r as above,

$$\begin{aligned} S - s &= \Sigma\omega_r D_r + \Sigma\omega_r d_r \\ &\geqq \Sigma\omega_r D_r \\ &\geqq \omega\Sigma D_r. \end{aligned}$$

---

*Cf. Pierpont, *Theory of Functions of Real Variables*, **1** (1905), § 498.

Therefore $\omega\sigma > \omega\Sigma D_r$,

and $\Sigma D_r < \sigma$.

43. Integrable Functions.

I. *If $f(x)$ is continuous in (a, b), it is integrable in (a, b).*

In the first place, we know that $f(x)$ is bounded in the interval, since it is continuous in (a, b) [cf. § 31. 1].

Next, we know that, to the arbitrary positive number ϵ, there corresponds a positive number η such that the oscillation of $f(x)$ is less than ϵ in all partial intervals less than or equal to η [cf. § 31. 1].

Now we wish to show that, given the arbitrary positive number ϵ, there is *a* mode of division such that $S - s < \epsilon$ [§ 42, II]. Starting with the given ϵ, we know that for $\epsilon/(b-a)$ there is a positive number η such that the oscillation of $f(x)$ is less than $\epsilon/(b-a)$ in all partial intervals less than or equal to η.

If we take a mode of division in which the partial intervals are less than or equal to this η, then for it we have

$$S - s < (b-a)\frac{\epsilon}{b-a} = \epsilon.$$

Therefore $f(x)$ is integrable in (a, b).

II. *If $f(x)$ is monotonic in (a, b), it is integrable in (a, b).**

In the first place, we note that the function, being given in the *closed* interval (a, b), and being monotonic, is also bounded. We shall take the case of a monotonic increasing function, so that we have

$$f(a) \leqq f(x_1) \leqq f(x_2) \ldots \leqq f(x_{n-1}) \leqq f(b)$$

for the mode of division given by

$$a, \quad x_1, \quad x_2, \ldots x_{n-1}, \quad b.$$

Thus we have

$$\left.\begin{aligned} S &= f(x_1)(x_1-a) + f(x_2)(x_2-x_1) \ldots + f(b)(b-x_{n-1}), \\ s &= f(a)\,(x_1-a) + f(x_1)(x_2-x_1) \ldots + f(x_{n-1})(b-x_{n-1}). \end{aligned}\right\}$$

Therefore, if all the partial intervals are less than or equal to η,

$$S - s \leqq \eta\,[f(b) - f(a)],$$

since $f(x_1) - f(a), \quad f(x_2) - f(x_1), \ldots f(b) - f(x_{n-1})$

are none of them negative.

---

*Since a function of bounded variation (cf. § 36. 2) is the difference of two monotonic functions, it follows from § 45, III that functions of bounded variation are integrable.

If we take $$\eta < \frac{\epsilon}{f(b)-f(a)},$$
it follows that $$S - s < \epsilon.$$

Thus $f(x)$ is integrable in (a, b).

The same proof applies to a monotonic decreasing function.

We have seen that a monotonic function, given in (a, b), can only have ordinary discontinuities, but these need not be finite in number (cf. § 34). We are thus led to consider other cases in which a bounded function is integrable, when discontinuities of the function occur in the given interval. A simple test of integrability is contained in the following theorem:

III. *A bounded function is integrable in* (a, b), *when all its points of discontinuity in* (a, b) *can be enclosed in a finite number** *of intervals the sum of which is less than any arbitrary positive number.*

Let ϵ be any positive number, as small as we please, and let the upper bound of $|f(x)|$ in (a, b) be A.

By our hypothesis we can enclose all the points of discontinuity of $f(x)$ in a finite number of intervals, the sum of which is less than $\epsilon/4A$.

The part of $S - s$ coming from these intervals is, at most, $2A$ multiplied by their sum.

On the other hand, $f(x)$ is continuous in all the remaining (closed) intervals.

We can, therefore, break up this part of (a, b) into a finite number of partial intervals such that the corresponding portion of $S - s < \frac{1}{2}\epsilon$ [cf. (I)].

Thus the combined mode of division for the whole of (a, b) is such that for it $S - s < \epsilon$.

Hence $f(x)$ is integrable in (a, b).

In particular, *a bounded function, with only a finite number of discontinuities in* (a, b), *is integrable in this interval.*

The discontinuities referred to in this Theorem III need not be ordinary discontinuities, but, as the function is bounded, they cannot be infinite discontinuities.

---

*It will be shown in Appendix II, § 10 that a bounded function is also integrable according to Riemann's definition of the integral, when the points of discontinuity can be enclosed in an infinite number of intervals, if the sum of the lengths of these intervals can be made as small as we please, and, in particular, when its points of discontinuity form a countably infinite set.

IV. *If a bounded function is integrable in each of the partial intervals* (a, α_1), (α_1, α_2), ... (α_{p-1}, b), *it is integrable in the whole interval* (a, b).

Since the function is integrable in each of these p intervals, there is a mode of division for each (*e.g.* α_{r-1}, α_r), such that $S-s$ for it is less than ϵ/p, where ϵ is any given positive number.

Then $S-s$ for the combined mode of division of the whole interval (a, b) is less than ϵ.

Therefore the function is integrable.

From the above results it is clear that *if a bounded function is such that the interval* (a, b) *can be broken up into a finite number of open partial intervals, in each of which the function is monotonic or continuous, then it is integrable in* (a, b).

V. *If the bounded function* $f(x)$ *is integrable in* (a, b), *then* $|f(x)|$ *is also integrable in* (a, b).

This follows at once, since $S-s$ for $|f(x)|$ is not greater than $S-s$ for $f(x)$ for the same mode of division.

It may be remarked that the converse does not hold.

E.g. let $f(x)=1$ for rational values of x in $(0, 1)$,

and $\quad f(x)=-1$ for irrational values of x in $(0, 1)$.

Then $|f(x)|$ is integrable, but $f(x)$ is not integrable, for it is obvious that the condition (II) of § 42 is not satisfied, as the oscillation is 2 in any interval, however small.

44. If the bounded function f(x) is integrable in (a, b), there are an infinite number of points in any partial interval of (a, b) at which f(x) is continuous.*

Let $\omega_1 > \omega_2 > \omega_3$ ... be an infinite sequence of positive numbers, such that

$$\lim_{n\to\infty} \omega_n = 0.$$

Let (α, β) be any interval contained in (a, b) such that $a \leqq \alpha < \beta < b$.

Then, by § 42, III, there is a mode of division of (a, b) such that the sum of the partial intervals in which the oscillation of $f(x)$ is greater than or equal to ω_1 is less than $(\beta-\alpha)$.

If we remove from (a, b) these partial intervals, the remainder must cover at least part of (α, β). We can thus choose within (α, β) a new interval (α_1, β_1) such that $(\beta_1-\alpha_1) < \frac{1}{2}(\beta-\alpha)$ and the oscillation in (α_1, β_1) is less than ω_1.

Proceeding in the same way, we obtain within (α_1, β_1) a new interval (α_2, β_2) such that $(\beta_2-\alpha_2) < \frac{1}{2}(\beta_1-\alpha_1)$ and the oscillation in (α_2, β_2) is less than ω_2. And so on.

---

*Cf. Pierpont, *loc. cit.*, § 508. A more general theorem is given in Appendix II, § 10.

Thus we find an infinite set of intervals $A_1, A_2, \ldots$, each contained *entirely within* the preceding, while the length of A_n tends to zero as $n \to \infty$, and the oscillation of $f(x)$ in A_n also tends to zero.

By the theorem of § 18, the set of intervals defines a point (*e.g.* c) which lies within all the intervals.

Let ϵ be any positive number, as small as we please.

Then we can choose in the sequence $\omega_1, \omega_2, \ldots$ a number ω_r less than ϵ. Let A_r be the corresponding interval (α_r, β_r), and η a positive number smaller than $(c-\alpha_r)$ and $(\beta_r - c)$.

Then $$|f(x)-f(c)| < \epsilon, \quad \text{when } |x-c| \leqq \eta,$$

and therefore we have shown that $f(x)$ is continuous at c.

Since this proof applies to any interval in (a, b), the interval (α, β) contains an infinite number of points at which $f(x)$ is continuous, for any part of (α, β), however small, contains a point of continuity.

45. Some Properties of the Definite Integral. We shall now establish some of the properties of $\int_a^b f(x)dx$, the integrand being bounded in (a, b) and integrable.

I. *If $f(x)$ is integrable in (a, b), it is also integrable in any interval (α, β) contained in (a, b).*

From § 42, I we know that to the arbitrary positive number ϵ there corresponds a positive number η such that the difference $S-s<\epsilon$ for every mode of division of (a, b) in which all the partial intervals are less than or equal to η.

We can choose a mode of division of this kind with (α, β) as ends of partial intervals.

Let Σ, σ be the sums for the mode of division of (α, β) included in the above.

Then we have $$0 \leqq \Sigma - \sigma \leqq S - s < \varepsilon.$$

Thus $f(x)$ is integrable in (α, β) [§ 42, II].

II. *If the value of the integrable function $f(x)$ is altered at a finite number of points of (a, b), the function $\phi(x)$ thus obtained is integrable in (a, b), and its integral is the same as that of $f(x)$.*

We can enclose the points to which reference is made in a finite number of intervals, the sum of which is less than $\epsilon/4A$, where ϵ is any given positive number, and A is the upper bound of $|\phi(x)|$ in (a, b).

The part of $S-s$ for $\phi(x)$, arising from these intervals, is at most $2A$ multiplied by their sum, *i.e.* it is less than $\frac{1}{2}\epsilon$.

On the other hand, $f(x)$ and $\phi(x)$, which is identical with $f(x)$ in the parts of (a, b) which are left, are integrable in each of these parts.

Thus we can obtain a mode of division for the whole of them which will contribute less than $\frac{1}{2}\epsilon$ to $S-s$, and, finally, we have a mode of division of (a, b) for which $S-s<\epsilon$.

Therefore $\phi(x)$ is integrable in (a, b).

Further,
$$\int_a^b \phi(x)\,dx = \int_a^b f(x)\,dx.$$

For we have seen in § 41 that $\int_a^b \phi(x)\,dx$ is the limit of

$$\phi(\xi_1)(x_1-a)+\phi(\xi_2)(x_2-x_1)+\ldots+\phi(\xi_n)(b-x_{n-1})$$

when the intervals (a, x_1), (x_1, x_2), ... (x_{n-1}, b) tend to zero, and $\xi_1, \xi_2, \ldots \xi_n$ are any values of x in these intervals.

We may put $f(\xi_1), f(\xi_2), \ldots f(\xi_n)$ for $\phi(\xi_1), \phi(\xi_2), \ldots \phi(\xi_n)$ in this sum, since in each interval there are points at which $\phi(x)$ and $f(x)$ are equal.

In this way we obtain a sum of the form $\lim \Sigma f(\xi_r)(x_r - x_{r-1})$, which is identical with $\int_a^b f(x)\,dx$.

III. It follows immediately from the definition of the integral, that *if $f(x)$ is integrable in (a, b), so also is $Cf(x)$, where C is any constant.*

Again, *if $f_1(x)$ and $f_2(x)$ are integrable in (a, b), their sum is also integrable.*

For, let (S, s), (S', s') and (Σ, σ) be the sums corresponding to the same mode of division for $f_1(x)$, $f_2(x)$ and $f_1(x)+f_2(x)$.

Then it is clear that

$$\Sigma-\sigma \leqq (S-s)+(S'-s'),$$

and the result follows.

Also it is easy to show that

$$\int_a^b Cf(x)\,dx = C\int_a^b f(x)\,dx,$$

and
$$\int_a^b \{f_1(x)+f_2(x)\}\,dx = \int_a^b f_1(x)\,dx + \int_a^b f_2(x)\,dx.$$

IV. *The product of two integrable functions $f_1(x)$, $f_2(x)$ is integrable.*

To begin with, let the functions $f_1(x)$, $f_2(x)$ be positive in (a, b).

Let M_r, m_r; M'_r, m'_r; $\mathbf{M}_r$, $\mathbf{m}_r$ be the upper and lower bounds of $f_1(x), f_2(x)$ and $f_1(x) f_2(x)$ in the partial interval (x_{r-1}, x_r).

Let (S, s), (S', s') and (Σ, σ) be the corresponding sums for a certain mode of division in which (x_{r-1}, x_r) is a partial interval.

Then it is clear that

$$\mathbf{M}_r - \mathbf{m}_r \leqq M_r M'_r - m_r m'_r = M_r(M'_r - m'_r) + m'_r(M_r - m_r).$$

A fortiori, $\quad \mathbf{M}_r - \mathbf{m}_r \leqq M(M'_r - m'_r) + M'(M_r - m_r)$,

where M, M' are the upper bounds of $f_1(x), f_2(x)$ in (a, b).

Multiplying this inequality by $(x_r - x_{r-1})$ and adding the corresponding results, we have

$$\Sigma - \sigma \leqq M(S' - s') + M'(S - s).$$

It follows that $\Sigma - \sigma$ tends to zero, and the product of $f_1(x), f_2(x)$ is integrable in (a, b).

If the two functions are not both positive throughout the interval, we can always add constants c_1 and c_2, so that $f_1(x) + c_1$, $f_2(x) + c_2$ remain positive in (a, b).

The product

$$(f_1(x) + c_1)(f_2(x) + c_2) = f_1(x) f_2(x) + c_1 f_2(x) + c_2 f_1(x) + c_1 c_2$$

is then integrable.

But $c_1 f_2(x) + c_2 f_1(x) + c_1 c_2$ is integrable.

It follows that $f_1(x) f_2(x)$ is integrable.

On combining these results, we see that *if* $f_1(x), f_2(x) \ldots f_n(x)$ *are integrable functions, every polynomial in*

$$f_1(x), \quad f_2(x) \ldots f_n(x)$$

*is also an integrable function.**

46. Properties of the Definite Integral (*continued*).

I. $$\int_a^b f(x)\,dx = -\int_b^a f(x)\,dx.$$

In the definition of the sums S and s, and of the definite integral $\int_a^b f(x)\,dx$, we assumed that a was less than b. This restriction is, however, unnecessary, and will now be removed.

If $a > b$, we take as before the set of points

$$a, \quad x_1, \quad x_2, \ldots x_{n-1}, \quad b,$$

and we deal with the sums

$$\left.\begin{aligned} S &= M_1(x_1 - a) + M_2(x_2 - x_1) + \ldots + M_n(b - x_{n-1}), \\ s &= m_1(x_1 - a) + m_2(x_2 - x_1) + \ldots + m_n(b - x_{n-1}). \end{aligned}\right\} \quad \ldots\ldots\ldots(1)$$

---

*This result can be extended to any continuous function of the n functions [cf. Hobson, *Theory of Functions of a Real Variable*, 1 (3rd ed., 1927), § 337 (6)].

The new sum S is equal in absolute value, but opposite in sign, to the sum obtained from

$$b, \quad x_{n-1}, \quad x_{n-2}, \ldots x_1, \quad a.$$

The existence of the bounds of S and s in (1) follows, and the definite integral is defined as the common value of these bounds, when they have a common value.

It is thus clear that, with this extension of the definition of § 41, we have
$$\int_a^b f(x)\,dx = -\int_b^a f(x)\,dx,^*$$
a, b being any points of an interval in which $f(x)$ is bounded and integrable.

II. *Let c be any point of an interval (a, b) in which $f(x)$ is bounded and integrable.*

Then
$$\int_a^b f(x)\,dx = \int_a^c f(x)\,dx + \int_c^b f(x)\,dx.$$

Consider a mode of division of (a, b) which has not c for a point of section. If we now introduce c as an additional point of section, the sum S is certainly not increased.

But the sums S for (a, c) and (c, b), given by this mode of division, are respectively not less than $\int_a^c f(x)\,dx$ and $\int_c^b f(x)\,dx$.

Thus every mode of division of (a, b) gives a sum S not less than
$$\int_a^c f(x)\,dx + \int_c^b f(x)\,dx.$$

It follows that
$$\int_a^b f(x)\,dx \geqq \int_a^c f(x)\,dx + \int_c^b f(x)\,dx.$$

If we consider the sum s, in the same way we find that every mode of division of (a, b) gives a sum s not greater than
$$\int_a^c f(x)\,dx + \int_c^b f(x)\,dx.$$

It follows that
$$\int_a^b f(x)\,dx \leqq \int_a^c f(x)\,dx + \int_c^b f(x)\,dx.$$

Thus we must have
$$\int_a^b f(x)\,dx = \int_a^c f(x)\,dx + \int_c^b f(x)\,dx.$$

---

*The results proved in §§ 42–45 are also applicable, in some cases with slight verbal alterations, to the Definite Integral thus generalised.

If c lies on (a, b) produced in either direction, it is easy to show, as above, that this result remains true, provided that $f(x)$ is integrable in (a, c) in the one case, and (c, b) in the other.

47. If $f(x) \geqq g(x)$, and both functions are integrable in (a, b), then $\int_a^b f(x)\,dx \geqq \int_a^b g(x)\,dx$.

Let $$\phi(x) = f(x) - g(x) \geqq 0.$$

Then $\phi(x)$ is integrable in (a, b), and obviously, from the sum s,

$$\int_a^b \phi(x)\,dx \geqq 0.$$

Therefore $$\int_a^b f(x)\,dx - \int_a^b g(x)\,dx \geqq 0.$$

Corollary I. *If $f(x)$ is integrable in (a, b), then*

$$\left|\int_a^b f(x)\,dx\right| \leqq \int_a^b |f(x)|\,dx.$$

We have seen in § 43 that if $f(x)$ is integrable in (a, b), so also is $|f(x)|$.

And $$-|f(x)| \leqq f(x) \leqq |f(x)|.$$

The result follows from the above theorem.

Corollary II. *Let $f(x)$ be integrable and never negative in (a, b). If $f(x)$ is continuous at c in (a, b) and $f(c) > 0$, then $\int_a^b f(x)\,dx > 0$.*

We have seen in § 44 that if $f(x)$ is integrable in (a, b), it must have points of continuity in the interval. What is assumed here is that at one of these points of continuity $f(x)$ is positive.

Let this point c be an internal point of the interval (a, b), and not an end-point. Then there is an interval (c', c''), where $a < c' < c < c'' < b$, such that $f(x) > k$ for every point of (c', c''), k being some positive number.

Thus, since $f(x) \geqq 0$ in (a, c'), $\int_a^{c'} f(x)\,dx \geqq 0$.

And, since $f(x) > k$ in (c', c''), $\int_{c'}^{c''} f(x)\,dx \geqq k(c'' - c') > 0$.

Also, since $f(x) \geqq 0$ in (c'', b), $\int_{c''}^{b} f(x)\,dx \geqq 0$.

Adding these results, we have $\int_a^b f(x)\,dx > 0$.

The changes in the argument when c is an end-point of (a, b) are slight.

COROLLARY III. *Let $f(x) \geqq g(x)$, and both be integrable in (a, b). At a point c in (a, b), let $f(x)$ and $g(x)$ both be continuous, and $f(c) > g(c)$. Then $\int_a^b f(x)\,dx > \int_a^b g(x)\,dx$.*

This follows at once from Corollary II by writing

$$\phi(x) = f(x) - g(x).$$

By the aid of the theorem proved in § 44, the following simpler result may be obtained:

If $f(x) > g(x)$, and both are integrable in (a, b), then

$$\int_a^b f(x)dx > \int_a^b g(x)dx.$$

For, if $f(x)$ and $g(x)$ are integrable in (a, b), we know that $f(x) - g(x)$ is integrable and has an infinite number of points of continuity in (a, b).

At any one of these points $f(x) - g(x)$ is positive, and the result follows from Corollary II.

48. The First Theorem of Mean Value. Let $\phi(x)$, $\psi(x)$ be two bounded functions, integrable in (a, b), and let $\psi(x)$ keep the same sign in this interval; *e.g.* let $\psi(x) \geqq 0$ in (a, b).

Also let M, m be the upper and lower bounds of $\phi(x)$ in (a, b).

Then we have, in (a, b),

$$m \leqq \phi(x) \leqq M,$$

and multiplying by the factor $\psi(x)$, which is not negative,

$$m\psi(x) \leqq \phi(x)\psi(x) \leqq M\psi(x).$$

It follows from § 47 that

$$m\int_a^b \psi(x)\,dx \leqq \int_a^b \phi(x)\psi(x)\,dx \leqq M\int_a^b \psi(x)\,dx,$$

since $\phi(x)\psi(x)$ is also integrable in (a, b).

Therefore $$\int_a^b \phi(x)\psi(x)\,dx = \mu\int_a^b \psi(x)\,dx$$

where μ is some number satisfying the relation $m \leqq \mu \leqq M$.

It is clear that the argument applies also to the case when

$$\psi(x) \leqq 0 \text{ in } (a, b).$$

If $\phi(x)$ is continuous in (a, b), we know that it takes the value μ for some value of x in the interval (cf. § 31).

We have thus established the important theorem:

If $\phi(x)$, $\psi(x)$ are two bounded functions, integrable in (a, b), $\phi(x)$

being continuous and $\psi(x)$ keeping the same sign in the interval, then

$$\int_a^b \phi(x)\psi(x)\,dx = \phi(\xi)\int_a^b \psi(x)\,dx,$$

where ξ is some definite value of x in $a \leqq x \leqq b$.

Further, *if $\phi(x)$ is not continuous in (a, b), we replace $\phi(\xi)$ by μ, where μ satisfies the relation $m \leqq \mu \leqq M$, and m, M are the bounds of $\phi(x)$ in (a, b).*

This is usually called the *First Theorem of Mean Value.*

As a particular case, when $\phi(x)$ is continuous,

$$\int_a^b \phi(x)dx = (b-a)\phi(\xi), \quad \text{where } a \leqq \xi \leqq b.$$

It will be seen from the corollaries to the theorem in § 47 that in certain cases we can replace $a \leqq \xi \leqq b$ by $a < \xi < b$.*

However, for most applications of the theorem, the more general statement in the text is sufficient.

49. The Integral considered as a Function of its Upper Limit. Let $f(x)$ be bounded and integrable in (a, b), and let

$$F(x) = \int_a^x f(x)\,dx,$$

where x is any point in (a, b).

Then if $(x+h)$ is also in the interval,

$$F(x+h) - F(x) = \int_x^{x+h} f(x)\,dx.$$

Thus
$$F(x+h) - F(x) = \mu h,$$

where $m \leqq \mu \leqq M$, the numbers M, m being the upper and lower bounds of $f(x)$ in $(x, x+h)$.

It follows that $F(x)$ *is a continuous function of x in* (a, b).

Further, if $f(x)$ is continuous in (a, b),

$$F(x+h) - F(x) = hf(\xi), \quad \text{where } x \lesseqgtr \xi \lesseqgtr x+h.$$

When h tends to zero, $f(\xi)$ has the limit $f(x)$.

Therefore
$$\lim_{h\to 0} \frac{F(x+h) - F(x)}{h} = f(x).$$

Thus *when $f(x)$ is continuous in (a, b), $\int_a^x f(x)dx$ is continuous in (a, b), and has a differential coefficient for every value of x in (a, b), this differential coefficient being equal to $f(x)$.*

---

*Cf. Pierpont, *loc. cit.*, pp. 367–8.

This is one of the most important theorems of the Calculus. It shows that every continuous function is the differential coefficient of a continuous function, usually called its primitive, or indefinite integral.

It also gives a means of evaluating definite integrals of continuous functions. For if $f(x)$ is continuous in (a, b) and

$$F(x)=\int_a^x f(x)\,dx,$$

we know that $\dfrac{d}{dx}F(x)=f(x)$. Suppose that, by some means or other, we have obtained a continuous function $\phi(x)$ such that

$$\frac{d}{dx}\phi(x)=f(x).$$

We must then have $F(x)=\phi(x)+C$, since $\dfrac{d}{dx}(F(x)-\phi(x))=0$ in (a, b).*

To determine the constant C, we use the fact that $F(x)$ vanishes at $x=a$.

Thus we have $$\int_a^x f(x)\,dx=\phi(x)-\phi(a).$$

50. 1. The Second Theorem of Mean Value. We now come to a theorem regarding the integral $\int_a^b \phi(x)\psi(x)\,dx$ of which frequent use will be made, especially in the more symmetrical form given in (III). The proof is simpler, when we begin with the special case taken in (I), where $\phi(x)$ is monotonic decreasing and never negative in (a, b).

I. *Let $\phi(x)$ be bounded, monotonic decreasing, and never negative in (a, b); and let $\psi(x)$ be bounded and integrable, and not change its sign more than a finite number of times in (a, b).*†

Then $$\int_a^b \phi(x)\psi(x)\,dx=\phi(a)\int_a^\xi \psi(x)\,dx,$$

where ξ is some definite value of x in $a \leqq x \leqq b$.

Since we are given that $\psi(x)$ does not change sign more than a definite number of times in (a, b), we can take

$$a=a_0,\ a_1,\ a_2,\ \dots\ a_{n-1},\ a_n=b,$$

such that $\psi(x)$ keeps the same sign in the partial intervals

$$(a, a_1),\ (a_1, a_2),\ \dots\ (a_{n-1}, a_n).$$

---

*Cf. Hardy, *loc. cit.* (5th ed., 1928), 228.

†This limitation will be removed in the proof of § 50. 2.

Then $\int_a^b \phi(x)\,\psi(x)\,dx = \sum_1^n \int_{a_{r-1}}^{a_r} \phi(x)\,\psi(x)\,dx.$

Now, by the First Theorem of Mean Value,

$$\int_{a_{r-1}}^{a_r} \phi(x)\,\psi(x)\,dx = \mu_r \int_{a_{r-1}}^{a_r} \psi(x)\,dx,$$

where
$$\phi(a_{r-1}) \leqq \mu_r \leqq \phi(a_r).$$

Therefore $\int_{a_{r-1}}^{a_r} \phi(x)\,\psi(x)\,dx = \mu_r\,[F(a_r) - F(a_{r-1})],$

where we have written $F(x) = \int_a^x \psi(x)\,dx.$

Thus we have $\int_a^b \phi(x)\,\psi(x)\,dx = \sum_1^n \mu_r\,[F(a_r) - F(a_{r-1})].$ ....(1)

Since $F(a)=0$, we may add on the term $F(a)\,\phi(a)$, and we rewrite (1) in the form

$$\int_a^b \phi(x)\,\psi(x)\,dx = (\phi(a) - \mu_1)\,F(a) + \sum_2^n (\mu_{r-1} - \mu_r)\,F(a_{r-1}) + \mu_n\,F(b). \quad (2)$$

But none of these multipliers of $F(a)$, $F(a_1)$, ... $F(b)$ are negative.

We may, therefore, replace the right-hand side of (2) by

$$\mathbf{M}[(\phi(a) - \mu_1) + (\mu_1 - \mu_2) + \ldots + (\mu_{n-1} - \mu_n) + \mu_n], \text{ ....(3)}$$

where $\mathbf{M}$ is some definite number between the greatest and least of $F(a)$, $F(a_1)$, ... $F(b)$, or coinciding with one or other.

Since $F(x) = \int_a^x \psi(x)\,dx$, we know that $F(x)$ is continuous in (a, b).

Therefore there is a number ξ satisfying $a \leqq x \leqq b$, such that

$$\mathbf{M} = F(\xi) \text{ [cf. § 31. 1]}.$$

It follows from (2) and (3), that

$$\int_a^b \phi(x)\,\psi(x)\,dx = \phi(a) \int_a^\xi \psi(x)\,dx,$$

where ξ is some definite value of x in $a \leqq x \leqq b$.

The corresponding theorem for the case when $\phi(x)$ is monotonic increasing and never negative in (a, b) is stated in (II). It can be proved in exactly the same way as (I), or deduced from it by the substitution $y = b - x$ in the integral

$$\int_a^b \phi(x)\,\psi(x)\,dx.$$

II. *Let* $\phi(x)$ *be bounded, monotonic increasing, and never negative in* (a, b); *and let* $\psi(x)$ *satisfy the same conditions as in* (I).

Then $$\int_a^b \phi(x)\psi(x)\,dx = \phi(b)\int_\xi^b \psi(x)\,dx,$$

where ξ is some definite value of x in $a \leqq x \leqq b$.

We now come to the more general case where $\phi(x)$ is monotonic, but not necessarily of the same sign in (a, b).

III. *Let $\phi(x)$ be bounded, and monotonic in (a, b); and let $\psi(x)$ satisfy the same conditions as in* (I).

Then $$\int_a^b \phi(x)\psi(x)\,dx = \phi(a)\int_a^\xi \psi(x)\,dx + \phi(b)\int_\xi^b \psi(x)\,dx,$$

where ξ is some definite value of x in $a \leqq x \leqq b$.

Let $\phi(x)$ be monotonic decreasing, and $f(x) = \phi(x) - \phi(b)$.

Then $f(x)$ is monotonic decreasing, and never negative in (a, b).

Using (I) we have

$$\int_a^b f(x)\psi(x)\,dx = f(a)\int_a^\xi \psi(x)\,dx,$$

where ξ is some definite value of x in $a \leqq x \leqq b$.

It follows that

$$\int_a^b \phi(x)\psi(x)\,dx = (\phi(a) - \phi(b))\int_a^\xi \psi(x)\,dx + \phi(b)\int_a^b \psi(x)\,dx.$$

Thus $$\int_a^b \phi(x)\psi(x)\,dx = \phi(a)\int_a^\xi \psi(x)\,dx + \phi(b)\int_\xi^b \psi(x)\,dx,$$

where ξ is some definite value of x in $a \leqq x \leqq b$.

If $\phi(x)$ is monotonic increasing, we put $f(x) = \phi(x) - \phi(a)$, and use (II).

The form of the Second Theorem of Mean Value given in (III) is the most useful and easily remembered.

Other modifications may be mentioned :

Since $\phi(x)$ is monotonic in (a, b), $\phi(a+0)$ and $\phi(b-0)$ exist. Also we may give $\phi(x)$ these values at $x = a$ and $x = b$ respectively, without changing the monotonic character of $\phi(x)$, or the value of the integral

$$\int_a^b \phi(x)\psi(x)\,dx.$$

We thus obtain the theorem :

IV. *Let $\phi(x)$ be bounded and monotonic in (a, b); and let $\psi(x)$ be bounded and integrable, and not change its sign more than a finite number of times in (a, b).*

Then $$\int_a^b \phi(x)\psi(x)\,dx = \phi(a+0)\int_a^\xi \psi(x)\,dx + \phi(b-0)\int_\xi^b \psi(x)\,dx,$$

*where ξ is some definite value of x in $a \leqq x \leqq b$.**

Also it is clear that we can in the same way replace $\phi(a+0)$ and $\phi(b-0)$, respectively, by any numbers A and B, provided that $A \leqq \phi(a+0)$ and $B \geqq \phi(b-0)$ in the case of the monotonic increasing function; and $A \geqq \phi(a+0)$, $B \leqq \phi(b-0)$ in the case of the monotonic decreasing function.

We thus obtain, *with the same limitation on $\phi(x)$ and $\psi(x)$ as before,*

V. $$\int_a^b \phi(x)\psi(x)\,dx = A\int_a^\xi \psi(x)\,dx + B\int_\xi^b \psi(x)\,dx,$$

where $A \leqq \phi(a+0)$ and $B \geqq \phi(b-0)$, if $\phi(x)$ is monotonic increasing, and $A \geqq \phi(a+0)$, $B \leqq \phi(b-0)$, if $\phi(x)$ is monotonic decreasing, ξ being some definite value of x in $a \leqq x \leqq b$.

The value of ξ in (I)-(V) need not, of course, be the same, and in (V) it will depend on the values chosen for A and B.

Theorems (I) and (II) are the earliest form of the Second Theorem of Mean Value, and are due to Bonnet,† by whom they were employed in the discussion of the Theory of Fourier's Series.

Theorem (III) was given by Weierstrass in his lectures and du Bois-Reymond,‡ independently of Bonnet.

50. 2. In the proof of the Second Theorem of Mean Value given in § 50. 1, it is assumed that the second function $\psi(x)$ does not change sign more than a finite number of times in the interval (a, b).

In this section, we show that this restriction is unnecessary.

It will be sufficient to prove (I), as the other results (II)-(V) follow directly from (I).

Let $\phi(x)$ be bounded, monotonic decreasing, and never negative in (a, b); and let $\psi(x)$ be bounded and integrable in (a, b).

Then $$\int_a^b \phi(x)\psi(x)\,dx = \phi(a)\int_a^\xi \psi(x)\,dx,$$

where ξ is some definite value of x in $a \leqq x \leqq b$.

Let the positive number ϵ be chosen, as small as we please.

---

*Corresponding results hold for (I) and (II):

$$e.g.\ \int_a^b \phi(x)\psi(x)\,dx = \phi(a+0)\int_a^\xi \psi(x)\,dx, \quad a \leqq \xi \leqq b,$$

takes the place of (I).

†*Mém. cour. Acad. roy. Bruxelles*, **23** (1850), 8; also *Journal de Math.*, **14** (1849), 249.

‡*Journal für Math.*, **69** (1869), 81; and **79** (1875), 42.

There is a mode of division of (a, b), say

$$a=x_0,\ x_1,\ x_2,\ \dots\ x_{n-1},\ x_n=b,$$

such that for it

$$\left.\begin{array}{l} \text{the sum } S \text{ for } \phi(x)\psi(x) < \int_a^b \phi(x)\psi(x)\,dx+\epsilon, \\ \text{and the sum } s \text{ for } \phi(x)\psi(x) > \int_a^b \phi(x)\psi(x)\,dx-\epsilon \end{array}\right\} \quad \dots\dots (1)$$

$$\left.\begin{array}{l} \text{also the sum } S \text{ for } \psi(x) < \int_a^b \psi(x)\,dx+\dfrac{\epsilon}{\phi(a)}, \\ \text{and the sum } s \text{ for } \psi(x) > \int_a^b \psi(x)\,dx-\dfrac{\epsilon}{\phi(a)} \end{array}\right\}; \quad \dots\dots (2)$$

Let $\sigma=\sum_0^{n-1}\phi(x_r)\psi(x_r)(x_{r+1}-x_r)$

$=\sum_0^{n-1}\phi(x_r)c_r$, where $c_r=\psi(x_r)(x_{r+1}-x_r)$,

$=d_0\phi(x_0)+\sum_1^{n-1}\phi(x_r)(d_r-d_{r-1})$, where $d_r=c_0+c_1+\dots+c_r$.

$=d_0\,[\phi(x_0)-\phi(x_1)]+d_1\,[\phi(x_1)-\phi(x_2)]+\dots$
$\qquad +d_{n-2}\,[\phi(x_{n-2})-\phi(x_{n-1})+d_{n-1}\phi(x_{n-1})]$.

None of the multipliers of $d_0, d_1, \dots d_{n-1}$ are negative.

Let d_p and d_q be the smallest and largest of $d_0, d_1, \dots d_{n-1}$.

Then we have

$$d_p\left\{\sum_0^{n-2}[\phi(x_r)-\phi(x_{r+1})]+\phi(x_{n-1})\right\}\leqq\sigma\leqq d_q\left\{\sum_0^{n-2}[\phi(x_r)-\phi(x_{r+1})]+\phi(x_{n-1})\right\},$$

i.e.

$$d_p\phi(a)\leqq\sigma\leqq d_q\phi(a). \quad \dots\dots (3)$$

Thus $\sigma=\mu\,\phi(a)$, where μ is some number satisfying $d_p\leqq\mu\leqq d_q$.

Now $d_p=\sum_0^p c_r=\sum_0^p\psi(x_r)(x_{r+1}-x_r)$.

Therefore the sum s for $\psi(x)$ for $(x_0, x_1, \dots x_{p+1})\leqq d_p\leqq$ the sum S for $\psi(x)$ for $(x_0, x_1, \dots x_{p+1})$.

And $\int_a^{x_{p+1}}\psi(x)dx$ lies between these numbers s and S.

Also $(S-s)$ for $\psi(x)$ for $(x_0, x_1, \dots x_{p+1})\leqq(S-s)$ for $\psi(x)$ for $(x_0, x_1, \dots x_n)<\dfrac{2\epsilon}{\phi(a)}$ by (2).

Therefore

$$\left.\begin{array}{l} d_p>\int_a^{x_{p+1}}\psi(x)\,dx-\dfrac{2\epsilon}{\phi(a)}, \\ d_q<\int_a^{x_{q+1}}\psi(x)\,dx+\dfrac{2\epsilon}{\phi(a)} \end{array}\right\} \quad \dots\dots (4)$$

and similarly (the second inequality above).

Therefore, by (3) and (4),

$$\phi(a)\int_a^{x_{p+1}}\psi(x)\,dx-2\epsilon<d_p\phi(a)\leqq\sigma\leqq d_q\phi(a)<\int_a^{x_{q+1}}\psi(x)\,dx+2\epsilon. \quad \dots\dots (5)$$

But the sum s for

$\phi(x)\psi(x)$ for $(x_0, x_1, \dots x_n)\leqq\sigma\leqq$ the sum S for $\phi(x)\psi(x)$ for $(x_0, x_1, \dots x_n)$.

Therefore by (1), $$\left| \sigma - \int_a^b \phi(x)\psi(x)\,dx \right| < \epsilon,$$

and $$\sigma - \epsilon < \int_a^b \phi(x)\psi(x)\,dx < \sigma + \epsilon.$$

Therefore by (5),

$$\phi(a)\int_a^{x_p+1} \psi(x)\,dx - 3\epsilon < \sigma - \epsilon < \int_a^b \phi(x)\psi(x)\,dx < \sigma + \epsilon <$$
$$\phi(a)\int_a^{x_q+1} \psi(x)\,dx + 3\epsilon. \quad \text{......................(6)}$$

Let M, m be the largest and smallest values of $\int_a^x \psi(x)\,dx$ in (a, b).

Then $\phi(a)\int_a^{x_p+1} \psi(x)\,dx \geqq m\phi(a)$ and $\phi(a)\int_a^{x_q+1} \psi(x)\,dx \leqq M\phi(a)$.

Thus we have from (6),

$$m\phi(a) - 3\epsilon < \int_a^b \phi(x)\psi(x)\,dx < M\phi(a) + 3\epsilon.$$

And it follows that

$$m\phi(a) \leqq \int_a^b \phi(x)\psi(x)\,dx \leqq M\phi(a).$$

Hence $$\int_a^b \phi(x)\psi(x)\,dx = \phi(a)\int_a^{\xi} \psi(x)\,dx,$$

where ξ is some definite value of x in $a \leqq x \leqq b$.

INFINITE INTEGRALS. INTEGRAND BOUNDED. INTERVAL INFINITE.

51. In the definition of the ordinary integral $\int_a^b f(x)\,dx$, and in the preceding sections of this chapter, we have supposed that the integrand is bounded in the interval of integration which extends from one given point a to another given point b. We proceed to extend this definition so as to include cases in which

(i) the interval increases without limit,

(ii) the integrand has a finite number of infinite discontinuities.*

I. Integrals to $+\infty$. $\int_a^{\infty} f(x)\,dx$.

Let $f(x)$ be bounded and integrable in the interval (a, b), where a is fixed and b is any number greater than a. We define the integral $\int_a^{\infty} f(x)\,dx$ as $\lim_{x\to\infty} \int_a^x f(x)\,dx$, when this limit exists.†

---

*For the definition of the term "infinite discontinuities," see § 33.

†It is more convenient to use this notation, but, if the presence of the variable x in the integrand offers difficulty, we may replace these integrals by

$$\int_a^x f(t)\,dt \quad \text{and} \quad \lim_{x\to\infty} \int_a^x f(t)\,dt.$$

We speak of $\int_a^\infty f(x)dx$ in this case as an *infinite integral,* and say that it *converges.*

On the other hand, when $\int_a^x f(x)dx$ tends to ∞ as $x\to\infty$, we say that the infinite integral $\int_a^\infty f(x)dx$ *diverges* to ∞, and there is a similar definition of *divergence* to $-\infty$ of $\int_a^\infty f(x)dx$.

Ex. 1. $$\int_0^\infty e^{-x}\,dx=1; \quad \int_1^\infty \frac{dx}{x^{\frac{3}{2}}}=2.$$

For $$\int_0^\infty e^{-x}\,dx=\lim_{x\to\infty}\int_0^x e^{-x}\,dx=\lim_{x\to\infty}(1-e^{-x})=1.$$

And $$\int_1^\infty \frac{dx}{x^{\frac{3}{2}}}=\lim_{x\to\infty}\int_1^x \frac{dx}{x^{\frac{3}{2}}}=\lim_{x\to\infty}2\left(1-\frac{1}{x^{\frac{1}{2}}}\right)=2.$$

Ex. 2. $$\int_0^\infty e^x\,dx=\infty\ ; \quad \int_1^\infty \frac{dx}{\sqrt{x}}=\infty.$$

For $$\int_0^\infty e^x\,dx=\lim_{x\to\infty}\int_0^x e^x\,dx=\lim_{x\to\infty}(e^x-1)=\infty.$$

And $$\int_1^\infty \frac{dx}{\sqrt{x}}=\lim_{x\to\infty}\int_1^x \frac{dx}{\sqrt{x}}=\lim_{x\to\infty}2(\sqrt{x}-1)=\infty.$$

Similarly $$\int_1^\infty \log\frac{1}{x}\,dx=-\infty; \quad \int_2^\infty \frac{dx}{1-x}=-\infty.$$

These integrals diverge to ∞ or $-\infty$, as the case may be.

Finally, when none of these alternatives occur, we say that the infinite integral $\int_a^\infty f(x)dx$ *oscillates* finitely or infinitely, as in §§ 16 and 25.

Ex. 3.
$$\int_a^\infty \sin x\,dx \text{ oscillates finitely.}$$
$$\int_a^\infty x\sin x\,dx \text{ oscillates infinitely.}$$

II. Integrals to $-\infty$. $\int_{-\infty}^b f(x)\,dx$.

When $f(x)$ is bounded and integrable in the interval (a, b), where b is fixed and a is any number less than b, we define the integral $\int_{-\infty}^b f(x)dx$ as $\lim_{x\to-\infty}\int_x^b f(x)\,dx$, when this limit exists.

We speak of $\int_{-\infty}^b f(x)dx$ as an infinite integral, and say that it converges.

The cases in which $\int_{-\infty}^b f(x)dx$ is said to diverge to ∞ or to $-\infty$, or to oscillate finitely or infinitely, are treated as before.

Ex. 1. $$\int_{-\infty}^0 e^x\,dx=1, \quad \int_{-\infty}^0 \frac{dx}{(1-2x)^2}=\frac{1}{2}.$$

Ex. 2. $\int_{-\infty}^{0} e^{-x}\,dx$ diverges to ∞.

$\int_{-\infty}^{0} \sinh x\,dx$ diverges to $-\infty$.

$\int_{-\infty}^{0} \sin x\,dx$ oscillates finitely.

$\int_{-\infty}^{0} x\sin x\,dx$ oscillates infinitely.

III. Integrals from $-\infty$ to ∞. $\int_{-\infty}^{\infty} f(x)\,dx$.

If the infinite integrals $\int_{-\infty}^{a} f(x)\,dx$ and $\int_{a}^{\infty} f(x)\,dx$ are both convergent, we say that the infinite integral $\int_{-\infty}^{\infty} f(x)\,dx$ is convergent and is equal to their sum.

Since
$$\int_{a}^{x} f(x)\,dx = \int_{a}^{\alpha} f(x)\,dx + \int_{\alpha}^{x} f(x)\,dx, \quad a<\alpha<x,$$
it follows that, if one of the two integrals $\int_{a}^{\infty} f(x)\,dx$ or $\int_{\alpha}^{\infty} f(x)\,dx$ converges, the other does.

Also
$$\int_{a}^{\infty} f(x)\,dx = \int_{a}^{\alpha} f(x)\,dx + \int_{\alpha}^{\infty} f(x)\,dx.$$

Similarly,
$$\int_{x}^{a} f(x)\,dx = \int_{x}^{\alpha} f(x)\,dx + \int_{\alpha}^{a} f(x)\,dx, \quad x<\alpha<a,$$
and, if one of the two integrals $\int_{-\infty}^{a} f(x)\,dx$ or $\int_{-\infty}^{\alpha} f(x)\,dx$ converges, the other does.

Also
$$\int_{-\infty}^{a} f(x)\,dx = \int_{-\infty}^{\alpha} f(x)\,dx + \int_{\alpha}^{a} f(x)\,dx.$$

Thus
$$\int_{-\infty}^{a} f(x)\,dx + \int_{a}^{\infty} f(x)\,dx = \int_{-\infty}^{\alpha} f(x)\,dx + \int_{\alpha}^{\infty} f(x)\,dx,$$
and the value of $\int_{-\infty}^{\infty} f(x)\,dx$ is independent of the point a used in the definition.

Ex.
$$\int_{-\infty}^{\infty} \frac{dx}{1+x^2} = \pi, \quad \int_{-\infty}^{\infty} e^{-x^2}\,dx = 2\int_{0}^{\infty} e^{-x^2}\,dx.$$

52. A necessary and sufficient condition for the convergence of $\int_{a}^{\infty} f(x)\,dx$.

Let
$$F(x) = \int_{a}^{x} f(x)\,dx.$$

The conditions under which $F(x)$ shall have a limit as $x \to \infty$

have been discussed in §§ 27 and 29. 1. In the case of the infinite integral we are thus able to say that:

I. *The integral* $\int_a^\infty f(x)\,dx$ *is convergent and has the value* I, *when, any positive number* ϵ *having been chosen, as small as we please, there is a positive number* X *such that*

$$\left| I - \int_a^x f(x)\,dx \right| < \epsilon, \text{ provided that } x \geqq X.$$

And further:

II. *A necessary and sufficient condition for the convergence of the integral* $\int_a^\infty f(x)dx$ *is that, when any positive number* ϵ *has been chosen, as small as we please, there shall be a positive number* X *such that*

$$\left| \int_{x'}^{x''} f(x)\,dx \right| < \epsilon$$

for all values of x', x'' *for which* $x'' > x' \geqq X$.

We have seen in § 51 that if $\int_a^\infty f(x)\,dx$ converges, then

$$\int_a^\infty f(x)\,dx = \int_a^\alpha f(x)\,dx + \int_\alpha^\infty f(x)\,dx, \quad a < \alpha.$$

It follows from (I) that, if $\int_a^\infty f(x)\,dx$ converges, to the arbitrary positive number ϵ there corresponds a positive number X such that

$$\left| \int_x^\infty f(x)\,dx \right| < \epsilon, \text{ when } x \geqq X.$$

Also, if this condition is satisfied, the integral converges.

These results, and the others given in §§ 53-58, can be extended immediately to the infinite integral

$$\int_{-\infty}^a f(x)\,dx.$$

53. $\int_a^\infty$ **f(x)dx. Integrand Positive.** If the integrand $f(x)$ is positive when $x > a$, it is clear that $\int_a^x f(x)\,dx$ is a monotonic increasing function of x. Thus $\int_a^\infty f(x)\,dx$ must either converge or diverge to ∞.

I. *It will converge if there is a positive number* A *such that* $\int_a^x f(x)\,dx < A$ *when* $x > a$, *and in this case* $\int_a^\infty f(x)\,dx \leqq A$.

It will diverge to ∞ if there is no such number.

These statements follow from the properties of monotonic functions (§ 34).

Further, there is an important "comparison test" for the convergence of integrals when the integrand is positive.

II. *Let $f(x)$, $g(x)$ be two functions which are positive, bounded and integrable in the arbitrary interval (a, b). Also let $g(x) \leqq f(x)$ when $x \geqq a$. Then, if $\int_a^\infty f(x)\,dx$ is convergent, it follows that $\int_a^\infty g(x)\,dx$ is convergent, and $\int_a^\infty g(x)\,dx \leqq \int_a^\infty f(x)\,dx$.*

For from § 47 we know that

$$\int_a^x g(x)\,dx \leqq \int_a^x f(x)\,dx, \text{ when } x>a.$$

Therefore $$\int_a^x g(x)\,dx < \int_a^\infty f(x)\,dx.$$

Then, from (I), $$\int_a^\infty g(x)\,dx \leqq \int_a^\infty f(x)\,dx.$$

III. *If $g(x) \geqq f(x)$, and $\int_a^\infty f(x)\,dx$ diverges, so also does*

$$\int_a^\infty g(x)\,dx.^*$$

This follows at once, since $\int_a^x g(x)\,dx \geqq \int_a^x f(x)\,dx$.

One of the most useful integrals for comparison is $\int_a^\infty \frac{dx}{x^n}$, where $a>0$.

We have $$\int_a^x \frac{dx}{x^n} = \frac{1}{1-n}\{x^{1-n} - a^{1-n}\}, \quad \text{when } n \neq 1,$$

and $$\int_a^x \frac{dx}{x} = \log x - \log a, \quad \text{when } n=1.$$

---

*Since the relative behaviour of the positive integrands $f(x)$ and $g(x)$ matters only as $x \to \infty$, these conditions may be expressed in terms of limits:

When $g(x)/f(x)$ has a limit as $x \to \infty$, $\int_a^\infty g(x)\,dx$ converges, if $\int_a^\infty f(x)\,dx$ converges.

When $g(x)/f(x)$ has a limit, not zero, or diverges, as $x \to \infty$, $\int_a^\infty g(x)\,dx$ diverges, if $\int_a^\infty f(x)\,dx$ diverges.

Thus, when $n>1$, $\lim_{x\to\infty}\int_a^x \frac{dx}{x^n}=\frac{a^{1-n}}{n-1}$,

$$i.e. \quad \int_a^\infty \frac{dx}{x^n}=\frac{a^{1-n}}{n-1}.$$

And, when $n\leqq 1$, $\lim_{x\to\infty}\int_a^x \frac{dx}{x^n}=\infty$,

$$i.e. \quad \int_a^\infty \frac{dx}{x^n} \text{ diverges.}$$

Ex. 1. $\int_a^\infty \frac{dx}{x\sqrt{(1+x^2)}}$ converges, since $\frac{1}{x\sqrt{(1+x^2)}}<\frac{1}{x^2}$, when $x\geqq a>0$.

2. $\int_2^\infty \frac{dx}{\sqrt{(x^2-1)}}$ diverges, since $\frac{1}{\sqrt{(x^2-1)}}>\frac{1}{x}$, when $x\geqq 2$.

3. $\int_a^\infty \frac{\sin^2 x}{x^2}\,dx$ converges, since $\frac{\sin^2 x}{x^2}\leqq\frac{1}{x^2}$, when $x\geqq a>0$.

54. Absolute Convergence. *The integral* $\int_a^\infty f(x)\,dx$ *is said to be* **absolutely** *convergent when* $f(x)$ *is bounded and integrable in the arbitrary interval* (a, b), *and* $\int_a^\infty |f(x)|\,dx$ *is convergent.*

Since $\left|\int_{x'}^{x''} f(x)\,dx\right| \leqq \int_{x'}^{x''} |f(x)|\,dx$, for $x''>x'\geqq a$
(cf. § 47, Cor. I),

it follows from § 52, II that if $\int_a^\infty |f(x)|\,dx$ converges, so also does $\int_a^\infty f(x)\,dx$.

But the converse is not true. *An infinite integral of this type may converge, and yet not converge* **absolutely**.

For example, consider the integral

$$\int_0^\infty \frac{\sin x}{x}\,dx.$$

The Second Theorem of Mean Value (§ 50. 1) shows that this integral converges.

For we have

$$\int_{x'}^{x''} \frac{\sin x}{x}\,dx=\frac{1}{x'}\int_{x'}^{\xi} \sin x\,dx+\frac{1}{x''}\int_{\xi}^{x''} \sin x\,dx,$$

where $0<x'\leqq\xi\leqq x''$.

But

$\left|\int_{x'}^{\xi} \sin x\,dx\right|$ and $\left|\int_{\xi}^{x''} \sin x\,dx\right|$ are each less than or equal to 2.

Therefore $$\left|\int_{x'}^{x''} \frac{\sin x}{x}\, dx\right| \leqq 2\left\{\frac{1}{x'}+\frac{1}{x''}\right\}$$
$$< \frac{4}{x'}.$$

Thus $$\left|\int_{x'}^{x''} \frac{\sin x}{x}\, dx\right| < \epsilon, \quad \text{when } x'' > x' \geqq X,$$

provided that $$X > \frac{4}{\epsilon}.$$

Therefore $\int_0^{\infty} \frac{\sin x}{x}\, dx$ converges, and we shall find in § 88 that its value is $\frac{1}{2}\pi$.

But the integral $\int_0^{\infty} \frac{|\sin x|}{x}\, dx$ diverges.

To prove this, it is only necessary to consider the integral
$$\int_0^{n\pi} \frac{|\sin x|}{x}\, dx,$$
where n is any positive integer.

We have $$\int_0^{n\pi} \frac{|\sin x|}{x}\, dx = \sum_1^n \int_{(r-1)\pi}^{r\pi} \frac{|\sin x|}{x}\, dx.$$

But $$\int_{(r-1)\pi}^{r\pi} \frac{|\sin x|}{x}\, dx = \int_0^{\pi} \frac{\sin y}{(r-1)\pi + y}\, dy,$$
on putting $x = (r-1)\pi + y$.

Therefore $$\int_{(r-1)\pi}^{r\pi} \frac{|\sin x|}{x}\, dx > \frac{1}{r\pi}\int_0^{\pi} \sin y\, dy$$
$$> \frac{2}{r\pi}.$$

Thus $$\int_0^{n\pi} \frac{|\sin x|}{x}\, dx > \frac{2}{\pi}\sum_1^n \frac{1}{r}.$$
But the series on the right hand diverges to ∞ as $n \to \infty$.

Therefore $$\lim_{n\to\infty} \int_0^{n\pi} \frac{|\sin x|}{x}\, dx = \infty.$$

But when $x > n\pi$,
$$\int_0^{x} \frac{|\sin x|}{x}\, dx > \int_0^{n\pi} \frac{|\sin x|}{x}\, dx.$$

Therefore $$\lim_{x\to\infty} \int_0^{x} \frac{|\sin x|}{x}\, dx = \infty.$$

When infinite integrals of this type converge, but do not converge absolutely, the convergence must be due to changes of sign in the integrand as $x \to \infty$.

55. The μ-Test for the Convergence of $\int_a^\infty f(x)\,dx$.

I. *Let $f(x)$ be bounded and integrable in the arbitrary interval (a, b) where $a > 0$. If there is a number μ greater than 1 such that $x^\mu f(x)$ is bounded when $x \geqq a$, then $\int_a^\infty f(x)\,dx$ converges absolutely.*

Here $|x^\mu f(x)| < A$, where A is some definite positive number and $x \geqq a$.

Thus
$$|f(x)| < \frac{A}{x^\mu}.$$

But we know that $\int_a^\infty \frac{dx}{x^\mu}$ converges.

It follows that $\int_a^\infty |f(x)|\,dx$ converges.

Therefore $\int_a^\infty f(x)\,dx$ converges, and the convergence is absolute.

II. *Let $f(x)$ be bounded and integrable in the arbitrary interval (a, b), where $a > 0$. If there is a number μ less than or equal to 1 such that $x^\mu f(x)$ has a positive lower bound when $x \geqq a$, then $\int_a^\infty f(x)\,dx$ diverges to ∞.*

Here we have, as before,
$$x^\mu f(x) \geqq A > 0, \text{ when } x \geqq a.$$

It follows that
$$\frac{A}{x^\mu} \leqq f(x).$$

But $\int_a^\infty \frac{dx}{x^\mu}$ diverges to ∞ when $\mu \leqq 1$.

It follows that $\int_a^\infty f(x)\,dx$ diverges to ∞.

III. *Let $f(x)$ be bounded and integrable in the arbitrary interval (a, b), where $a > 0$. If there is a number μ less than or equal to 1 such that $x^\mu f(x)$ has a negative upper bound when $x \geqq a$, then $\int_a^\infty f(x)\,dx$ diverges to $-\infty$.*

This follows from (II), for in this case
$$-x^\mu f(x)$$
must have a positive lower bound when $x \geqq a$.

But, if $\lim_{x\to\infty} (x^\mu f(x))$ exists, it follows that $x^\mu f(x)$ is bounded in $x \geqq a$; also, by properly choosing the positive number X, $x^\mu f(x)$ will either have a positive lower bound, when this limit is positive, or a negative upper bound, when this limit is negative provided that $x \geqq X$.

Thus, from (I)-(III), the following theorem can be immediately deduced:

Let $f(x)$ be bounded and integrable in the arbitrary interval (a, b), where $a > 0$.

If there is a number μ greater than 1 such that $\lim_{x\to\infty} (x^\mu f(x))$ exists, then $\int_a^\infty f(x)\,dx$ converges.

If there is a number μ less than or equal to 1 such that $\lim_{x\to\infty} (x^\mu f(x))$ exists and is not zero, then $\int_a^\infty f(x)\,dx$ diverges; and the same is true if $x^\mu f(x)$ diverges to $+\infty$, or to $-\infty$, as $x\to\infty$.

We shall make very frequent use of this test, and refer to it as the "μ-test." It is clear that we are simply comparing the integral $\int_a^\infty f(x)\,dx$ with the integral $\int_a^\infty \frac{dx}{x^\mu}$, and deducing the convergence or divergence of the former from that of the latter.

Ex. 1. $\int_0^\infty \frac{x^2}{(a^2+x^2)^2}dx$ converges, since $\lim_{x\to\infty}\left(x^2 \times \frac{x^2}{(a^2+x^2)^2}\right)=1.$

2. $\int_0^\infty \frac{x^3}{(a^2+x^2)^2}dx$ diverges, since $\lim_{x\to\infty}\left(x \times \frac{x^3}{(a^2+x^2)^2}\right)=1.$

3. $\int_0^\infty \frac{x^{\frac{3}{2}}}{b^2x^2+c^2}dx$ diverges, since $\lim_{x\to\infty}\left(x^{\frac{1}{2}} \times \frac{x^{\frac{3}{2}}}{b^2x^2+c^2}\right)=\frac{1}{b^2}.$

It should be noticed that the theorems of this section do not apply to the integral $\int_0^\infty \frac{\sin x}{x}dx$.

56. Further Tests for the Convergence of $\int_a^\infty f(x)dx$.

I. *If $\phi(x)$ is bounded when $x \geqq a$, and integrable in the arbitrary interval (a, b), and $\int_a^\infty \psi(x)\,dx$ converges absolutely, then $\int_a^\infty \phi(x)\psi(x)\,dx$ is absolutely convergent.*

For we have $|\phi(x)| < A$, where A is some definite positive number and $x \geqq a$.

Also $$\int_{x'}^{x''} |\phi(x)|\,|\psi(x)|\,dx < A\int_{x'}^{x''} |\psi(x)|\,dx,$$
when $x'' > x' > a$.

Since we are given that $\int_a^\infty |\psi(x)|\,dx$ converges, the result follows.

Ex. 1. $\int_a^\infty \frac{\sin x}{x^{1+n}}\,dx$, $\int_a^\infty \frac{\cos x}{x^{1+n}}\,dx$ converge absolutely, when n and a are positive.

2. $\int_a^\infty e^{-ax}\cos bx\,dx$ converges absolutely, when a is positive.

3. $\int_0^\infty \frac{\cos mx}{a^2+x^2}\,dx$ converges absolutely.

II. *Let $\phi(x)$ be monotonic and bounded when $x \geqq a$. Let $\psi(x)$ be bounded and integrable in the arbitrary interval (a, b), and not change sign more than a finite number of times in the interval. Also let $\int_a^\infty \psi(x)\,dx$ converge.*

Then $\int_a^\infty \phi(x)\psi(x)\,dx$ converges.

This follows from the Second Theorem of Mean Value, since
$$\int_{x'}^{x''} \phi(x)\psi(x)\,dx = \phi(x')\int_{x'}^{\xi} \psi(x)\,dx + \phi(x'')\int_{\xi}^{x''} \psi(x)\,dx,$$
where $a < x' \leqq \xi \leqq x''$.

But $|\phi(x')|$ and $|\phi(x'')|$ are each less than some definite positive number A.

Also we can choose X so that
$$\left|\int_{x'}^{\xi} \psi(x)\,dx\right| \quad\text{and}\quad \left|\int_{\xi}^{x''} \psi(x)\,dx\right|$$
are each less than $\epsilon/2A$, when $x'' > x' \geqq X$, and ϵ is any given positive number, as small as we please.

It follows that
$$\left|\int_{x'}^{x''} \phi(x)\psi(x)\,dx\right| < \epsilon, \text{ when } x'' > x' \geqq X,$$
and the given integral converges.

Ex. 1. $\int_0^\infty e^{-x}\frac{\sin x}{x}\,dx$ converges.

2. $\int_a^\infty (1-e^{-x})\frac{\cos x}{x}\,dx$ converges when $a > 0$.

III. *Let $\phi(x)$ be monotonic and bounded when $x \geqq a$, and*

$$\lim_{x\to\infty} \phi(x) = 0.$$

Let $\psi(x)$ be bounded and integrable in the arbitrary interval (a, b), and not change sign more than a finite number of times in the interval. Also let $\int_a^x \psi(x)\,dx$ be bounded when $x > a$.

Then $\int_a^\infty \phi(x)\psi(x)\,dx$ is convergent.

As above, in (II), we know that

$$\int_{x'}^{x''} \phi(x)\psi(x)\,dx = \phi(x')\int_{x'}^{\xi} \psi(x)\,dx + \phi(x'')\int_{\xi}^{x''} \psi(x)\,dx,$$

where $a < x' \leqq \xi \leqq x''$.

But $\left|\int_a^x \psi(x)\,dx\right| < A$, when $x > a$, where A is some definite positive number.

And
$$\left|\int_{x'}^{\xi} \psi(x)\,dx\right| \leqq \left|\int_a^{\xi} \psi(x)\,dx\right| + \left|\int_a^{x'} \psi(x)\,dx\right|$$
$$< 2A.$$

Similarly
$$\left|\int_{\xi}^{x''} \psi(x)\,dx\right| < 2A.$$

Also
$$\lim_{x\to\infty} \phi(x) = 0.$$

Therefore, if ϵ is any positive number, as small as we please, there will be a positive number X such that

$$|\phi(x)| < \frac{\epsilon}{4A}, \quad \text{when } x \geqq X.$$

It follows that

$$\left|\int_{x'}^{x''} \phi(x)\psi(x)\,dx\right| < \epsilon, \quad \text{when } x'' > x' \geqq X,$$

and $\int_a^\infty \phi(x)\psi(x)\,dx$ converges.

Ex. 1. $\int_a^\infty \frac{\sin x}{x^n}\,dx$, $\int_a^\infty \frac{\cos x}{x^n}\,dx$ converge, when n and a are positive.

2. $\int_1^\infty \frac{x}{1+x^2}\sin x\,dx$ converges.

3. $\int_0^\infty \frac{\cos ax - \cos bx}{x}\,dx$ converges.

The Mean Value Theorems for the Infinite Integral.

57. The First Theorem of Mean Value.

Let $\phi(x)$ be bounded when $x \geqq a$, and integrable in the arbitrary interval (a, b). Let $\psi(x)$ keep the same sign in $x \geqq a$, and $\int_a^\infty \psi(x)dx$ converge.

Then
$$\int_a^\infty \phi(x)\psi(x)dx = \mu \int_a^\infty \psi(x)dx,$$
where $m \leqq \mu \leqq M$, the upper and lower bounds of $\phi(x)$ in $x \geqq a$ being M and m.

We have
$$m \leqq \phi(x) \leqq M, \quad \text{when } x \geqq a,$$
and, if $\psi(x) \geqq 0$,
$$m\psi(x) \leqq \phi(x)\psi(x) \leqq M\psi(x).$$

Therefore $m \int_a^x \psi(x)dx \leqq \int_a^x \phi(x)\psi(x)dx \leqq M \int_a^x \psi(x)dx, \quad \text{when } x \geqq a.$

But, by § 56, I, $\int_a^\infty \phi(x)\psi(x)dx$ converges, and we are given that $\int_a^\infty \psi(x)dx$ converges.

Thus we have from these inequalities
$$m \int_a^\infty \psi(x)dx \leqq \int_a^\infty \phi(x)\psi(x)dx \leqq M \int_a^\infty \psi(x)dx.$$

In other words,
$$\int_a^\infty \phi(x)\psi(x)dx = \mu \int_a^\infty \psi(x)dx,$$
where $m \leqq \mu \leqq M$.

58. The Second Theorem of Mean Value.

LEMMA. *Let $\int_a^\infty f(x)dx$ be a convergent integral, and $F(x) = \int_x^\infty f(x)dx \, (x \geqq a)$. Then $F(x)$ is continuous when $x \geqq a$, and bounded in the interval (a, ∞). Also it takes at least once in that interval every value between its upper and lower bounds, these being included.*

The continuity of $F(x)$ follows from the equation
$$F(x+h) - F(x) = -\int_x^{x+h} f(x)dx.$$

Further, $\lim_{x\to\infty} F(x)$ exists and is zero.

It follows from § 32 that $F(x)$ is bounded in the interval (a, ∞), as defined in that section, and, if M, m are its upper and lower bounds, it takes at least once in (a, ∞) the values M and m and every value between M, m.

Let $\phi(x)$ be bounded and monotonic when $x \geqq a$.

Let $\psi(x)$ be bounded and integrable in the arbitrary interval (a, b), and not change sign more than a finite number of times in the interval. Also let $\int_a^\infty \psi(x)dx$ converge.

Then
$$\int_a^\infty \phi(x)\psi(x)dx = \phi(a+0)\int_a^\xi \psi(x)dx + \phi(\infty)\int_\xi^\infty \psi(x)dx,$$
*where $a \leqq \xi \leqq \infty$.**

*Cf. Pierpont, *loc. cit.*, § 654.

Suppose $\phi(x)$ to be monotonic increasing.

We apply the Second Theorem of Mean Value to the arbitrary interval (a, b).

Then we have

$$\int_a^b \phi(x)\psi(x)dx = \phi(a+0)\int_a^\xi \psi(x)dx + \phi(b-0)\int_\xi^b \psi(x)dx,$$

where $a \leqq \xi \leqq b$.

Add to both sides $$B = \phi(\infty)\int_b^\infty \psi(x)dx,$$

observing that $\phi(\infty)$ exists, since $\phi(x)$ is monotonic increasing in $x \geqq a$ and does not exceed some definite number (§ 34).

Also $\lim_{b\to\infty} B = 0$ and $\int_a^\infty \phi(x)\psi(x)dx$ converges [§ 56, II].

Then $B + \int_a^b \phi(x)\psi(x)dx$

$$= \phi(a+0)\int_a^\xi \psi(x)dx + \phi(b-0)\int_\xi^b \psi(x)dx + \phi(\infty)\int_b^\infty \psi(x)dx$$

$$= \phi(a+0)\left[\int_a^\infty \psi(x)dx - \int_\xi^\infty \psi(x)dx\right] + \phi(b-0)\left[\int_\xi^\infty \psi(x)dx - \int_b^\infty \psi(x)dx\right] + \phi(\infty)\int_b^\infty \psi(x)dx$$

$$= \phi(a+0)\int_a^\infty \psi(x)dx + U + V, \quad \text{...(1)}$$

where $$U = \{\phi(b-0) - \phi(a+0)\}\int_\xi^\infty \psi(x)dx,$$

and $$V = \{\phi(\infty) - \phi(b-0)\}\int_b^\infty \psi(x)dx.$$

Now we know from the above Lemma that $\int_x^\infty \psi(x)dx$ is bounded in (a, ∞).

Let M, m be its upper and lower bounds.

Then $$m \leqq \int_\xi^\infty \psi(x)dx \leqq M,$$

and $$m \leqq \int_b^\infty \psi(x)dx \leqq M.$$

Therefore $\{\phi(b-0) - \phi(a+0)\}m \leqq U \leqq \{\phi(b-0) - \phi(a+0)\}M,$

$\{\phi(\infty) \quad - \phi(b-0)\}m \leqq V \leqq \{\phi(\infty) - \phi(b-0)\}M.$

Adding these, we see that

$$\{\phi(\infty) - \phi(a+0)\}m \leqq U + V \leqq \{\phi(\infty) - \phi(a+0)\}M.$$

Therefore $U + V = \mu\{\phi(\infty) - \phi(a+0)\}$, where $m \leqq \mu \leqq M$.

Insert this value for $U+V$ in (1), and proceed to the limit when $b \to \infty$.

Then $$\int_a^\infty \phi(x)\psi(x)dx = \phi(a+0)\int_a^\infty \psi(x)dx + \mu'\{\phi(\infty) - \phi(a+0)\},$$

where $\mu' = \lim_{b\to\infty} \mu$.

This limit must exist, since the other terms in (1) have limits when $b \to \infty$.

Also, since $$m \leqq \mu \leqq M,$$
it follows that $$m \leqq \mu' \leqq M.$$

But $\int_x^\infty \psi(x)dx$ takes the value μ' at least once in the interval (a, ∞).

Thus we may put $\mu' = \int_\xi^\infty \psi(x)dx$, where $a \leqq \xi' \leqq \infty$.

Therefore we have finally

$$\int_a^\infty \phi(x)\psi(x)dx = \phi(a+0)\int_a^{\xi'} \psi(x)dx + \phi(\infty)\int_{\xi'}^\infty \psi(x)dx,$$

where $a \leqq \xi' \leqq \infty$.

It is clear that we might have used the other forms (III) and (V), § 50. 1, of the Second Theorem of Mean Value and obtained corresponding results.

INFINITE INTEGRALS. INTEGRAND INFINITE.

59. $\int_a^b f(x)dx$. In the preceding sections we have dealt with the infinite integrals $\int_a^\infty f(x)\,dx$, $\int_{-\infty}^a f(x)\,dx$ and $\int_{-\infty}^\infty f(x)\,dx$, when the integrand $f(x)$ is bounded in any arbitrary interval, however large.

A further extension of the definition of the integral is required so as to include the case in which $f(x)$ has a finite number of infinite discontinuities (cf. § 33) in the interval of integration.

First we take the case when a is the only point of infinite discontinuity in (a, b). The integrand $f(x)$ is supposed bounded and integrable in the arbitrary interval $(a+\xi,\ b)$, where $a < a+\xi < b$.

On this understanding, *if the integral* $\int_{a+\xi}^b f(x)\,dx$ *has a limit as* $\xi \to 0$, *we define the infinite integral* $\int_a^b f(x)\,dx$ *as* $\lim_{\xi\to 0}\int_{a+\xi}^b f(x)\,dx$.

Similarly, *when the point b is the only point of infinite discontinuity in (a, b), and $f(x)$ is bounded and integrable in the arbitrary interval $(a, b-\xi)$, where $a < b-\xi < b$, we define the infinite integral* $\int_a^b f(x)\,dx$ as $\lim_{\xi\to 0}\int_a^{b-\xi} f(x)\,dx$, *when this limit exists.*

Again, *when a and b are both points of infinite discontinuity, we define the infinite integral* $\int_a^b f(x)\,dx$ *as the sum of the infinite integrals* $\int_a^c f(x)\,dx$ *and* $\int_c^b f(x)\,dx$, *when these integrals exist, as defined above, c being a point between a and b.*

This definition is independent of the position of c between a and b, since we have

$$\int_a^c f(x)\,dx = \int_a^{c'} f(x)\,dx + \int_{c'}^c f(x)\,dx,$$

where $a < c' < c$ (cf. § 51, III).

Finally, *let there be a finite number of points of infinite discontinuity in the interval* (a, b). *Let these points be* $x_1, x_2, \dots x_n$, *where* $a \leqq x_1 < x_2, \dots < x_n \leqq b$. *We define the infinite integral* $\int_a^b f(x)\,dx$ *by the equation*

$$\int_a^b f(x)\,dx = \int_a^{x_1} f(x)\,dx + \int_{x_1}^{x_2} f(x)\,dx + \dots + \int_{x_n}^b f(x)\,dx,$$

when the integrals on the right-hand exist, according to the definitions just given.

It should be noticed that with this definition there are only to be a finite number of points of infinite discontinuity, and $f(x)$ is to be bounded in any partial interval of (a, b), which has not one of these points as an interval point or an end-point.

This definition was extended by du Bois-Reymond, Dini and Harnack to certain cases in which the integrand has an infinite number of points of infinite discontinuity, but the case given in the text is amply sufficient for our purpose. The modern treatment of the integral has rendered further generalisation of Riemann's discussion chiefly of historical interest.

It is convenient to speak of the infinite integrals of this and the succeeding section as *convergent*, as we did when one or other of the limits of integration was infinite, and the terms *divergent* and *oscillatory* are employed as before.

Some writers use the term *proper integral* for the ordinary integral $\int_a^b f(x)dx$, when $f(x)$ is bounded and integrable in the interval (a, b), and *improper integral* for the case when it has points of infinite discontinuity in (a, b), reserving the term *infinite integral* for

$$\int_a^\infty f(x)dx, \quad \int_{-\infty}^a f(x)dx \text{ or } \int_{-\infty}^\infty f(x)dx.$$

French mathematicians refer to both as *intégrales généralisées*; Germans refer to both as *uneigentliche Integrale*, to distinguish them from *eigentliche Integrale* or ordinary integrals.

60. $\int_a^\infty f(x)\,dx$. $\int_{-\infty}^a f(x)\,dx$. $\int_{-\infty}^\infty f(x)\,dx$.

Let $f(x)$ have infinite discontinuities at a finite number of points in any interval, however large.

For example, let there be infinite discontinuities only at x_1, x_2, ... x_n in $x \geqq a$, $f(x)$ being bounded in any interval (c, b), where $c > x_n$.

Let $$a \leqq x_1 < x_2, \ldots < x_n < b.$$

Then we have, as above (§ 59),

$$\int_a^b f(x)\,dx = \int_a^{x_1} f(x)\,dx + \int_{x_1}^{x_2} f(x)\,dx + \ldots + \int_{x_n}^{c} f(x)\,dx + \int_c^b f(x)\,dx,$$

where $x_n < c < b$, provided that the integrals on the right-hand side exist.

It will be noticed that the last integral $\int_c^b f(x)\,dx$ is an ordinary integral, $f(x)$ being bounded and integrable in (c, b).

If the integral $\int_c^\infty f(x)\,dx$ *also converges, we define the infinite integral* $\int_a^\infty f(x)\,dx$ *by the equation:*

$$\int_a^\infty f(x)\,dx = \int_a^{x_1} f(x)\,dx + \int_{x_1}^{x_2} f(x)\,dx + \ldots + \int_{x_n}^{c} f(x)\,dx + \int_c^\infty f(x)\,dx.$$

It is clear that this definition is independent of the position of c, since we have

$$\int_{x_n}^{c} f(x)\,dx + \int_c^\infty f(x)\,dx = \int_{x_n}^{c'} f(x)\,dx + \int_{c'}^\infty f(x)\,dx,$$

where $x_n < c < c'$.

Also we may write the above in the form

$$\int_a^\infty f(x)\,dx = \int_a^{x_1} f(x)\,dx + \int_{x_1}^{x_2} f(x)\,dx + \ldots + \int_{x_n}^\infty f(x)\,dx.$$

The verbal alterations required in the definition of $\int_{-\infty}^{a} f(x)\,dx$ are obvious, and we define $\int_{-\infty}^{\infty} f(x)\,dx$, as before, as the sum of the integrals $\int_{-\infty}^{a} f(x)\,dx$ and $\int_a^\infty f(x)\,dx$.

It is easy to show that this definition is independent of the position of the point a.

61. Tests for Convergence of $\int_a^b f(x)\,dx$. It is clear that we need only discuss the case when there is a point of infinite discontinuity at an end of the interval of integration.

If $x=a$ is the only point of infinite discontinuity, we have

$$\int_a^b f(x)\,dx = \lim_{\xi \to 0} \int_{a+\xi}^b f(x)\,dx,$$

when this limit exists.

It follows at once, from the definition, that:

I. *The integral* $\int_a^b f(x)\,dx$ *is convergent and has the value I when, any positive number* ϵ *having been chosen, as small as we please, there is a positive number* η *such that*

$$\left| I - \int_{a+\xi}^b f(x)\,dx \right| < \epsilon, \text{ provided that } 0 < \xi \leqq \eta.$$

And further:

II. *A necessary and sufficient condition for the convergence of the interval* $\int_a^b f(x)\,dx$ *is that, if any positive number* ϵ *has been chosen, as small as we please, there shall be a positive number* η *such that*

$$\left| \int_{a+\xi''}^{a+\xi'} f(x)\,dx \right| < \epsilon, \text{ when } 0 < \xi'' < \xi' \leqq \eta.$$

Also, if this infinite integral $\int_a^b f(x)\,dx$ converges, we have

$$\int_a^b f(x)\,dx = \int_a^x f(x)\,dx + \int_x^b f(x)\,dx, \quad a < x < b.$$

It follows from (I) that, if $\int_a^b f(x)\,dx$ converges, to the arbitrary positive number ϵ, there corresponds a positive number η such that

$$\left| \int_a^x f(x)\,dx \right| < \epsilon, \text{ when } 0 < (x-a) \leqq \eta.$$

Absolute Convergence. *The infinite integral* $\int_a^b f(x)\,dx$ *is said to be* **absolutely** *convergent, if* $f(x)$ *is bounded and integrable in the arbitrary interval* $(a+\xi, b)$, *where* $0 < \xi < b-a$, *and* $\int_a^b |f(x)|\,dx$ *converges.*

It follows from (II) that absolute convergence carries with it ordinary convergence. But the converse is not true. An *infinite integral of this kind may converge, but not converge* **absolutely,*** as the following example shows.

---

*Cf. § 43, V; § 47, Cor. I; and § 54.

An example of such an integral is suggested at once by § 54.

It is clear that $$\int_0 \frac{\sin 1/x}{x}\,dx$$

converges, but not absolutely, for this integral is reduced to

$$\int_1^\infty \frac{\sin x}{x}\,dx$$

by substituting $1/x$ for x.

Again, it is clear that $\int_a^b \frac{dx}{(x-a)^n}$ converges, if $0<n<1$.

For we have

$$\int_{a+\xi}^b \frac{dx}{(x-a)^n}=\frac{1}{1-n}\{(b-a)^{1-n}-\xi^{1-n}\}.$$

Therefore $\lim\limits_{\xi\to 0}\int_{a+\xi}^b \frac{dx}{(x-a)^n}=\frac{(b-a)^{1-n}}{1-n}$, when $0<n<1$.

Also the integral diverges when $n\geqq 1$.

From this we obtain results which correspond to those of § 55.

III. *Let $f(x)$ be bounded and integrable in the arbitrary interval $(a+\xi, b)$, where $0<\xi<b-a$. If there is a number μ between 0 and 1 such that $(x-a)^\mu f(x)$ is bounded when $a<x\leqq b$, then $\int_a^b f(x)\,dx$ converges absolutely.*

Again,

IV. *Let $f(x)$ be bounded and integrable in the arbitrary interval $(a+\xi, b)$, where $0<\xi<b-a$. If there is a number μ greater than or equal to 1 such that $(x-a)^\mu f(x)$ has a positive lower bound when $a<x\leqq b$, or a negative upper bound, then $\int_a^b f(x)\,dx$ diverges to $+\infty$ in the first case, and to $-\infty$ in the second case.*

And finally,

V. *Let $f(x)$ be bounded and integrable in the arbitrary interval $(a+\xi, b)$, where $0<\xi<b-a$.*

If there is a number μ between 0 and 1 such that $\lim\limits_{x\to a+0}(x-a)^\mu f(x)$ exists, then $\int_a^b f(x)\,dx$ converges absolutely.

If there is a number μ greater than or equal to 1 such that $\lim\limits_{x\to a+0}(x-a)^\mu\, f(x)$ exists and is not zero, then $\int_a^b f(x)\,dx$ diverges; and the same is true if $(x-a)^\mu f(x)$ tends to $+\infty$, or to $-\infty$, as $x\to a+0$.

We shall speak of this test as the μ-test for the infinite integral $\int_a^b f(x)\,dx$, when $x=a$ is a point of infinite discontinuity. It is clear that in applying this test we are simply asking ourselves the order of the infinity that occurs in the integrand.

The results can be readily adapted to the case when the upper limit b is a point of infinite discontinuity.

Also, it is easy to show that

VI. *If* $\phi(x)$ *is bounded and integrable in* (a, b), *and* $\int_a^b \psi(x)\,dx$ *converges absolutely, then* $\int_a^b \phi(x)\psi(x)\,dx$ *is absolutely convergent.* (Cf. § 56, I.)

The tests given in (III)-(VI) will cover most of the cases which we shall meet. But it would not be difficult to develop in detail the results which correspond to the other tests obtained for the convergence of the infinite integral $\int_a^\infty f(x)\,dx$.

No special discussion is required for the integral $\int_a^b f(x)\,dx$, when a certain number of points of infinite discontinuity occur in (a, b), or for $\int_a^\infty f(x)\,dx$, $\int_{-\infty}^a f(x)\,dx$, and $\int_{-\infty}^\infty f(x)\,dx$, as defined in § 60. These integrals all reduce to the sum of integrals of the types for which we have already obtained the required criteria.

We add some examples illustrating the points to which we have referred.

Ex. 1. Prove that $\int_0^1 \frac{dx}{(1+x)\sqrt{x}}$ converges and that $\int_0^1 \frac{dx}{x(1+x)}$ diverges.

(i) Let
$$f(x)=\frac{1}{(1+x)\sqrt{x}}.$$

Then
$$\lim_{x\to 0}\sqrt{x}\,f(x)=1.$$

The μ-test thus establishes the convergence of $\int_0^1 \frac{dx}{(1+x)\sqrt{x}}$.

(ii) Let
$$f(x)=\frac{1}{x(1+x)}.$$

Then
$$\lim_{x\to 0} x f(x)=1.$$

Therefore the integral diverges by the same test.

Ex. 2. Prove that $\int_0^{\frac{1}{2}\pi} \frac{\sin x}{x^{1+n}}\,dx$ converges, when $0<n<1$.

The integral is an ordinary finite integral if $n\leqq 0$.

Also
$$\lim_{x\to 0} x^n \left(\frac{\sin x}{x^{1+n}}\right) = 1.$$

Therefore the integral converges when $0 < n < 1$ It diverges when $n \leqq 1$.

Ex. 3. Prove that $\int_0^1 \frac{dx}{\sqrt{\{x(1-x)\}}}$ converges.

The integrand has infinities at $x=0$ and $x=1$.

We have thus to examine the convergence of the two infinite integrals
$$\int_0^a \frac{dx}{\sqrt{\{x(1-x)\}}}, \quad \int_a^1 \frac{dx}{\sqrt{\{x(1-x)\}}},$$
where a is some number between 0 and 1.

The μ-test is sufficient in each case.
$$\int_0^a \frac{dx}{\sqrt{\{x(1-x)\}}} \text{ converges, since } \lim_{x\to 0} \{x^{\frac{1}{2}} f(x)\} = 1,$$
$$\int_a^1 \frac{dx}{\sqrt{\{x(1-x)\}}} \text{ converges, since } \lim_{x\to 1-0} ((1-x)^{\frac{1}{2}} f(x)) = 1,$$
where we have written $f(x) = \dfrac{1}{\sqrt{\{x(1-x)\}}}$.

Ex. 4. Show that $\int_0^{\frac{1}{2}\pi} \log \sin x \, dx$ converges and is equal to $-\frac{1}{2}\pi \log 2$.

The only infinity is at $x=0$, and the convergence of the integral follows from the μ-test.

Further,
$$\begin{aligned}\int_0^{\pi} \log \sin x \, dx &= 2\int_0^{\frac{1}{2}\pi} \log \sin 2x \, dx \\ &= \pi \log 2 + 2\int_0^{\frac{1}{2}\pi} \log \sin x \, dx + 2\int_0^{\frac{1}{2}\pi} \log \cos x \, dx \\ &= \pi \log 2 + 4\int_0^{\frac{1}{2}\pi} \log \sin x \, dx.\end{aligned}$$

But
$$\int_0^{\pi} \log \sin x \, dx = 2\int_0^{\frac{1}{2}\pi} \log \sin x \, dx.$$

Therefore
$$\int_0^{\frac{1}{2}\pi} \log \sin x \, dx = -\tfrac{1}{2}\pi \log 2.$$

From this result it is easy to show that the convergent integrals
$$\int_0^{\pi} \log(1-\cos x)dx \quad \text{and} \quad \int_0^{\pi} \log(1+\cos x)dx$$
are equal to $-\pi \log 2$.

Ex. 5. Show that $\int_0^{\frac{1}{2}\pi} \cos 2nx \log \sin x \, dx$ converges and is equal to $-\dfrac{\pi}{4n}$, when n is a positive integer.

The only infinity is at $x=0$ and the convergence of the integral follows from the μ test (or from the last example).

Further, on integrating by parts, we see that

$$\int_0^{\frac{1}{2}\pi} \cos 2nx \log \sin x \, dx = -\frac{1}{2n}\int_0^{\frac{1}{2}\pi} \frac{\sin 2nx \cos x}{\sin x} dx$$

$$= -\frac{1}{4n}\int_0^{\frac{1}{2}\pi} \frac{\sin (2n+1)x + \sin (2n-1)x}{\sin x} dx$$

But
$$\frac{\sin (2m+1)x}{\sin x} = 1 + 2\sum_1^m \cos 2rx.$$

It follows that

$$\int_0^{\frac{1}{2}\pi} \cos 2nx \log \sin x \, dx = -\frac{\pi}{4n}.$$

From this we obtain at once

$$\int_0^{\frac{1}{2}\pi} \cos 2nx \log \cos x \, dx = -\frac{\pi}{4n} \cos n\pi.$$

$$\int_0^{\pi} \cos nx \log 2(1 - \cos x) dx = -\frac{\pi}{n},$$

and
$$\int_0^{\pi} \cos nx \log 2(1 + \cos x) dx = -\frac{\pi}{n} \cos n\pi.$$

Ex. 6. Discuss the convergence or divergence of the Gamma Function integral $\int_0^\infty e^{-x} x^{n-1} \, dx$.

(i) Let $n \geqq 1$.

Then the integrand is bounded in $0 < x \leqq a$, where a is arbitrary, and we need only consider the convergence of $\int_a^\infty e^{-x} x^{n-1} \, dx$.

The μ-test of § 55 establishes that this integral converges, since the order of e^x is greater than any given power of x.

Or we might procced as follows:

Since $e^x = 1 + x + \dfrac{x^2}{2!} + \cdots$,

when $x > 0$,
$$e^x > \frac{x^r}{r!} \ (r = \text{any positive integer}),$$

and
$$e^{-x} x^{n-1} < \frac{r!}{x^{r-n+1}}.$$

But whatever n may be, we can choose r so that $r - n + 1 > 1$.

It follows that, whatever n may be,

$$\int_a^\infty e^{-x} x^{n-1} \, dx \text{ converges.}$$

(ii) Let $0 < n < 1$.

In this case $e^{-x} x^{n-1}$ has an infinity at $x = 0$.

The μ-test shows that $\int_0^1 e^{-x} x^{n-1} \, dx$ converges, and we have just shown that $\int_1^\infty e^{-x} x^{n-1} \, dx$ converges.

Therefore $\int_0^\infty e^{-x} x^{n-1} \, dx$ converges.

(iii) Let $n \leqq 0$.

In this case $e^{-x}x^{n-1}$ has an infinity at $x=0$, and the μ-test shows that $\int_0^1 e^{-x}x^{n-1}\,dx$ diverges to $+\infty$.

Ex. 7. Discuss the integral $\int_0^1 x^{n-1}\log x\,dx$.*

Since $\lim_{x\to 0}(x^r \log x)=0$, when $r>0$, the integral is an ordinary integral, when $n>1$.

Also we know that

$$\int_x^1 \log x\,dx = \Big[x(\log x - 1)\Big]_x^1 = x(1-\log x) - 1.$$

It follows that $\int_0^1 \log x\,dx = \lim_{x\to 0} x\{(1-\log x)-1\} = -1$.

Again, $\lim_{x\to 0}(x^\mu \times x^{n-1}\log x) = \lim_{x\to 0}(x^{\mu+n-1}\log x) = 0$, if $\mu > 1-n$.

And when $0<n<1$, we can choose a positive number μ less than 1 which satisfies this condition.

Therefore $\int_0^1 x^{n-1}\log x\,dx$ converges, when $0<n\leqq 1$.

Finally, we have

$$\lim_{x\to 0}(x\times x^{n-1}\,|\log x|) = \lim_{x\to 0} x^n\,|\log x| = \infty, \quad \text{when } n\leqq 0.$$

Therefore $\int_0^1 x^{n-1}\log x\,dx$ diverges, when $n\leqq 0$.

REFERENCES.

De la Vallée Poussin, *loc. cit.*, I (5e éd., 1923), Ch. VI; 2 (4e éd., 1922), Ch. I.

Dini, *Lezioni di Analisi Infinitesimale*, 2 (Pisa, 1909), 1a Parte, Cap. I and VII.

Goursat, *loc. cit.*, 1 (4e éd., 1923), Ch. IV and V.

Hobson, *loc. cit.*, 1 (3rd ed., 1927), Ch. VI.

Kowalewski, *Grundzüge der Differential- u. Integralrechnung* (Leipzig, 1909), Kap. XIV.

Lebesgue, *Leçons sur l'Intégration*, (2e éd., Paris, 1928), Ch. I and II.

Osgood, *loc. cit.*, 1 (4 Aufl., 1923), Kap. III.

Pierpont, *loc. cit.*, 1 (1905), Ch. XII-XV.

Stolz, *Grundzüge der Differential- u. Integralrechnung*, 1 (Leipzig, 1893), Absch. X.

And

Brunel, "Bestimmte Integrale," *Enc. d. math. Wiss.*, Bd. II, Tl. I (Leipzig, 1899).

Montel-Rosenthal, "Neuere Untersuchungen über Funktionen reeller Veränderlichen: II C 9 *b*, Integration und Differentiation," *Enc. d. math. Wiss.*, Bd. II, Tl. III, 2 (Leipzig, 1923).

---

*This integral can be reduced to the Gamma Function integral, and its convergence or divergence follows from Ex. 6. Also see below, Ex. 5, p. 134.

EXAMPLES ON CHAPTER IV.

1. Show that the following integrals converge:

$$\int_0^\infty \frac{\sin x}{1+\cos x+e^x}\,dx, \quad \int_1^\infty \frac{(x-1)\sqrt{x}}{1+x+x^3+\sin x}\,dx, \quad \int_0^\infty e^{-x^2}\,dx,$$

$$\int_0^\infty e^{-a^2x^2}\cosh bx\,dx, \quad \int_0^1 \frac{\log x}{1+x}\,dx, \quad \int_0^1 \frac{\log x}{1-x^2}\,dx.$$

2. Discuss the convergence or divergence of the following integrals:

$$\int_a^b \frac{dx}{(x-a)\sqrt{(b-x)}}, \quad \int_0^c \frac{x^{a-1}}{x+1}\,dx, \quad \int_0^c \frac{x^{a-1}}{x-1}\,dx, \text{ where } 0<c<1,$$

$$\int_0^\infty \frac{x^{a-1}}{x+1}\,dx, \quad \int_0^\infty \frac{x^{a-1}}{x-1}\,dx, \quad \int_0^{\frac{\pi}{2}} \sin^m\theta\cos^n\theta\,d\theta.$$

3. Show that the following integrals are absolutely convergent:

$$\int_0^b \sin\frac{1}{x}\frac{dx}{\sqrt{x}}, \quad \int_0^\infty e^{-a^2x^2}\cos bx\,dx, \quad \int_0^\infty e^{-a^2x^2}x^m\sin nx\,dx \quad (m>0),$$

and

$$\int_a^\infty \frac{P(x)}{Q(x)}\,dx,$$

where $P(x)$ is a polynomial of the mth degree, and $Q(x)$ a polynomial of the nth degree, $n \geqq m+2$, and a is a number greater than the largest root of $Q(x)=0$.

4. Let $f(x)$ be defined in the interval $0<x\leqq 1$ as follows:

$$f(x)=2,\ \tfrac{1}{2}<x\leqq 1,\quad f(x)=-3,\ \tfrac{1}{3}<x\leqq\tfrac{1}{2},$$
$$f(x)=4,\ \tfrac{1}{4}<x\leqq\tfrac{1}{3},\quad f(x)=-5,\ \tfrac{1}{5}<x\leqq\tfrac{1}{4},$$

and so on, the values being alternately positive and negative.

Show that the infinite integral $\int_0^1 f(x)\,dx$ converges, but not absolutely.

5. Using the substitution $x=e^{-u}$, show that

$$\int_0^1 x^{m-1}(\log x)^n dx$$

converges, provided that $m>0$ and $n>-1$.

And by means of a similar substitution, show that

$$\int_1^\infty x^{m-1}(\log x)^n dx$$

converges, provided that $m<0$ and $n>-1$.

6. Show that $\displaystyle\int_a^\infty \frac{dx}{x(\log x)^{1+\mu}}$ converges when $\mu>0$ and that it diverges when $\mu\leqq 0$, the lower limit a of the integral being some number greater than unity.

Deduce that if there is a number $\mu>0$, such that $\lim\limits_{x\to\infty}\{x(\log x)^{1+\mu}f(x)\}$ exists, then $\int_a^\infty f(x)\,dx$ converges, and give a corresponding test for the divergence of this integral, $f(x)$ being bounded and integrable in any arbitrary interval (a, b), where $b>a$.

Show that $$\int_2^\infty \frac{\cos x}{(x+\sin^2 x)(\log x)^2}\,dx \text{ converges,}$$

and $$\int_2^\infty \frac{dx}{(x+\sin^2 x)\log x} \text{ diverges.}$$

7. On integrating $\int_1^x \cos x \log x\,dx$ by parts, we obtain

$$\int_1^x \cos x \log x\,dx = \sin x \log x - \int_1^x \frac{\sin x}{x}\,dx.$$

Deduce that $\int_1^\infty \cos x \log x\,dx$ oscillates infinitely.

Also show that $\int_0^1 \cos x \log x\,dx$ converges, and is equal to $-\int_0^1 \frac{\sin x}{x}\,dx$.

8. On integrating $\int_{x'}^{x''} \cos x^2\,dx$ by parts, we obtain

$$\int_{x'}^{x''} \cos x^2 dx = \frac{1}{2x''}\sin x''^2 - \frac{1}{2x'}\sin x'^2 + \frac{1}{2}\int_{x'}^{x''} \frac{\sin x^2}{x^2}\,dx,$$

where $x'' > x' > 0$.

Deduce the convergence of $\int_0^\infty \cos x^2 dx$.

9. Let $f(x)$ and $g(x)$ be bounded and integrable in (a, b), except at a certain number of points of infinite discontinuity, these points being different for the two functions.

Prove that $\int_a^b f(x)g(x)dx$ converges, if $\int_a^b |f(x)|\,dx$ and $\int_a^b |g(x)|\,dx$ converge.

10. Let $f(x)$ be monotonic when $x \geqq a$, and $\lim\limits_{x\to\infty} f(x) = 0$.

Then the series $$f(a) + f(a+1) + f(a+2) + \ldots$$
is convergent or divergent according as $\int_a^\infty f(x)dx$ converges or diverges.

Prove that for all values of the positive integer n,

$$2\sqrt{(n+1)} - 2 < \frac{1}{\sqrt{1}} + \frac{1}{\sqrt{2}} \ldots + \frac{1}{\sqrt{n}} < 2\sqrt{n} - 1.$$

Also show that $$\frac{1}{2\sqrt{1}} + \frac{1}{3\sqrt{2}} + \frac{1}{4\sqrt{3}} \ldots$$
converges to a value between $\frac{1}{2}(\pi+1)$ and $\frac{1}{2}\pi$.

11. (i) From the relation $\dfrac{\sin 2nx}{\sin x} = 2\sum\limits_1^n \cos(2r-1)x$,

show that $$\int_0^{\frac{1}{2}\pi} \frac{\sin 2nx}{\sin x}\,dx = 2\sum_1^n \frac{(-1)^{r-1}}{2r-1}$$

Deduce that $$\lim_{n\to\infty} \int_0^{\frac{1}{2}\pi} \frac{\sin 2nx}{\sin x}\,dx = \frac{\pi}{2}.$$

(ii) By integration by parts, show that

$$\lim_{n\to\infty} \int_0^{\frac{1}{2}\pi} \sin 2nx \left(\frac{1}{\sin x} - \frac{1}{x}\right) dx = 0.$$

(iii) From the above, prove that $\int_0^\infty \frac{\sin x}{x} = \frac{\pi}{2}$.

12 (i) Prove that if $u_n = \int_0^{\frac{1}{2}\pi} \sin 2nx \cot x \, dx$

and
$$v_n = \int_0^{\frac{1}{2}\pi} \frac{\sin 2nx}{x} dx,$$

then
$$u_n = \tfrac{1}{2}\pi, \quad \text{and} \quad v = \lim_{n\to\infty} v_n = \int_0^\infty \frac{\sin x}{x} dx.$$

(ii) By integration by parts, show that $\lim_{n\to\infty} (v_n - u_n) = 0$.

(iii) From the above, prove that $\int_0^\infty \frac{\sin x}{x} dx = \frac{\pi}{2}$.

CHAPTER V

THE THEORY OF INFINITE SERIES, WHOSE TERMS ARE FUNCTIONS OF A SINGLE VARIABLE

62. We shall now consider some of the properties of series whose terms are functions of x.

We denote such a series by

$$u_1(x) + u_2(x) + u_3(x) + \dots,$$

and the terms of the series are supposed to be given for values of x in some interval, *e.g.* (a, b).*

When we speak of the sum of the infinite series

$$u_1(x) + u_2(x) + u_3(x) + \dots$$

it is to be understood:†

(i) that we settle for what value of x we wish the sum of the series;

(ii) that we then insert this value of x in the different terms of the series;

(iii) that we then find the sum—$s_n(x)$—of the first n terms; and

(iv) that we then find the limit of this sum as $n \to \infty$, keeping x all the time at the value settled upon.

On this understanding, *the series*

$$u_1(x) + u_2(x) + u_3(x) + \dots$$

is said to be convergent for the value x, and to have $f(x)$ for its sum,

---

*As mentioned in § 24, when we say that x *lies in the interval* (a, b) we mean that $a \leqq x \leqq b$. In some of the results of this chapter the ends of the interval are excluded from the range of x. When this is so, the fact that we are dealing with the *open* interval $(a < x < b)$ will be stated.

†Cf. Baker, *Nature*, **59** (1899), 319.

if, this value of x having been first inserted in the different terms of the series, and any positive number ϵ having been chosen, as small as we please, there is a positive integer ν such that

$$|f(x)-s_n(x)|<\epsilon, \text{ when } n \geqq \nu.$$

Further,

A necessary and sufficient condition for convergence is that, if any positive number ϵ has been chosen, as small as we please, there shall be a positive integer ν such that

$$|s_{n+p}(x)-s_n(x)|<\epsilon, \text{ when } n \geqq \nu,$$

for every positive integer p.

A similar convention exists when we are dealing with other limiting processes. In the definition of the differential coefficient of $f(x)$ it is understood that we first agree for what value of x we wish to know $f'(x)$; that we then calculate $f(x)$ and $f(x+h)$ for this value of x; then obtain the value of $\dfrac{f(x+h)-f(x)}{h}$; and finally take the limit of this fraction as $h \to 0$.

Again, in the case of the definite integral $\int_a^b f(x, a)\,dx$, it is understood that we insert in $f(x, a)$ the particular value of a for which we wish the integral before we proceed to the summation and limit involved in the integration.

We shall write, as before (§ 19),

$$f(x)-s_n(x)=R_n(x),$$

where $f(x)$ is the sum of the series, and we shall call $R_n(x)$ *the remainder after n terms.*

As we have seen in § 19, $R_n(x)$ is the sum of the series

$$u_{n+1}(x)+u_{n+2}(x)+u_{n+3}(x)+\dots.$$

Also we shall write

$${}_pR_n(x)=s_{n+p}(x)-s_n(x),$$

and call this a *partial remainder.*

With this notation, the two conditions for convergence are

(i) $$|R_n(x)|<\epsilon, \text{ when } n \geqq \nu;$$

(ii) $$|{}_pR_n(x)|<\epsilon, \text{ when } n \geqq \nu,$$

for every positive integer p.*

A series may converge for every value of x in the open interval $a<x<b$ and not for the end-points a or b.

---

* When there is no ambiguity it will sometimes be convenient to omit the x in $s_n(x)$, $R_n(x)$, ${}_pR_n(x)$ and write s_n, R_n and ${}_pR_n$.

E.g. the series $$1+x+x^2+\dots$$

converges and has $\dfrac{1}{1-x}$ for its sum, when $-1<x<1$.

When $x=1$, it diverges to $+\infty$; when $x=-1$, it oscillates finitely.

63. The Sum of a Series whose Terms are Continuous Functions of x may be discontinuous. Until Abel* pointed out that the periodic function of x given by the series

$$2(\sin x - \tfrac{1}{2}\sin 2x + \tfrac{1}{3}\sin 3x - \dots),$$

which represents x in the interval $-\pi<x<\pi$, is discontinuous at the points $x=(2r+1)\pi$, r being any integer, it was supposed that a function defined by a convergent series of functions, continuous in a given interval, must itself be continuous in that interval. Indeed Cauchy† distinctly stated that this was the case, and later writers on Fourier's Series have sometimes tried to escape the difficulty by asserting that the sums of these trigonometrical series, at the critical values of x, passed continuously from the values just before those at the points of discontinuity to those just after.‡

This mistaken view of the sum of such series was due to two different errors. The first consisted in the assumption that, as n increases, the curves $y=s_n(x)$ must approach more and more nearly to the curve $y=f(x)$, when the sum of the series is $f(x)$ an ordinary function capable of graphical representation. These curves $y=s_n(x)$ we shall call the approximation curves for the series, but we shall see that cases may arise where the approximation curves, even for large values of n, differ very considerably from the curve $y=f(x)$.

It is true that, in a certain sense, the curves

$$\text{(i) } y=s_n(x) \quad \text{and} \quad \text{(ii) } y=f(x)$$

approach towards coincidence; but the sense is that, if we choose any particular value of x in the interval, and the arbitrary small positive number ϵ, there will be a positive integer ν such that, for this value of x, the absolute value of the difference of the ordinates of the curves (i) and (ii) will be less than ϵ when $n \geqq \nu$.

---

*Abel, *Journal für Math.*, 1 (1826), 316.

†Cauchy, *Cours d'Analyse* (1821), 1re Partie, p. 131. Also *Œuvres de Cauchy*, (Sér. 2), T. III, p. 120.

‡Cf. Sachse, *loc. cit.*; Donkin, *Acoustics* (1870), 53.

Still this is not the same thing as saying that the curves coincide geometrically. They do not, in fact, lie near to each other in the neighbourhood of a point of discontinuity of $f(x)$; and they may not do so, even where $f(x)$ is continuous.

The following examples and diagrams illustrate these points:

Ex. 1. Consider the series

$$\frac{x}{x+1}+\frac{x}{(x+1)(2x+1)}+\dots, \quad x \geqq 0.$$

Here
$$u_n(x)=\frac{1}{(n-1)x+1}-\frac{1}{nx+1},$$

and
$$s_n(x)=1-\frac{1}{nx+1}.$$

Thus, when $x>0$, $\quad \lim_{n\to\infty} s_n(x)=1;$

when $x=0$, $\quad \lim_{n\to\infty} s_n(x)=0$, since $s_n(0)=0$.

The curve $y=f(x)$, when $x\geqq 0$, consists of the part of the line $y=1$ for which $x>0$, and the origin. The sum of the series is discontinuous at $x=0$.

Now examine the approximation curves

$$y=s_n(x)=1-\frac{1}{nx+1}.$$

This equation may be written

$$(y-1)\left(x+\frac{1}{n}\right)=-\frac{1}{n}.$$

As n increases, this rectangular hyperbola (cf. Fig. 10) approaches more and more closely to the lines $y=1$, $x=0$. If we reasoned from the shape of the

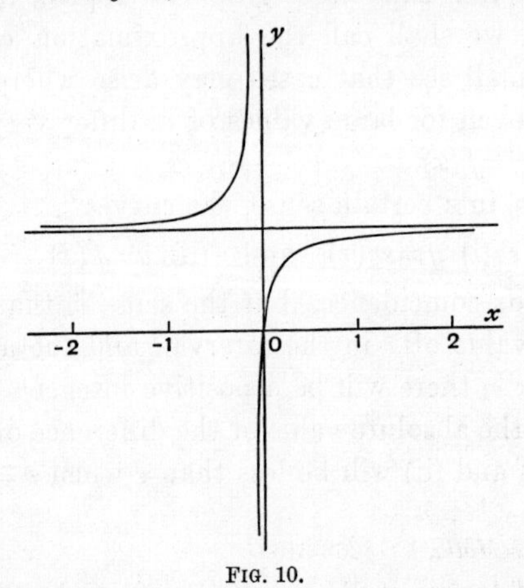

FIG. 10.

approximate curves, we should expect to find that part of the axis of y for which $0<y<1$ appearing as a portion of the curve $y=f(x)$ when $x\geqq 0$.

As $s_n(x)$ is certainly continuous, when the terms of the series are continuous, the approximation curves will always differ very materially from the curve $y=f(x)$, when the sum of the series is discontinuous.

Ex. 2. Consider the series

$$u_1(x)+u_2(x)+u_3(x)+\dots, \quad x \geqq 0,$$

where
$$u_n(x)=\frac{nx}{1+n^2x^2}-\frac{(n-1)x}{1+(n-1)^2x^2}.$$

In this case
$$s_n(x)=\frac{nx}{1+n^2x^2},$$

and
$$\lim_{n\to\infty} s_n(x)=0 \text{ for all values of } x.$$

Thus the sum of this series is continuous for all values of x, but we shall see that the approximation curves differ very materially from the curve $y=f(x)$ in the neighbourhood of the origin.

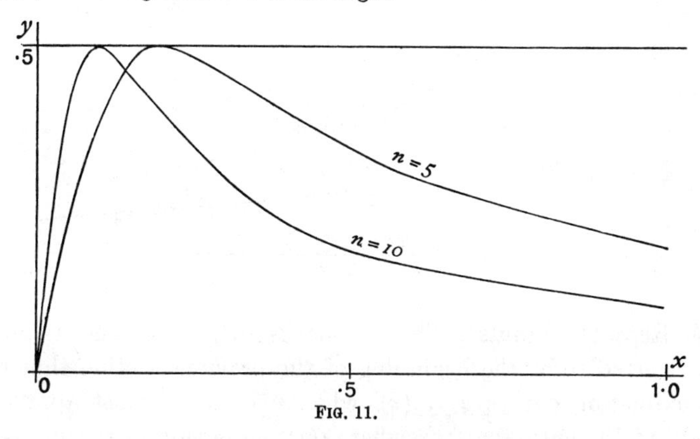

FIG. 11.

The curve
$$y=s_n(x)=\frac{nx}{1+n^2x^2}$$

has a maximum at $(1/n, \frac{1}{2})$ and a minimum at $(-1/n, -\frac{1}{2})$ (cf. Fig. 11). The points on the axis of x just below the maximum and minimum move in towards the origin as n increases. And if we reasoned from the shape of the curves $y=s_n(x)$, we should expect to find the part of the axis of y from $-\frac{1}{2}$ to $\frac{1}{2}$ appearing as a portion of the curve $y=f(x)$.

Ex. 3. Consider the series

$$u_1(x)+u_2(x)+u_3(x)+\dots, \quad x \geqq 0,$$

where
$$u_n(x)=\frac{n^2x}{1+n^3x^2}-\frac{(n-1)^2x}{1+(n-1)^3x^2}.$$

Here
$$s_n(x)=\frac{n^2x}{1+n^3x^2},$$

and
$$\lim_{n\to\infty} s_n(x)=0 \text{ for all values of } x.$$

The sum of the series is again continuous, but the approximation curves (cf. Fig. 12), which have a maximum at $(1/\sqrt{n^3}, \frac{1}{2}\sqrt{n})$ and a minimum at $(-1/\sqrt{n^3}, -\frac{1}{2}\sqrt{n})$, differ very greatly from the curve $y=f(x)$ in the neighbourhood of the origin. Indeed they would suggest that the whole of the axis of y should appear as part of $y=f(x)$.

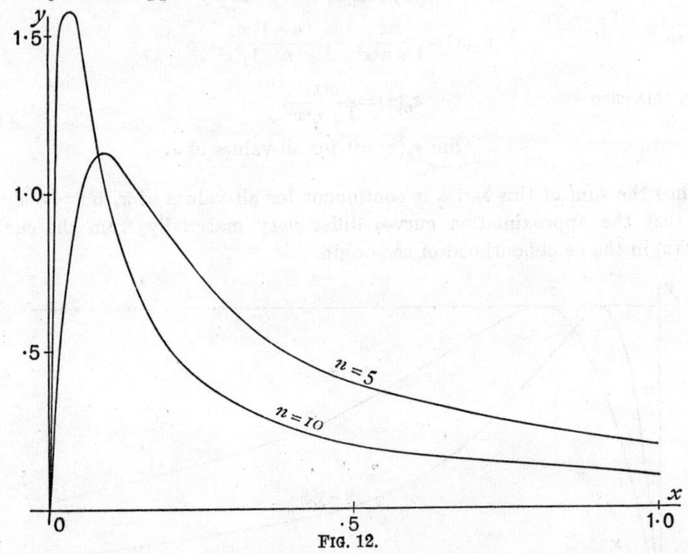

FIG. 12.

64. Repeated Limits. These remarks dispose of the assumption referred to at the beginning of the previous section that the approximation curves $y=s_n(x)$, when n is large, must approach closely to the curve $y=f(x)$, where $f(x)$ is the sum of the series.

The second error alluded to above arose from neglect of the convention implied in the definition of the sum of an infinite series whose terms are functions of x. The proper method of finding the sum has been set out in § 62, but the mathematicians to whom reference is now made proceeded in quite a different manner. In finding the sum for a value of x, say x_0, at which a discontinuity occurs, they replaced x by a function of n, which converges to x_0 as n increases. Then they took the limit when $n \to \infty$ of $s_n(x)$ in its new form. In this method x and n approach their limits concurrently, and the value of this limit may quite well differ from the actual sum for $x=x_0$. Indeed, by choosing the function of n suitably, it can be made to take any value between $f(x_0+0)$ and $f(x_0-0)$, while in some cases it goes outside this interval (cf. Ch. IX, p. 293).

For instance, in the series of § 63, Ex. 1,

$$\frac{x}{x+1}+\frac{x}{(x+1)(2x+1)}+\dots, \quad x \geqq 0,$$

we have seen that $x=0$ is a point of discontinuity.

If we put $x=p/n$ where p is positive, in the expression for $s_n(x)$, and then let $n\to\infty$, p remaining fixed, we can make $\lim\limits_{n\to\infty} s_n(p/n)$ take any value between 0 and 1, according to our choice of p. For we have $s_n(p/n)=\frac{p}{p+1}$, which is independent of n, and

$$\lim_{n\to\infty} s_n(p/n)=\frac{p}{p+1},$$

which passes from 0 to 1 as p increases from 0 to ∞.

It will be seen that the matter at issue was partly a question of words and the misunderstanding of a definition. The confusion can also be traced, in some cases, to ignorance of the care which must be exercised in any operation involving repeated limits, for we are really dealing here with two limiting processes.

If the series is convergent and its sum is $f(x)$, then

$$f(x)=\lim_{n\to\infty} s_n(x),$$

and the limit of $f(x)$ as x tends to x_0, assuming that there is such a limit, is given by

$$\lim_{x\to x_0} f(x)=\lim_{x\to x_0}\ [\lim_{n\to\infty} s_n(x)]. \quad \dots\dots\dots\dots\dots\dots(1)$$

If we may use the curve as an illustration, this is the ordinate of the point towards which we move as we proceed along the curve $y=f(x)$, the abscissa getting nearer and nearer to x_0, but not quite reaching x_0. According as x approaches x_0 from the right or left, the limit given in (1) will be $f(x_0+0)$ or $f(x_0-0)$.

Now $f(x_0)$, the sum of the series for $x=x_0$, is, by definition,

$$\lim_{n\to\infty} [s_n(x_0)],$$

and since we are now dealing with a definite number of continuous functions, $s_n(x)$ is a continuous function of x in the interval with which we are concerned.

Thus $$s_n(x_0)=\lim_{x\to x_0} s_n(x).$$

Therefore the sum of the series for $x=x_0$ may be written

$$\lim_{n\to\infty}\ [\lim_{x\to x_0} s_n(x)]. \qquad \text{.........................(2)}$$

The two expressions in (1) and (2) need not be the same. They are so only when $f(x)$ is continuous at x_0.

35. Uniform Convergence.* When the question of changing the order of two limiting processes arises, the principle of uniform convergence, which we shall now explain for the case of infinite series whose terms are functions of x, is fundamental. What is involved in this principle will be seen most clearly by returning to the series

$$\frac{x}{x+1}+\frac{x}{(x+1)(2x+1)}+\ldots,\ x\geqq 0.$$

In this series $$s_n(x)=1-\frac{1}{nx+1},$$

and $$\lim_{n\to\infty} s_n(x)=1 \text{ when } x>0.$$

Also $$R_n(x)=\frac{1}{nx+1},\ x>0, \text{ and } R_n(0)=0.$$

If the arbitrary positive number ϵ is chosen, less than unity, and some positive x is taken, it is clear that $1/(nx+1)<\epsilon$ for a positive n, only if

$$n>\frac{\frac{1}{\epsilon}-1}{x}.$$

E.g. let $\epsilon=\dfrac{1}{10^3+1}$.

If $x=0{\cdot}1, 0{\cdot}01, 0{\cdot}001, \ldots, 10^{-p}$, respectively, $1/(nx+1)<\epsilon$ only when

$$n>10^4,\ 10^5,\ 10^6, \ldots 10^{p+3}.$$

And when $\epsilon=\dfrac{1}{10^q+1}$ and $x=10^{-p}$, n must be greater than 10^{p+q} if

$$1/(nx+1)<\epsilon.$$

As we approach the origin we have to take more and more terms of the series to make the sum of n terms differ from the sum of the series by less than a given number. When $x=10^{-8}$, the first million terms do not contribute 1 per cent. of the sum.

The inequality $$n>\frac{\frac{1}{\epsilon}-1}{x}$$

shows that when n is any given positive number less than unity,

---

*A simple treatment of uniform convergence will be found in a paper by Osgood, *Bull. Amer. Math. Soc.*, 3 (1896).

and x approaches nearer and nearer to zero, the smallest positive integer which will make $R_n(x)$, $R_{n+1}(x)$, ... all less than ϵ increases without limit.

There is no positive integer ν which will make $R_\nu(x)$, $R_{\nu+1}(x)$, ... all less than this ϵ in $x \geqq 0$, *the same* ν *serving for all values of* x *in this range.*

On the other hand there is a positive integer ν which will satisfy this condition, if the range of x is given by $x \geqq a$, where a is some definite positive number.

Such a value of ν would be the integer next above $\left(\frac{1}{\epsilon} - 1\right)\Big/a$.

Our series is said to *converge uniformly* in $x \geqq a$, but it does not *converge uniformly* in $x \geqq 0$.

We turn now to the series

$$u_1(x) + u_2(x) + u_3(x) + \dots,$$

and define uniform convergence* in an interval as follows:

Let the series $\qquad u_1(x) + u_2(x) + u_3(x) + \dots$

converge for all values of x *in the interval* $a \leqq x \leqq b$ *and its sum be* $f(x)$. *It is said to* **converge uniformly in that interval,** *if, any positive number* ϵ *having been chosen, as small as we please, there is a positive integer* ν *such that,* **for all values of x in the interval,**

$$|f(x) - s_n(x)| < \epsilon, \quad \textit{when } n \geqq \nu.\dagger$$

It is true that, if the series converges, $|R_n(x)| < \epsilon$ for each x in (a, b) when $n \geqq \nu$.

The additional point in the definition of uniform convergence is that, any positive number ϵ having been chosen, as small as we please, *the same value of* ν *is to serve for all the values of* x *in the interval.*

For this integer ν we must have

$$|R_\nu(x)|, \quad |R_{\nu+1}(x)|, \dots$$

all less than ϵ, no matter where x lies in (a, b).

---

*The property of uniform convergence was discovered independently by Stokes (cf. *Trans. Phil. Soc. Camb.*, 8 (1847), 533) and Seidel (cf. *Abh. Ak. Wiss. München*, 5 (1848), 381). See also Hardy, *Proc. Phil. Soc. Camb.*, 19 (1920), 148.

†We can also have uniform convergence in the *open* interval $a < x < b$, or the half-open intervals $a < x \leqq b$, $a \leqq x < b$; but, when the terms are continuous in the closed interval, uniform convergence in the *open* interval carries with it uniform convergence in the *closed* interval (cf. § 68).

The series does not converge uniformly in (a, b) if we know that for some positive number (say ϵ_0) there is no positive integer ν which will make

$$|R_\nu(x)|, \quad |R_{\nu+1}(x)|, \ldots$$

all less than ϵ_0 for every x in (a, b).

It will be seen that the series

$$\frac{x}{x+1} + \frac{x}{(x+1)(2x+1)} + \ldots$$

converges uniformly in any interval $a \leqq x \leqq b$, where a, b are any given positive numbers.

It may be said to converge infinitely slowly as x tends to zero, in the sense that, as we get nearer and still nearer to the origin, we cannot fix a limit to the number of terms which we must take to make $|R_n(x)| < \epsilon$. It is this property of *infinitely slow convergence* at a point (*e.g.* x_0) which prevents a series converging uniformly in an interval $(x_0 - \delta, x_0 + \delta)$ including that point.

Further, the above series converges uniformly in the *infinite interval* $x \geqq a$, where a is any given positive number.

It is sometimes necessary to distinguish between uniform convergence in an *infinite interval* and uniform convergence in a *fixed interval, which may be as large as we please.*

The exponential series is convergent for all values of x, but it does not converge uniformly in the infinite interval $x \geqq 0$.

For in this series $R_n(x)$ is greater than $x^n/n!$, when x is positive.

Thus, if the series were uniformly convergent in $x \geqq 0$, $x^n/n!$ would need to be less than ϵ when $n \geqq \nu$, the same ν serving for all values of x in the interval.

But it is clear that we need only take x greater than $(\nu!\epsilon)^{\frac{1}{\nu}}$ to make $R_n(x)$ greater than ϵ for n equal to ν.

However, the exponential series is uniformly convergent in the interval $(0, b)$, where b is fixed, but may be fixed as large as we please.

For take c greater than b. We know that the series converges for $x = c$.

Therefore $\qquad R_n(c) < \epsilon$, when $n \geqq \nu$.

But $\qquad R_n(x) < R_n(c)$, when $0 \leqq x \leqq b < c$.

Therefore $R_n(x) < \epsilon$, when $n \geqq \nu$, the same ν serving for all values of x in $(0, b)$.

From the uniform convergence of the exponential series in the interval $(0, b)$, it follows that the series also converges uniformly in the interval $(-b, b)$, where in both cases b is fixed, but may be fixed as large as we please.

Ex. 1. Prove that the series

$$1+x+x^2+\ldots$$

converges uniformly to $1/(1-x)$ in $0 \leqq x \leqq x_0 < 1$.

Ex. 2. Prove that the series

$$(1-x)+x(1-x)+x^2(1-x)+\ldots$$

converges uniformly to 1 in $0 \leqq x \leqq x_0 < 1$.

Ex. 3. Prove that the series

$$(1-x)^2+x(1-x)^2+x^2(1-x)^2+\ldots$$

converges uniformly to $(1-x)$ in $0 \leqq x \leqq 1$.

Ex. 4. Prove that the series

$$\frac{1}{1+x^2}-\frac{1}{2+x^2}+\frac{1}{3+x^2}-\ldots$$

converges uniformly in the infinite interval $x \geqq 0$.

Ex. 5. Prove that the series

$$\frac{x}{1.2}+\frac{x}{2.3}+\frac{x}{3.4}+\ldots$$

converges uniformly in the interval $(0, b)$, where b is fixed, but may be fixed as large as we please, and that it does not converge uniformly in the infinite interval $x \geqq 0$.

66. A necessary and sufficient condition for Uniform Convergence. When the sum $f(x)$ is known, the above definition often gives a convenient means of deciding whether the convergence is uniform or not.

When the sum is not known, the following test, corresponding to the general principle of convergence (§ 15), is more suitable.

Let

$$u_1(x)+u_2(x)+u_3(x)+\ldots$$

be an infinite series, whose terms are given in the interval (a, b). *A necessary and sufficient condition for the uniform convergence of the series in this interval is that, if any positive number* ϵ *has been chosen, as small as we please, there shall be a positive integer* ν *such that,* for all values of x in the interval, $|{}_pR_n(x)| < \epsilon$, *when* $n \geqq \nu$, *for every positive integer* p.

(i) *The condition is necessary.*

Let the positive number ϵ be chosen, as small as we please. Then take $\frac{1}{2}\epsilon$.

Since the series is uniformly convergent, there is a positive integer ν, such that

$$|f(x)-s_n(x)|<\tfrac{1}{2}\epsilon, \quad \text{when } n \geqq \nu,$$

the same ν serving for all values of x in (a, b), $f(x)$ being the sum of the series.

Let n'', n' be any two positive integers such that $n''>n' \geqq \nu$.

Then
$$\begin{aligned}|s_{n''}(x)-s_{n'}(x)| &\leqq |s_{n''}(x)-f(x)|+|f(x)-s_{n'}(x)| \\ &<\tfrac{1}{2}\epsilon+\tfrac{1}{2}\epsilon \\ &<\epsilon.\end{aligned}$$

Thus $|s_{n+p}(x)-s_n(x)|<\epsilon$, when $n \geqq \nu$, for every positive integer p the same ν serving for all values of x in (a, b).

(ii) *The condition is sufficient.*

We know that the series converges, when this condition is satisfied.

Let its sum be $f(x)$.

Again let the arbitrary positive number ϵ be chosen. Then there is a positive integer ν such that

$|s_{n+p}(x)-s_n(x)|<\tfrac{1}{2}\epsilon$, when $n \geqq \nu$, for every positive integer p, the same ν serving for all values of x in (a, b).

Thus
$$s_\nu(x)-\tfrac{1}{2}\epsilon<s_{\nu+p}(x)<s_\nu(x)+\tfrac{1}{2}\epsilon.$$

Also
$$\lim_{p\to\infty} s_{\nu+p}(x)=f(x).$$

Therefore
$$s_\nu(x)-\tfrac{1}{2}\epsilon \leqq f(x) \leqq s_\nu(x)+\tfrac{1}{2}\epsilon.$$

But
$$|s_n(x)-f(x)| \leqq |s_n(x)-s_\nu(x)|+|s_\nu(x)-f(x)|.$$

It follows that, when n is greater than or equal to the value ν specified above,

$$\begin{aligned}|s_n(x)-f(x)| &< \tfrac{1}{2}\epsilon+\tfrac{1}{2}\epsilon \\ &<\epsilon,\end{aligned}$$

and this holds for all values of x in (a, b).

Thus the series converges uniformly in this interval.

67. 1. Weierstrass's M-Test for Uniform Convergence. The following simple test for uniform convergence is due to Weierstrass:

The series $\quad u_1(x)+u_2(x)+u_3(x)+\dots$

will converge uniformly in (a, b), *if there is a convergent series of positive constants*

$$M_1+M_2+M_3+\dots,$$

such that, no matter what value x may have in (a, b),

$$|u_n(x)| \leqq M_n \textit{ for every positive integer } n.$$

Since the series $M_1+M_2+M_3+\dots$ is convergent, with the usual notation,

$$M_{n+1}+M_{n+2}+\dots+M_{n+p}<\epsilon,$$

when $n \geqq \nu$, for every positive integer p.

But $\quad |{}_pR_n(x)| \leqq |u_{n+1}(x)|+|u_{n+2}(x)|+\dots+|u_{n+p}(x)|.$

Thus $\quad |{}_pR_n(x)| \leqq M_{n+1}+M_{n+2}+\dots+M_{n+p}$

$$<\epsilon, \text{ when } n \geqq \nu, \text{ for every positive integer } p,$$

the inequality holding for all values of x in (a, b).

Thus the given series is uniformly convergent in (a, b).

For example, we know that the series

$$1+2a+3a^2+\dots$$

is convergent, when a is any given positive number less than unity.

It follows that the series

$$1+2x+3x^2+\dots$$

is uniformly convergent in the interval $(-a, a)$.

Ex. 1. Show that the series

$$x\cos\theta+x^2\cos 2\theta+x^3\cos 3\theta+\dots$$

is uniformly convergent for any interval (x_0, x_1), where $-1<x_0<x_1<1$ and θ is any given number.

Ex. 2. Show that the series

$$x\cos\theta+x^2\cos 2\theta+x^3\cos 3\theta+\dots$$

and

$$x\cos\theta+\frac{x^2}{2}\cos 2\theta+\frac{x^3}{3}\cos 3\theta+\dots$$

are uniformly convergent for all values of θ, when $|x|$ is any given positive number less than unity.

67. 2. Further Tests for Uniform Convergence. In the M-Test the series converges absolutely and uniformly. But absolute convergence is not required in the following tests, usually called Abel's Test and Dirichlet's Test.

I. Abel's Test. *Let the series*

$$u_1(x)+u_2(x)+u_3(x) \quad \dots\dots\dots\dots(1)$$

converge uniformly in (a, b) and the sequence

$$v_1(x), \quad v_2(x), \quad v_3(x), \quad \dots\dots\dots\dots(2)$$

*be monotonic for every (fixed) x in (a, b) and uniformly bounded.**

*A function $f_n(x)$ is said to be *uniformly bounded* in an interval, when there is a positive number K, independent of x and n, such that $|f_n(x)|<K$, for every value of x in the interval, and every positive integer n.

Then the series

$$u_1(x)v_1(x)+u_2(x)v_2(x)+u_3(x)v_3(x)+\ldots \quad \ldots\ldots\ldots\ldots(3)$$

is uniformly convergent in (a, b).

Let ${}_pR_n(x)$ be the partial remainder for the series (3) and $s_n(x)$, ${}_pr_n(x)$ the sum of n terms and the partial remainder for the series (1).

$$\begin{aligned}\text{Then}\quad {}_pR_n(x) &= u_{n+1}(x)v_{n+1}(x)+u_{n+2}(x)v_{n+2}(x)+\cdots\\ &\quad +u_{n+p}(x)v_{n+p}(x)\\ &= {}_1r_n(x)v_{n+1}(x)+[{}_2r_n(x)-{}_1r_n(x)]\,v_{n+2}(x)+\cdots\\ &\quad +[{}_pr_n(x)-{}_{p-1}r_n(x)]\,v_{n+p}(x)\\ &= {}_1r_n(x)\,[v_{n+1}(x)-v_{n+2}(x)]+\cdots\\ &\quad +{}_{p-1}r_n(x)\,[v_{n+p-1}(x)-v_{n+p}(x)]+{}_pr_n(x)v_{n+p}(x)\ldots(4)\end{aligned}$$

Now it is known that

$$[v_{n+1}(x)-v_{n+2}(x)],\quad [v_{n+2}(x)-v_{n+3}(x)],\quad [v_{n+p-1}(x)-v_{n+p}(x)]$$

all have the same sign, x being fixed; and that there is a positive number K such that $|v_n(x)|<K$ for all values of x in (a, b), and every positive integer n.

Also, since the series (1) converges uniformly, when the arbitrary positive number ϵ is chosen, there is a positive integer ν, such that

$$|{}_1r_n(x)|,\quad |{}_2r_n(x)|,\ \ldots\ |{}_pr_n(x)|$$

are each less than $\epsilon/3K$ when $n \geqq \nu$, the same ν serving for all values of x in (a, b).

It follows from (4) that

$$\begin{aligned}\left|{}_pR_n(x)\right| &< \frac{\epsilon}{3K}\left|v_{n+1}(x)-v_{n+p}(x)\right| + \frac{\epsilon}{3K}\left|v_{n+p}(x)\right|\\ &< \tfrac{2}{3}\epsilon+\tfrac{1}{3}\epsilon\\ &< \epsilon,\ \text{when } n \geqq \nu,\end{aligned}$$

the same ν serving for all values of x in (a, b).

Thus the series (3) is uniformly convergent in (a, b).

Ex. 1. Let $a_0+a_1+a_2+\ldots$ be a convergent series of constants and $v_n(x)=x^n$. Then $\sum_0^\infty a_n v_n(x)$ converges uniformly in $0 \leqq x \leqq 1$.

Ex. 2. Let $a_1+a_2+\ldots$ be a convergent series of constants and $v_n(x)=\dfrac{1}{n^x}$. Then $\sum_0^\infty a_n v_n(x)$ converges uniformly in $x \geqq 0$.

Ex. 3. Let $a_0+a_1+a_2+\ldots$ be a convergent series of constants and α_0, α_1, $\alpha_2, \ldots$ be a monotonic ascending sequence of positive numbers. Then the series

$$a_0e^{-\alpha_0x}+a_1e^{-\alpha_1x}+a_2e^{-\alpha_2x}$$

converges uniformly in $x \geqq 0$.

II. Dirichlet's Test. *Let*

$$s_n(x) = u_1(x) + u_2(x) + \dots u_n(x).$$

Then the series

$$u_1(x)v_1(x) + u_2(x)v_2(x) + u_3(x)v_3(x) + \dots$$

converges uniformly in (a, b) *provided that*

(i) $s_n(x)$ *is uniformly bounded in* (a, b)*

and

(ii) $v_1(x)$, $v_2(x)$, $v_3(x)$, ... *is a monotonic sequence converging uniformly to zero in* (a, b).

With the same notation as above,

$$\begin{aligned} {}_pR_n(x) &= u_{n+1}(x)v_{n+1}(x) + \dots + u_{n+p}(x)v_{n+p}(x) \\ &= [s_{n+1}(x) - s_n(x)]\,v_{n+1}(x) + [s_{n+2}(x) - s_{n+1}(x)]\,v_{n+2}(x) + \dots \\ &\qquad + [s_{n+p}(x) - s_{n+p-1}(x)]\,v_{n+p}(x) \\ &= s_{n+1}(x)\,[v_{n+1}(x) - v_{n+2}(x)] + \dots \\ &\qquad + s_{n+p-1}(x)\,[v_{n+p-1}(x) - v_{n+p}(x)] \\ &\qquad + s_{n+p}(x)v_{n+p}(x) - s_n(x)v_{n+1}(x). \end{aligned}$$

Then we have at once

$$|{}_pR_n(x)| < K\{|v_{n+1}(x) - v_{n+p}(x)| + |v_{n+p}(x)| + |v_{n+1}(x)|\}.$$

But the sequence $v_1(x)$, $v_2(x)$, ... converges uniformly to zero.

Therefore we know that, however small the arbitrary positive number ϵ may be, there is a positive integer ν, such that

$$\left| v_n(x) \right| < \frac{\epsilon}{3K}, \text{ when } n \geqq \nu,$$

the same ν serving for all values of x in the interval.

And $v_1(x)$, $v_2(x)$, etc., are all of the same sign.

Therefore

$$|{}_pR_n(x)| < \epsilon, \text{ when } n \geqq \nu,$$

the same ν serving for all values of x in (a, b), and the series

$$\sum_1^\infty u_n(x)v_n(x)$$

converges uniformly in (a, b).

Ex. 1. The series $\dfrac{1}{1+x^2} - \dfrac{1}{2+x^2} + \dfrac{1}{3+x^2} - \dots$

converges uniformly when $x \gtreqless 0$.

---

*Cf. footnote on p. 149.

Ex. 2. The series
$$\sin x + \tfrac{1}{2}\sin 2x + \tfrac{1}{3}\sin 3x + \dots$$
$$\cos x + \tfrac{1}{2}\cos 2x + \tfrac{1}{3}\cos 3x + \dots$$
converge uniformly in (a, b), when $0 < a < b < 2\pi$.

Ex. 3. The series
$$\sin x - \tfrac{1}{2}\sin 2x + \tfrac{1}{3}\sin 3x \dots$$
$$\cos x - \tfrac{1}{2}\cos 2x + \tfrac{1}{3}\cos 3x \dots$$
converge uniformly in $(-a, a)$, where $0 < a < \pi$.

Ex. 4. The series
$$\sin x + \tfrac{1}{3}\sin 3x + \tfrac{1}{5}\sin 5x + \dots$$
$$\tfrac{1}{2}\sin 2x + \tfrac{1}{4}\sin 4x + \tfrac{1}{6}\sin 6x + \dots$$
$$\cos x + \tfrac{1}{3}\cos 3x + \tfrac{1}{5}\cos 5x + \dots$$
$$\tfrac{1}{2}\cos 2x + \tfrac{1}{4}\cos 4x + \tfrac{1}{6}\cos 6x + \dots$$
converge uniformly in (a, b), when $0 < a < b < \pi$.

Ex. 5. The series $\sum_1^\infty a_n \sin nx$ and $\sum_1^\infty a_n \cos nx$

converge uniformly in (x_0, x_1) when $0 < x_0 < x_1 < 2\pi$, provided that the constants $a_1, a_2, \dots$ form a monotonic sequence and $\lim_{n\to\infty} a_n = 0$.

68. Uniform Convergence of Series whose Terms are Continuous Functions of x. In the previous sections dealing with uniform convergence the terms of the series have not been assumed continuous in the given interval. We shall now prove some properties of these series when this condition is added.

I. *Uniform convergence implies continuity in the sum.*

If the terms of the series
$$u_1(x) + u_2(x) + u_3(x) + \dots$$
are continuous in (a, b), and the series converges uniformly to $f(x)$ in this interval, then $f(x)$ is a continuous function of x in (a, b).

Since the series converges uniformly, we know that, however small the positive number ϵ may be, there is a positive integer ν, such that
$$|f(x) - s_n(x)| < \tfrac{1}{3}\epsilon, \text{ when } n \geqq \nu,$$
the same ν serving for all values of x in (a, b).

Choosing such a value of n, we have
$$f(x) = s_n(x) + R_n(x),$$
where $|R_n(x)| < \tfrac{1}{3}\epsilon$, for all values of x in (a, b).

Since $s_n(x)$ is the sum of n continuous functions, it is also continuous in (a, b).

Thus we know from §31 that there is a positive number η such that
$$|s_n(x') - s_n(x)| < \tfrac{1}{3}\epsilon,$$
when x, x' are any two values of x in the interval (a, b) for which $|x' - x| \leqq \eta$.

But $$f(x')=s_n(x')+R_n(x'),$$
where $$|R_n(x')|<\tfrac{1}{3}\epsilon.$$
Also $$f(x')-f(x)=[s_n(x')-s_n(x)]+R_n(x')-R_n(x).$$
Thus $$\begin{aligned}|f(x')-f(x)| &\leqq |s_n(x')-s_n(x)|+|R_n(x')|+|R_n(x)|.\\ &<\tfrac{1}{3}\epsilon+\tfrac{1}{3}\epsilon+\tfrac{1}{3}\epsilon\\ &<\epsilon, \text{ when } |x'-x|\leqq\eta.\end{aligned}$$

Therefore $f(x)$ is continuous in (a, b).

II. *If a series, whose terms are continuous functions, has a discontinuous sum, it cannot be uniformly convergent in an interval which contains a point of discontinuity.*

For if the series were uniformly convergent, we have just seen that its sum must be continuous in the interval of uniform convergence.

III. *Uniform convergence is thus a sufficient condition for the continuity of the sum of a series of continuous functions. It is not a necessary condition; since different non-uniformly convergent series are known, which represent continuous functions in the interval of non-uniform convergence.*

For example, the series discussed in Ex. 2 and Ex. 3 of § 63 are uniformly convergent in $x\geqq a>0$, for in both cases

$$|R_n(x)|<\frac{1}{nx}\leqq\frac{1}{na}, \text{ when } x\geqq a>0.$$

Thus $|R_n(x)|<\epsilon$, when $n>1/a\epsilon$, which is independent of x.

But the interval of uniform convergence does not extend up to and include $x=0$, even though the sum is continuous for all values of x.

This is clear in Ex. 2, where $R_n(x)=\dfrac{nx}{1+n^2x^2}$, for if it is asserted that $|R_n(x)|<\epsilon$, when $n\geqq\nu$, the same ν serving for all values of x in $x\geqq 0$, the statement is shown to be untrue by pointing out that for $x=1/\nu$, $R_\nu(x)=\frac{1}{2}$, and thus $|R_n(x)|\nless\epsilon$, when $n\geqq\nu$, right through the interval, if $\epsilon<\frac{1}{2}$.

Similarly in Ex. 3, where $R_n(x)=\dfrac{n^2x}{1+n^3x^2}$, if it is asserted that $|R_n(x)|<\epsilon$, when $n\geqq\nu$, the same ν serving for all values of x in $x\geqq 0$, we need only point out that for $x=1/\sqrt{\nu^3}$, $R_\nu(x)=\frac{1}{2}\sqrt{\nu^3}$. Thus $|R_n(x)|\nless\epsilon$, when $n\geqq\nu$, right through the interval, if $\epsilon<\frac{1}{2}$.

There is, in both cases, a positive integer ν for which $R_n(1/m) < \epsilon < \frac{1}{2}$, when $n \geqq \nu$, but this integer is greater than $1/m$.

Thus it is clear that the convergence becomes infinitely slow as $x \to 0$.

IV. *If the terms of the series are continuous in the closed interval* (a, b), *and the series converges uniformly in* $a < x < b$, *then it must converge for* $x = a$ *and* $x = b$, *and the uniformity of the convergence will hold for the closed interval* (a, b).

Since the series is uniformly convergent in the open interval $a < x < b$, we have, with the usual notation,

$$|s_m(x) - s_n(x)| < \tfrac{1}{3}\epsilon, \text{ when } m > n \geqq \nu, \quad \text{..............(1)}$$

the same ν serving for every x in this open interval.

Let m, n be any two positive integers satisfying this relation.

Since the terms of the series are continuous in the closed interval (a, b), there are positive numbers η_1 and η_2, say, such that

$$|s_m(x) - s_m(a)| > \tfrac{1}{3}\epsilon, \text{ when } 0 \leqq (x - a) \leqq \eta_1,$$

and

$$|s_n(x) - s_n(a)| < \tfrac{1}{3}\epsilon, \text{ when } 0 \leqq (x - a) \leqq \eta_2.$$

Choose a positive number η not greater than η_1 or η_2, and let

$$0 \leqq (x - a) \leqq \eta.$$

Then

$$\begin{aligned} &|s_m(a) - s_n(a)| \\ &\leqq |s_m(a) - s_m(x)| + |s_m(x) - s_n(x)| + |s_n(x) - s_n(a)| \\ &< \tfrac{1}{3}\epsilon + \tfrac{1}{3}\epsilon + \tfrac{1}{3}\epsilon \\ &< \epsilon, \quad \text{when } m > n \geqq \epsilon. \quad \text{..(2)} \end{aligned}$$

A similar argument shows that

$$|s_m(b) - s_n(b)| < \epsilon, \quad \text{when } m > n \geqq \nu. \quad \text{..............(3)}$$

From (2) and (3) we see that the series converges for $x = a$ and $x = b$, and, combining (1), (2) and (3), we see that the condition for uniform convergence in the closed interval (a, b) is satisfied.

If the terms of the series

$$u_1(x) + u_2(x) + u_3(x) + \ldots$$

are continuous in (a, b), and the series converges uniformly in every interval (α, β), where $a < \alpha < \beta < b$, the series need not converge for $x = a$ or $x = b$.

E.g. the series $$1 + 2x + 3x^2 + \ldots$$

converges uniformly in $(-\alpha, \alpha)$, where $\alpha < 1$, but it does not converge for

$$x = -1 \text{ or } x = 1.$$

However we shall see that in the case of the Power Series, if it converges for $x = a$ or $x = b$, the uniform convergence in (α, β) extends up to a or b, as the case may be. (Cf. § 72.)

But this property is not true in general.

The series of continuous functions

$$u_1(x)+u_2(x)+u_3(x)+\dots$$

may converge uniformly for every interval (α, β) within (a, b), and converge for $x=a$ or $x=b$, while the range of uniform convergence does not extend up to and include the point a or b.

E.g. the series $$x+\tfrac{1}{3}x^3-\tfrac{1}{2}x^2+\tfrac{1}{5}x^5+\tfrac{1}{7}x^7-\tfrac{1}{4}x^4+\dots \quad\text{.........................(1)}$$
formed from the logarithmic series

$$x-\tfrac{1}{2}x^2+\tfrac{1}{3}x^3-\dots \quad\text{.........................(2)}$$

by taking two consecutive positive terms and then one negative term, is convergent when $-1<x\leqq 1$, and its sum, when $x=1$, is $\frac{3}{2}\log 2$.*

Further, the series (2) is absolutely convergent when $|x|<1$, and therefore the sum is not altered by taking the terms in any other order. (Cf. § 22.)

It follows that when $|x|<1$ the sum of (1) is $\log(1+x)$, and when $x=1$ its sum is $\frac{3}{2}\log 2$.

Hence (1) is discontinuous at $x=1$ and therefore the interval of uniform convergence does not extend up to and include that point.

69. Uniform Convergence and the Approximation Curves. Let a series of continuous functions be uniformly convergent in (a, b).

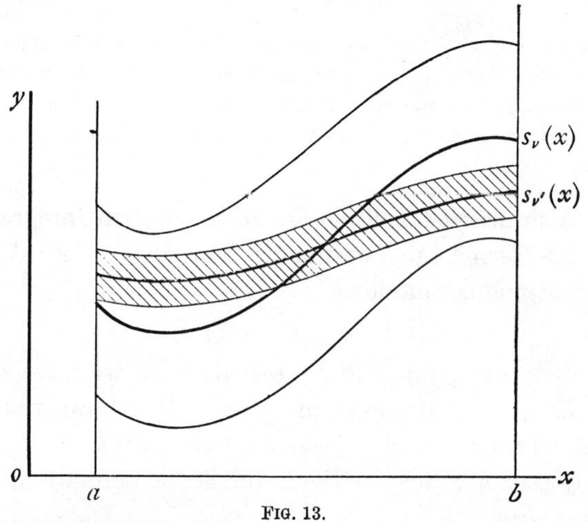

FIG. 13.

Then we have, as before,

$$|s_m(x)-s_n(x)|<\epsilon, \text{ when } m>n\geqq\nu,$$

the same ν serving for all values of x in the interval.

In particular, $$|s_m(x)-s_\nu(x)|<\epsilon, \text{ when } m>\nu,$$

and we shall suppose ν the smallest positive integer which will satisfy this condition for the given ϵ and every x in the interval.

Plot the curve $y=s_\nu(x)$ and the two parallel curves $y=s_\nu(x)\pm\epsilon$, forming a strip σ of breadth 2ϵ, whose central line is $y=s_\nu(x)$. (Fig. 13.)

*Cf. Hobson, *Plane Trigonometry* (7th ed., 1928), 251.

All the approximation curves $y=s_m(x)$, $m>\nu$, lie in this strip, and the curve $y=f(x)$, where $f(x)$ is the sum of the series, also lies within the strip, or at most reaches its boundaries. (Cf. § 66 (ii).)

Next choose ϵ' less than ϵ, and let the corresponding smallest positive integer satisfying the condition for uniform convergence be ν'. Then ν' is greater than or equal to ν. The new curve $y=s_{\nu'}(x)$ thus lies in the first strip, and the new strip σ' of breadth $2\epsilon'$, formed as before, if it goes outside the first strip in any part, can have this portion blotted out, for we are concerned only with the region in which the approximation curves may lie as m increases from the value ν.

In this way, if we take the set of positive numbers

$$\epsilon>\epsilon'>\epsilon''\ldots, \quad \text{where } \lim_{\kappa\to\infty} \epsilon^{(\kappa)}=0,$$

and the corresponding positive integers

$$\nu \leqq \nu' \leqq \nu'' \ldots,$$

we obtain the set of strips $\qquad \sigma, \sigma', \sigma'', \ldots$

Any strip lies within, or at most reaches, the boundary of the preceding one, and their breadth tends to zero as their number increases.

Further, the curve $y=f(x)$ lies within, or at most reaches, the boundary of the strips.

This construction, therefore, not only establishes the continuity of the sum of the series of continuous functions, in an interval of uniform convergence, but it shows that the approximation curves, as the number of the terms increase, may be used as a guide to the shape of the curve for the sum right through the interval.*

70. 1. A sufficient Condition for Term by Term Integration of a Series whose Terms are Continuous Functions of x When the series of continuous functions

$$u_1(x)+u_2(x)+u_3(x)+\ldots$$

is uniformly convergent in the interval (a, b), we have seen that its sum, $f(x)$, is continuous in (a, b). It follows that $f(x)$ is integrable between x_0 and x_1, when $a \leqq x_0 < x_1 \leqq b$.

But it does not follow, without further examination, that the series of integrals

$$\int_{x_0}^{x_1} u_1(x)\,dx+\int_{x_0}^{x_1} u_2(x)\,dx+\int_{x_0}^{x_1} u_3(x)\,dx+\ldots$$

is convergent, and, even if it be convergent, it does not follow, without proof, that its sum is $\int_{x_0}^{x_1} f(x)\,dx$.

---

*Of course the argument of this section applies only to such functions as can be graphically represented.

The geometrical treatment of the approximation curves in § 69 suggests that this result will be true, when the given series is uniformly convergent, arguing from the areas of the respective curves.

We shall now state the theorem more precisely and give its demonstration :

Let the functions $u_1(x)$, $u_2(x)$, $u_3(x)$, ... *be continuous in* (a, b), *and let the series*

$$u_1(x)+u_2(x)+u_3(x)+\dots$$

be uniformly convergent in (a, b) *and have* $f(x)$ *for its sum.*

Then

$$\int_{x_0}^{x_1} f(x)\,dx=\int_{x_0}^{x_1} u_1(x)\,dx+\int_{x_0}^{x_1} u_2(x)\,dx+\int_{x_0}^{x_1} u_3(x)\,dx+\dots,$$

where $a \leqq x_0 < x_1 \leqq b$.

Let the arbitrary positive number ϵ be chosen.

Since the series is uniformly convergent, we may put

$$f(x)=s_n(x)+R_n(x),$$

where $$|R_n(x)|<\frac{\epsilon}{b-a}, \text{ when } n \geqq \nu,$$

the same ν serving for all values of x in (a, b).

Also $f(x)$ and $s_n(x)$ are continuous in (a, b) and therefore integrable.

Thus we have

$$\int_{x_0}^{x_1} f(x)\,dx=\int_{x_0}^{x_1} s_n(x)\,dx+\int_{x_0}^{x_1} R_n(x)\,dx,$$

where $a \leqq x_0 < x_1 \leqq b$.

Therefore $$\left|\int_{x_0}^{x_1} f(x)\,dx-\int_{x_0}^{a_1} s_n(x)\,dx\right|=\left|\int_{x_0}^{x_1} R_n(x)\,dx\right|$$

$$<\epsilon\frac{x_1-x_0}{b-a}$$

$$<\epsilon, \text{ when } n \geqq \nu.$$

But $$\int_{x_0}^{x_1} s_n(x)\,dx=\sum_1^n\int_{x_0}^{x_1} u_r(x)\,dx.$$

Therefore $$\left|\int_{x_0}^{x_1} f(x)\,dx-\sum_1^n\int_{x_0}^{x_1} u_r(x)\,dx\right|<\epsilon, \text{ when } n \geqq \nu.$$

Thus the series of integrals is convergent and its sum is $\int_{x_0}^{x_1} f(x)\,dx$.

COROLLARY I. *Let* $u_1(x), u_2(x), u_3(x), \ldots$ *be continuous in* (a, b) *and the series*

$$u_1(x)+u_2(x)+u_3(x)+\ldots$$

converge uniformly to $f(x)$ *in* (a, b).

Then the series of integrals

$$\int_{x_0}^{x} u_1(x)\,dx+\int_{x_0}^{x} u_2(x)\,dx+\int_{x_0}^{x} u_3(x)\,dx+\ldots$$

converges uniformly to $\int_{x_0}^{x} f(x)dx$ in (a, b), *when* $a \leqq x_0 < x \leqq b$.

This follows at once from the argument above.

COROLLARY II. *Let* $u_1(x), u_2(x), u_3(x), \ldots$ *be continuous in* (a, b) *and the series*

$$u_1(x)+u_2(x)+u_3(x)+\ldots$$

converge uniformly to $f(x)$ *in* (a, b).

Also let $g(x)$ *be bounded and integrable in* (a, b).

Then

$$\int_{x_0}^{x} f(x)g(x)\,dx=\sum_1^{\infty}\int_{x_0}^{x} u_n(x)g(x)\,dx,$$

where $a \leqq x_0 < x \leqq b$, *and the convergence of the series of integrals is uniform in* (a, b).

Let the arbitrary positive number ϵ be chosen, and let M be the upper bound of $|g(x)|$ in (a, b).

Since the series $\sum_1^{\infty} u_n(x)$ converges uniformly to $f(x)$ in (a, b), we may put

$$f(x)=s_n(x)+R_n(x),$$

where

$$|R_n(x)|<\frac{\epsilon}{M(b-a)}, \text{ when } n \geqq \nu,$$

the same ν serving for all values of x in (a, b).

Therefore we have

$$\int_{x_0}^{x} f(x)g(x)\,dx=\int_{x_0}^{x} s_n(x)g(x)dx+\int_{x_0}^{x} R_n(x)\,g(x)\,dx,$$

where

$$a \leqq x_0 < x \leqq b.$$

Thus $\left|\int_{x_0}^{x} f(x)g(x)\,dx-\int_{x_0}^{x} s_n(x)g(x)\,dx\right|=\left|\int_{x_0}^{x} R_n(x)g(x)\,dx\right|.$

And $\left|\int_{x_0}^{x} f(x)g(x)\,dx-\sum_1^{n}\int_{x_0}^{x} u_r(x)g(x)\,dx\right|$

$$<\frac{\epsilon}{M(b-a)}\times M(x-x_0)$$

$$<\epsilon, \text{ when } n \geqq \nu,$$

which proves our theorem.

It is clear that these integrations can be repeated as many times as we wish.

Ex. 1. To prove that

$$\int_0^\pi \log(1-2y\cos x+y^2)\,dx=0, \text{ when } |y|<1,^*$$
$$=\pi\log y^2, \text{ when } |y|>1.$$

We know that

$$\frac{y-\cos x}{1-2y\cos x+y^2}=-\cos x-y\cos 2x-y^2\cos 3x+\dots, \quad \dots\dots\dots\dots(1)$$

when $|y|<1$.

Also the series (1) converges uniformly for any interval of y within $(-1, 1)$ (§ 67. 1).

Therefore $\displaystyle\int_0^y \frac{y-\cos x}{1-2y\cos x+y^2}\,dy=-\sum_1^\infty \cos nx\int_0^y y^{n-1}\,dy$, when $|y|<1$.

Therefore $\displaystyle\tfrac{1}{2}\log(1-2y\cos x+y^2)=-\sum_1^\infty \frac{\cos nx}{n}y^n$, when $|y|<1$.† ....(2)

But the series (2) converges uniformly for all values of x, when $|y|$ is some positive number less than unity (§ 67. 1).

Thus $\displaystyle\tfrac{1}{2}\int_0^\pi \log(1-2y\cos x+y^2)\,dx=-\sum_1^\infty \frac{y^n}{n}\int_0^\pi \cos nx\,dx$, when $|y|<1$.

Therefore $\displaystyle\int_0^\pi \log(1-2y\cos x+y^2)\,dx=0$, when $|y|<1$.

But $\displaystyle\int_0^\pi \log(1-2y\cos x+y^2)\,dx=\int_0^\pi\left[\log y^2+\log\left(1-\frac{2}{y}\cos x+\frac{1}{y^2}\right)\right]dx.$

Therefore $\displaystyle\int_0^\pi \log(1-2y\cos x+y^2)\,dx=\pi\log y^2$, when $|y|>1$.

Ex. 2. Prove that, if m is a positive integer,

$$\int_0^\pi \cos mx\log(1-2y\cos x+y^2)\,dx=-\pi\frac{y^m}{m} \text{ or } -\pi\frac{y^{-m}}{m},$$

according as $|y|<1$ or $|y|>1$.‡

---

*It follows from Ex. 4, p. 131, that we may replace the symbols $<$, $>$ by $\leqq$ and $\geqq$ respectively.

†If x is not zero or an even multiple of π, the series on the right-hand of (2) converges when $y=1$.

It follows from Abel's Theorem on the Power Series (§ 72, VII) that

$$\tfrac{1}{2}\log 2(1-\cos x)=-\sum_1^\infty\frac{\cos nx}{n}, \text{ when } x\neq 0 \text{ or } 2r\pi.$$

Again, if x is not an odd multiple of π, the series on the right-hand of (2) converges when $y=-1$.

Then, as above, we have

$$\tfrac{1}{2}\log 2(1+\cos x)=\sum_1^\infty(-1)^{n-1}\frac{\cos nx}{n}, \text{ when } x\neq(2r+1)\pi.$$

‡It follows from Ex. 5, p. 131 that these results hold also for $|y|=1$.

70. 2. The following extension of the theorem in the preceding section is sometimes useful.

Let $u_1(x), u_2(x) \ldots$ *be functions integrable in* (a, b) *and let* $s_n(x) = \sum_1^n u_r(x)$ *and* $f(x) = \sum_1^\infty u_r(x)$.

Thus if (i) $f(x)$ *is integrable in* (a, b), *and* (ii) *the series converges uniformly to* $f(x)$ *in* (a, c), *where* c *is any number between* a *and* b, *and* (iii) $s_n(x)$ *is uniformly bounded in* (a, b),

$$\int_a^b f(x)dx = \sum_1^\infty \int_a^b u_n(x)dx.$$

Since the sum $s_n(x)$ is uniformly bounded in (a, b), there is a positive number K (independent of x and n), such that

$$|s_n(x)| < K$$

for every x in (a, b) and every positive integer n.

Let the arbitrary positive number ϵ be chosen, and let c be taken so that $b - c < \dfrac{\epsilon}{4K}$.

Then $$\int_a^b f(x)dx - \int_a^b s_n(x)dx = \int_a^c (f(x) - s_n(x))dx + \int_c^b f(x)dx - \int_c^b s_n(x)dx. \quad (1)$$

And $$\left|\int_a^b f(x)dx - \int_a^b s_n(x)dx\right| \leqq \int_a^c |f(x) - s_n(x)|dx + \int_c^b |f(x)|dx + \int_c^b |s_n(x)|dx. \quad (2)$$

But $$|s_n(x)| < K.$$

Therefore $$|f(x)| = |\lim_{n\to\infty} s_n(x)| \leqq K.$$

But the series converges uniformly in (a, c).

Thus there is a positive integer ν, such that

$$|f(x) - s_n(x)| < \frac{\epsilon}{2(b-a)}, \text{ when } n \geqq \nu,$$

the same ν serving for all values of x in (a, c).

It follows from (2), that

$$\left|\int_a^b f(x)dx - \sum_1^n \int_a^b u_r(x)dx\right| < \frac{\epsilon}{2(b-a)}(c-a) + 2(b-c)K,$$
$$< \tfrac{1}{2}\epsilon + \tfrac{1}{2}\epsilon,$$
$$< \epsilon,$$

when $n \geqq \nu$.

Thus, under the conditions stated in the theorem,

$$\int_a^b f(x)dx = \sum_1^\infty \int_a^b u_r(x)dx.$$

This may be extended as follows:

Let the integrable function $f(x)$ *be the sum of the series of integrable functions* $u_1(x), u_2(x), \ldots$ *and let this series converge uniformly in* $a \leqq x \leqq b$, *except for a finite number of sub-intervals, the sum of whose lengths can be made less than any given number. Also let* $s_n(x) = \sum_1^n u_r(x)$ *be uniformly bounded in* (a, b).

Then $$\int_a^b f(x)\,dx = \sum_1^\infty \int_a^b u_r(x)\,dx.*$$

71. A sufficient Condition for Term by Term Differentiation.

If the series $$u_1(x) + u_2(x) + u_3(x) + \ldots$$

converges in (a, b) *and each of its terms has a differential coefficient, continuous in* (a, b), *and if the series of differential coefficients*

$$u_1'(x) + u_2'(x) + u_3'(x) + \ldots$$

converges uniformly in (a, b), *then* $f(x)$, *the sum of the original series, has a differential coefficient at every point of* (a, b), *and*

$$f'(x) = u_1'(x) + u_2'(x) + u_3'(x) + \ldots .$$

Let $$\phi(x) = u_1'(x) + u_2'(x) + u_3'(x) + \ldots .$$

Since this series of continuous functions converges uniformly in (a, b), we can integrate it term by term.

Thus we have

$$\int_{x_0}^{x_1} \phi(x)\,dx = \int_{x_0}^{x_1} u_1'(x)\,dx + \int_{x_0}^{x_1} u_2'(x)\,dx + \int_{x_0}^{x_1} u_3'(x)\,dx + \ldots,$$

where $a \leqq x_0 < x_1 \leqq b$.

Therefore $$\int_{x_0}^{x_1} \phi(x)\,dx = [u_1(x_1) - u_1(x_0)] + [u_2(x_1) - u_2(x_0)] + \ldots .$$

But $$f(x_1) = u_1(x_1) + u_2(x_1) + u_3(x_1) + \ldots$$

and $$f(x_0) = u_1(x_0) + u_2(x_0) + u_3(x_0) + \ldots .$$

Therefore $$\int_{x_0}^{x_1} \phi(x)\,dx = f(x_1) - f(x_0).$$

Now put $$x_0 = x \quad \text{and} \quad x_1 = x + \Delta x.$$

Then, by the First Theorem of Mean Value,

$$\phi(\xi)\Delta x = f(x + \Delta x) - f(x),$$

where $x \leqq \xi \leqq x + \Delta x$.

Therefore $$\phi(\xi) = \frac{f(x + \Delta x) - f(x)}{\Delta x}.$$

---

*Cf. *The Mathematical Gazette*, 13 (1927), 438. In this paper by the author on "Term by Term Integration of Infinite Series" further information on this subject will be found; and a proof is given of the theorem due to Arzela (1885) that when the series of integrable functions $u_1(x)$, $u_2(x)$, ... converges to the integrable function $f(x)$, and the sum $\sum_1^n u_r(x)$ is uniformly bounded in (a, b), then

$$\int_a^b f(x)\,dx = \sum_1^\infty \int_a^b u_r(x)\,dx.$$

But $$\lim_{\Delta x\to 0} \phi(\xi)=\phi(x),$$
since $\phi(x)$ is continuous in (a, b).

Therefore $f(x)$ has a differential coefficient $f'(x)$ in (a, b),

and $$f'(x)=\phi(x)$$
$$=u_1'(x)+u_2'(x)+u_3'(x)+\ldots .$$

It must be remembered that the conditions for continuity, and for term by term differentiation and integration, which we have obtained are only *sufficient* conditions. They are not *necessary* conditions. We have imposed more restrictions on the functions than are required. But no other conditions of equal simplicity have yet been found, and for that reason these theorems are of importance.

It should also be noted that in these sections we have again been dealing with repeated limits (cf. § 64), and we have found that in certain cases the order in which the limits are taken may be reversed without altering the result.

In term by term integration, we have been led to the equality, in certain cases, of

$$\int_{x_0}^{x_1} \lim_{n\to\infty} s_n(x)\,dx \quad \text{and} \quad \lim_{n\to\infty}\int_{x_0}^{x_1} s_n(x)\,dx.$$

Similarly in term by term differentiation we have found that, in certain cases,

$$\lim_{h\to 0}\left[\lim_{n\to\infty}\left(\frac{s_n(x+h)-s_n(x)}{h}\right)\right] \quad \text{and} \quad \lim_{n\to\infty}\left[\lim_{h\to 0}\left(\frac{s_n(x+h)-s_n(x)}{h}\right)\right]$$

are equal.

72. The Power Series. The properties of the Power Series
$$a_0+a_1x+a_2x^2+\ldots$$
are so important, and it offers so simple an illustration of the results we have just obtained, that a separate discussion of this series will now be given.

I. *If the series* $$a_0+a_1x+a_2x^2+\ldots$$
is convergent for $x=x_0$, *it is absolutely convergent for every value of* x *such that* $|x|<|x_0|$.

Since the series is convergent for $x=x_0$, there is a positive number M such

that $$|a_n x_0^n|<M, \text{ when } n \geqq 0.$$

But $$|a_n x^n|=|a_n x_0^n|\times\left|\frac{x}{x_0}\right|^n.$$

Therefore if $\left|\frac{x}{x_0}\right| = c < 1$, the terms of the series

$$|a_0| + |a_1 x| + |a_2 x^2| + \ldots$$

are less than the corresponding terms of the convergent series

$$M\{1 + c + c^2 + \ldots\},$$

and our theorem follows.

II. *If the series does not converge for $x = x_0$, it does not converge for any value of x such that $|x| > |x_0|$.*

This follows from (I), since if the series converges for a value of x, such that $|x| > |x_0|$, it must converge for $x = x_0$.

III. It follows from (I) and (II) that only the following three cases can occur:

(i) The series converges for $x = 0$ and no other value of x.

$$E.g. \quad 1 + 1!\,x + 2!\,x^2 + 3!\,x^3 + \ldots,$$
$$1 + x + 2^2 x^2 + 3^3 x^3 + \ldots.$$

(ii) The series converges for all values of x.

$$E.g. \quad 1 + x + \frac{x^2}{2!} + \ldots.$$

(iii) There is some positive number ρ such that, when $|x| < \rho$, the series converges, and, when $|x| > \rho$, the series does not converge.

$$E.g. \quad x - \tfrac{1}{2}x^2 + \tfrac{1}{3}x^3 - \ldots.$$

The interval $-\rho < x < \rho$ is called the *interval of convergence* of the series. Also it is convenient to say that, in the first case, the interval is zero, and, in the second, infinite. It will be seen that the interval of convergence of the following three series is $(-1, 1)$:

$$1 + x + x^2 + \ldots,$$
$$1 + \frac{x}{1} + \frac{x^2}{2} + \ldots,$$
$$1 + \frac{x}{1^2} + \frac{x^2}{2^2} + \ldots.$$

But it should be noticed that the first of these does not converge at the ends of the interval; the second converges at one of the ends; and the third converges at both.

In the Power Series there cannot be first an interval of convergence, then an interval where the convergence fails, and then a return to convergence.

Also the interval of convergence is symmetrical with regard to the origin. We shall denote its ends by L', L. The series need not converge at L' or L, but it may do so; and it must converge within $L'L$.

IV. *If the series converges for a value of $x \gtrless 0$, then the sequence*

$$|a_1|, \quad |a_2|^{\frac{1}{2}}, \quad |a_3|^{\frac{1}{3}}, \ldots |a_n|^{\frac{1}{n}} \ldots$$

is bounded above: and if $\overline{\lim_{n\to\infty}} |a_n|^{\frac{1}{n}} = \mu > 0$, the interval of convergence is $-\frac{1}{\mu} < x < \frac{1}{\mu}$. If $\overline{\lim_{n\to\infty}} |a_n|^{\frac{1}{n}} = 0$, the series converges for all values of x.

We are given that the series converges for some value of $x \gtrless 0$.

Then, as in (I), there is a positive number M, which we may take greater than unity, such that $|a_n x^n| < M$, for all values of n.

Thus $|a_n|^{\frac{1}{n}} < \frac{M}{|x|}$, for all values of n, and the given sequence is bounded above and below.

By § 17. 2, $\overline{\lim}_{n\to\infty} |a_n|^{\frac{1}{n}}$ exists.

Now let
$$\overline{\lim}_{n\to\infty} |a_n|^{\frac{1}{n}} = \mu > 0.$$

Take any x for which $|x| < 1/\mu$, and choose a definite point x_0 between $|x|$ and $1/\mu$.

Then $\mu < 1/x_0$, and, from the properties of the upper limit of indetermination, there is a positive integer ν, such that

$$|a_n|^{\frac{1}{n}} < \frac{1}{x_0}, \text{ when } n \geqq \nu.$$

Therefore
$$|a_n x_0^n| < 1, \text{ when } n \geqq \nu.$$

And
$$|a_n x^n| = |a_n x_0^n| \times \left|\frac{x}{x_0}\right|^n$$
$$< \left|\frac{x}{x_0}\right|^n < 1, \text{ when } n \geqq \nu.$$

Thus the series $\Sigma a_n x^n$ converges absolutely when $|x| < 1/\mu$.

Again, take any x for which $|x| > 1/\mu$.

Then $|a_n|^{\frac{1}{n}} > \frac{1}{|x|}$, for an infinite number of values of n.

Thus $|a_n x^n| > 1$ for an infinite number of values of n.

And the series $\Sigma a_n x^n$ cannot converge when $|x| > 1/\mu$.

It follows that *when* $\mu > 0$, *the interval of convergence of the series is*

$$-1/\mu < x < 1/\mu.$$

Finally let $\overline{\lim}_{n\to\infty} |a_n|^{\frac{1}{n}} = 0$.

Take any value of x other than zero.

Then, by the properties of the upper limit of indetermination, there is a positive integer ν such that

$$|a_n|^{\frac{1}{n}} < \frac{1}{2|x|}, \text{ when } n \geqq \nu.$$

Thus
$$|a_n x^n| < \frac{1}{2^n}, \text{ when } n \geqq \nu,$$

and in this case the series converges for all values of x.

Returning to the notation of (III), we now show that

V. *The series is absolutely convergent in the open interval* $-\rho < x < \rho$.

VI. *The series is absolutely and uniformly convergent in the closed interval* $-\rho+\delta \leqq x \leqq \rho-\delta$, *where* δ *is any assigned positive number less than* ρ.

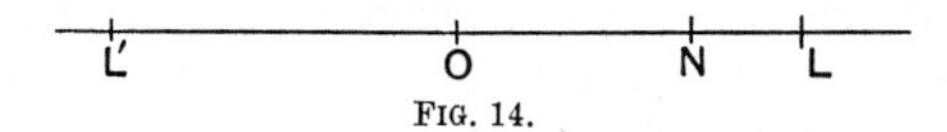

FIG. 14.

To prove (V), we have only to remark that if N is a point x_0, where $-\rho < x_0 < \rho$, between N and the nearer boundary of the interval of convergence, there are values of x for which the series converges, and thus by (I) it converges absolutely for $x=x_0$.

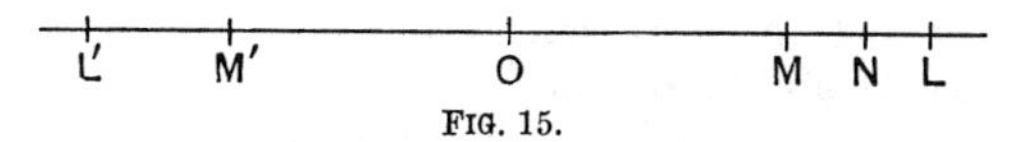

FIG. 15.

To prove (VI), let M', M correspond to $x=-\rho+\delta$ and $x=\rho-\delta$ respectively. We now choose a point N (say x_0) between M and the nearer boundary L. The series converges absolutely for $x=x_0$, by (V).

Thus, with the usual notation,

$$|a_n x_0^n| + |a_{n+1} x_0^{n+1}| + \ldots < \epsilon, \text{ when } n \geqq \nu.$$

But

$$|a_n x^n| + |a_{n+1} x^{n+1}| + \ldots$$

is less than the above for every point in $M'M$, including the ends M', M.

It follows that our series is absolutely and uniformly convergent in the closed interval (M', M).* And the sum of the series is continuous in this closed interval.

It remains to examine the behaviour of the series at the ends of the interval of convergence, and we shall now prove Abel's Theorem:†

VII. *If the series converges for either of the ends of the interval of convergence, the interval of uniform convergence extends up to and includes that point, and the continuity of* $f(x)$, *the sum of the series, extends up to and includes that point.*

This follows at once from Abel's test for uniform convergence given in § 67. 2.

Let the series converge for the end ρ of the interval of convergence.

Then in Abel's test, take $u_n = a_n \rho^n$ and $v_n = \left(\dfrac{x}{\rho}\right)^n$. (Cf. Ex. 1, p. 150.)

We thus establish that the series

$$a_0 + a_1 x + a_2 x^2 + \ldots$$

converges uniformly in this case in $0 \leqq x \leqq \rho$.

But we know from (VI) that the series is uniformly convergent in

$$-\rho+\delta \leqq x \leqq 0,$$

when δ is any positive number less than ρ.

---

*When the interval of convergence extends to infinity, the series will be absolutely convergent for every value of x, but it need not be uniformly convergent in the *infinite interval.* However, it will be uniformly convergent in any interval $(-b, b)$, where b is fixed, but may be fixed as large as we please.

E.g. the exponential series converges uniformly in any fixed interval, which may be arbitrarily great, but not in an infinite interval [cf. § 65].

† *Journal für Math.*, 1 (1826), 311.

It follows that the series is uniformly convergent in the interval

$$-\rho+\delta \leqq x \leqq \rho.$$

And that $f(x)$, the sum of the series, is continuous in this closed interval.

In particular, when the series converges at $x=\rho$,

$$\lim_{x \to \rho - 0} f(x)=f(\rho)=a_0+a_1\rho+a_2\rho^2+\dots.$$

In the case of the logarithmic series,

$$\log(1+x)=x-\tfrac{1}{2}x^2+\tfrac{1}{3}x^3-\dots,$$

the interval of convergence is $-1<x<1$.

Further, when $x=1$, the series converges.

It follows from Abel's Theorem that

$$\lim_{x\to 1}\log(1+x)=1-\tfrac{1}{2}+\tfrac{1}{3}-\dots,$$

$$\text{i.e.}\quad \log 2=1-\tfrac{1}{2}+\tfrac{1}{3}-\dots$$

Similarly, in the Binomial Series,

$$(1+x)^m=1+mx+\frac{m(m-1)}{2!}x^2+\dots,$$

when $-1<x<1$.

And it is known* that $1+m+\dfrac{m(m-1)}{2!}+\dots$

is conditionally convergent when $-1<m<0$, and absolutely convergent when $m>0$.

Hence $$\lim_{x\to 1}(1+x)^m=1+m+\frac{m(m-1)}{2!}+\dots,$$

$$\text{i.e.}\quad 2^m=1+m+\frac{m(m-1)}{2!}+\dots,$$

in both these cases.

On the other hand, if we put $x=1$ in the series for $(1+x)^{-1}$, we get a series which does not converge. The uniformity of the convergence of the series

$$1-x+x^2-\dots$$

is for the interval $-l\leqq x\leqq l$, where l is any given positive number less than 1.

VIII. *The intervals of convergence of the series*

$$a_0+a_1x+a_2x^2+a_3x^3+\dots$$

and

$$a_1+2a_2x+3a_3x^2+\dots$$

are the same.†

---

*Cf. Chrystal, *Algebra*, 2 (2nd ed., 1900), 131.

†If we know that $\lim_{n\to\infty}\left|\dfrac{a_{n+1}}{a_n}\right|$ exists, this result follows immediately from the ratio-test for convergence, since in both series

$$\lim_{n\to\infty}\left|\frac{u_{n+1}}{u_n}\right|<1, \text{ when } |x|<\lim_{n\to\infty}\left|\frac{a_n}{a_{n+1}}\right|.$$

From (IV), it will be seen that we need only prove that

$$\overline{\lim_{n\to\infty}} \, |a_n|^{\frac{1}{n}} \quad \text{and} \quad \overline{\lim_{n\to\infty}} \, |na_n|^{\frac{1}{n}}$$

are the same. This is a special case of the theorem established in § 17. 3, since $\overline{\lim_{n\to\infty}} \, n^{\frac{1}{n}}=1$.

Or we may proceed as follows:

We are given that

$$|a_0|+|a_1x|+|a_2x^2|+\dots$$

converges when $|x|<\rho$.

Take x_0 so that $|x|<x_0<\rho$.

Then $$\frac{1}{x_0}+\frac{2}{x_0}\left|\frac{x}{x_0}\right|+\frac{3}{x_0}\left|\frac{x}{x_0}\right|^2+\dots$$

is convergent, because the ratio of the nth term to the preceding has for its limit $\left|\frac{x}{x_0}\right|$, which is less than unity.

If we multiply the different terms of this series by the factors

$$|a_1x_0|, \quad |a_2x_0^2|, \quad |a_3x_0^3| \dots,$$

which form a bounded sequence (by V), it is clear that the series which we thus obtain, namely

$$|a_1|+2|a_2x|+3|a_3x^2|+\dots$$

is convergent when $|x|<\rho$.

We have yet to show that this last series diverges when $|x|>\rho$.

But if it converges when $x=|x_0|$, where $|x_0|>\rho$, the same would hold for the series

$$|a_1x_0|+2|a_2x_0^2|+3|a_3x_0^3|+\dots$$

and also for the series

$$|a_1x_0|+|a_2x_0^2|+|a_3x_0^3|+\dots,$$

since the terms of the latter are not greater than those of the former.

But this is impossible, since we are given that the interval of convergence of the original series is $-\rho<x<\rho$.

It follows that the series

$$a_0+a_1x+a_2x^2+a_3x^3+\dots$$

and the series $$a_1+2a_2x+3a_3x^2+\dots,$$

obtained by differentiating the first term by term, have the same interval of convergence.

IX. *Term by term differentiation and integration of the Power Series.*

Let the power series

$$a_0+a_1x+a_2x^2+\dots$$

have $-\rho<x<\rho$ for its interval of convergence.

Let its sum be $f(x)$ in this interval.

It follows from (VI) and § 70 that

$$\int_{x_0}^{x} f(x)dx=a_0(x-x_0)+\sum_{1}^{\infty}\frac{a_n}{n+1}(x^{n+1}-x_0^{n+1}), \text{ when } -\rho<x_0<x<\rho.$$

Also from (VIII) and § 71 we see that

$$f'(x)=a_1+2a_2x+3a_3x^2+\dots,$$

where x is any point in the open interval $-\rho<x<\rho$, and these integrations and differentiations may be repeated any number of times.

73. Extensions of Abel's Theorem on the Power Series.

I. We have seen in § 72 that if the series

$$a_0+a_1+a_2+\dots$$

converges, the Power Series

$$a_0+a_1x+a_2x^2+\dots$$

is uniformly convergent, when $0\leqq x\leqq 1$; and that, if

$$f(x)=a_0+a_1x+a_2x^2+\dots,$$

$$\lim_{x\to 1-0} f(x)=a_0+a_1+a_2+\dots.$$

The above theorem of Abel's is a special case of the following:

Let the series $\sum_0^\infty a_n$ *converge, and* $\alpha_0, \alpha_1, \alpha_2, \dots$ *be a sequence of positive numbers such that* $0\leqq\alpha_0<\alpha_1<\alpha_2\dots$. *Then the series* $\sum_0^\infty a_n e^{-\alpha_n t}$ *is uniformly convergent, when* $t\geqq 0$, *and if* $f(t)=\sum_0^\infty a_n e^{-\alpha_n t}$, *we have* $\lim_{t\to+0} f(t)=\sum_0^\infty a_n$.*

This results immediately from Abel's test of § 67. 2 (cf. Ex. 3, p. 150).

II. In Abel's Theorem and its extension stated above, the series $\sum_0^\infty a_n$ are supposed convergent. We proceed to prove Bromwich's Theorem dealing with series which need not converge.† In this discussion we shall adopt the following notation:

$$s_n=a_0+a_1+a_2+\dots+a_n,$$

$$S_n=s_0+s_1+s_2+\dots+s_n,$$

and we write σ_n for the Arithmetic Mean of the first n terms of the sequence $s_0, s_1, s_2, \dots$.

Thus
$$\sigma_n=\frac{s_0+s_1+s_2+\dots+s_{n-1}}{n}=\frac{S_{n-1}}{n}.$$

It can be shown (cf. § 102) that, if the series $\sum_0^\infty a_n$ converges and its sum is s, then, with the above notation, $\lim_{n\to\infty}\sigma_n=s$. But the converse does not hold.

---

*If $a_0, a_1, \dots$ are functions of x and the series $\sum_0^\infty a_n$ converges uniformly to $F(x)$ in a given interval, it follows from Abel's Test of § 67. 2 that $\lim_{t\to+0}\sum_0^\infty a_n e^{-\alpha_n t}$ converges uniformly to $F(x)$ in this interval.

†*Math. Annalen*, 65 (1908), 350. See also a paper by C. N. Moore in *Bull. Amer. Math. Soc.* 25 (1919), 258.

The sequence of Arithmetic Means may converge, while the sum $\sum_0^\infty a_n$ fails to converge.*

Bromwich's Theorem. *Let the sequence of Arithmetic Means σ_n for the series $\sum_0^\infty a_n$ converge to σ. Also let u_n be a function of t with the following properties, when $t>0$:*

(α) $\sum_p^q n\,|\,\Delta^2 u_n\,| < K$† $\left(\begin{array}{l}\textit{p, q any positive integers; K, a positive}\\ \textit{number independent of p, q and t}\end{array}\right).$

(β) $\lim_{n\to\infty} nu_n = 0,$

(γ) $\lim_{t\to+0} u_n = 1.$

Then the series $\sum_0^\infty a_n u_n$ converges when $t>0$, and $\lim_{t\to+0}\sum_0^\infty a_n u_n = \sigma$.

We have

$$\begin{aligned} S_0 &= s_0 = a_0,\\ S_1 - 2S_0 &= s_1 - s_0 = a_1,\\ S_2 - 2S_1 + S_0 &= s_2 - s_1 = a_2,\\ S_3 - 2S_2 + S_1 &= s_3 - s_2 = a_3, \text{ etc.}\end{aligned}$$

Thus

$$\sum_0^n a_n u_n = S_0 u_0 + (S_1 - 2S_0)u_1 + (S_2 - 2S_1 + S_0)u_2 + \ldots + (S_n - 2S_{n-1} + S_{n-2})u_n.$$

Therefore

$$\sum_0^n a_n u_n = S_0\Delta^2 u_0 + S_1\Delta^2 u_1 + \ldots + S_n\Delta^2 u_n + 2S_n u_{n+1} - S_n u_{n+2} - S_{n-1}u_{n+1}. \quad (1)$$

But the sequence of Arithmetic Means

$$\sigma_1, \sigma_2, \sigma_3, \ldots$$

converges, and $\lim_{n\to\infty}\sigma_n = \sigma$.

It follows that there is a number C, not less than $|\sigma|$, such that $|S_n| < (n+1)C$ for every integer n.

Also from (β) it is clear that

$$\lim_{n\to\infty}(S_n u_{n+1}) = \lim_{n\to\infty}(S_n u_{n+2}) = \lim_{n\to\infty}(S_{n-1}u_{n+1}) = 0. \quad \ldots\ldots\ldots\ldots(2)$$

---

*If $a_n = (-1)^n$, $n \geqq 0$, it is obvious that $\lim_{n\to\infty}\sigma_n = \frac{1}{2}$, and the series $\sum_0^\infty a_n$ is not convergent. But see the Hardy-Landau Theorem, § 102, II. When $\lim_{n\to\infty}\sigma_n = \sigma$, the series $\sum_0^\infty a_n$ is often said to be "*summable* $(C, 1)$" and its sum $(C, 1)$ is said to be σ. For a discussion of this method of treating series, due to Cesàro, reference may be made to Whittaker and Watson's *Course of Modern Analysis* (5th ed., 1928), p. 155.

Knopp, *loc. cit.*, English transl., Ch. XIII.

Hobson, *loc. cit.*, 2 (2nd ed., 1926), Ch. I.

Also see below, §§ 101–103, 108.

† $\Delta^2 u_n$ is written for $(u_n - 2u_{n+1} + u_{n+2})$. Since all the terms in the series $\sum_0^\infty n\,|\,\Delta^2 u_n\,|$ are positive this condition (α) implies the convergence of this series.

Further, the series $\sum_0^\infty (n+1)\,|\,\Delta^2 u_n\,|$ converges, since, from (a), the series $\sum_0^\infty n\,|\,\Delta^2 u_n\,|$ converges.

Also
$$|\,S_n \Delta^2 u_n\,| < C(n+1)\,|\,\Delta^2 u_n|.$$

Therefore the series $\sum_0^\infty S_n \Delta^2 u_n$ converges absolutely.

It follows from (1) and (2) that

$$\sum_0^\infty a_n u_n = \sum_0^\infty S_n \Delta^2 u_n. \quad \text{.................................(3)}$$

Taking the special case $a_0 = 1, \quad a_1 = a_2 = \ldots = 0,$

we have
$$S_n = n+1 \quad \text{and} \quad u_0 = \sum_0^\infty (n+1)\Delta^2 u_n. \quad \text{.....................(4)}$$

Thus, from (3) and (4),

$$\sum_0^\infty a_n u_n - \sigma u_0 = \sum_0^\infty (S_n - (n+1)\sigma)\,\Delta^2 u_n. \quad \text{.....................(5)}$$

Now
$$\lim_{n\to\infty} \frac{S_n}{n+1} = \sigma.$$

Therefore, to the arbitrary positive number ϵ, there corresponds a positive integer ν such that

$$\left|\frac{S_n}{n+1} - \sigma\right| < \frac{\epsilon}{4K}, \text{ when } n \geqq \nu.$$

Thus
$$|\,S_n - (n+1)\sigma\,| < \frac{\epsilon}{4K}(n+1), \text{ when } n \geqq \nu.$$

Also $|\,S_n| < (n+1)C$, for every positive integer, and $|\sigma| \leqq C$.

It follows, from these inequalities and (5), that

$$|\sum_0^\infty a_n u_n - \sigma u_0| \leqq |\sum_0^{\nu-1} (S_n - (n+1)\sigma)\Delta^2 u_n| + |\sum_\nu^\infty (S_n - (n+1)\sigma)\Delta^2 u_n|.$$

$$< 2C\sum_0^{\nu-1}(n+1)\,|\,\Delta^2 u_n| + \frac{\epsilon}{4K}\sum_\nu^\infty (n+1)\,|\,\Delta^2 u_n|. \quad \text{...............(6)}$$

But
$$\sum_\nu^\infty (n+1)\,|\,\Delta^2 u_n\,| < 2\sum_\nu^\infty n\,|\,\Delta^2 u_n|$$
$$< 2K, \text{ by } (a). \quad \text{...........................(7)}$$

And
$$\lim_{t\to+0} \Delta^2 u_n = 0, \quad \text{since} \quad \lim_{t\to+0} u_n = 1, \text{ by } (\gamma).$$

It follows that, ν being fixed, there is a positive number η such that

$$|\,\Delta^2 u_n\,| < \frac{\epsilon}{2\nu(\nu+1)C}, \text{ when } 0 < t \leqq \eta \text{ and } n \leqq \nu - 1. \quad \text{...........(8)}$$

Thus from (6), (7) and (8), we see that

$$|\sum_0^\infty a_n u_n - \sigma u_0\,| < \tfrac{1}{2}\epsilon + \tfrac{1}{2}\epsilon < \epsilon, \text{ when } 0 < t \leqq \eta.$$

Therefore
$$\lim_{t\to+0} (\sum_0^\infty a_n u_n - \sigma u_0) = 0.$$

And, finally,
$$\lim_{t\to+0} \sum_0^\infty a_n u_n = \sigma,$$

since, from (γ),
$$\lim_{t\to+0} u_0 = 1.$$

III. *Let the sequence of Arithmetic Means for the series* $\sum_0^\infty a_n$ *converge to* σ, *and let* u_n *be* e^{-nt} (*or* e^{-n^2t}). *Then the series* $\sum_0^\infty a_n u_n$ *converges, when* $t > 0$, *and* $\lim_{t\to+0} \sum_0^\infty a_n u_n = \sigma$.

This follows at once from Bromwich's Theorem above, if e^{-nt} (or e^{-n^2t}) satisfy the conditions (α), (β) and (γ) of that theorem.

It is obvious that (β) and (γ) are satisfied, so it only remains to establish that (α) is satisfied.

(i) Let
$$u_n = e^{-nt}, \quad t > 0.$$
Then
$$\Delta^2 u_n = u_n - 2u_{n+1} + u_{n+2}$$
$$= e^{-nt}(1 - e^{-t})^2.$$

Therefore $\Delta^2 u_n$ is positive.

Also
$$\sum_1^n n|\Delta^2 u_n| = \sum_1^n n\Delta^2 u_n$$
$$= nu_{n+2} - (n+1)u_{n+1} + u_1.$$

Therefore
$$\sum_1^\infty n|\Delta^2 u_n| = e^{-t}, \text{ and the condition } (\alpha) \text{ is satisfied.}$$

(ii) Let $u_n = e^{-n^2t}$, $t > 0$.

In this case
$$\Delta^2 u_n = e^{-(n+\theta)^2 t}(4(n+\theta)^2 t^2 - 2t),$$
where $0 < \theta < 2$.*

Therefore the sign of $\Delta^2 u_n$ depends upon that of $(4(n+\theta)^2 t^2 - 2t)$.

It follows that it is positive or negative according as
$$2(n+\theta)t - \sqrt{(2t)} \gtrless 0.$$

Also
$$(n+\theta) > \frac{1}{\sqrt{(2t)}}, \text{ when } n > \frac{1}{\sqrt{(2t)}},$$
and
$$(n+\theta) < \frac{1}{\sqrt{(2t)}}, \text{ when } n+2 < \frac{1}{\sqrt{(2t)}}.$$

Therefore $\Delta^2 u_n$ cannot change sign more than three times for any positive value of t.

But it follows at once from the equation
$$n\Delta^2 u_n = ne^{-(n+\theta)^2 t}(4(n+\theta)^2 t^2 - 2t)$$
that a positive number, independent of t, can be assigned such that $n|\Delta^2 u_n|$ is less than this number for all values of n.

Hence K can be chosen so that the condition (α) is satisfied, provided that the sum of any sequence of terms, all of the same sign, that we can choose from $\sum n\Delta^2 u_n$, is less in absolute value than some fixed positive number for all values of t.

Let $\sum_r^s n\Delta^2 n_n$ be the sum of such a set of consecutive terms.

---

*This follows from the fact that
$$\frac{f(x+2h) - 2f(x+h) + f(x)}{h^2} = f''(x+\theta h),$$
where $0<\theta<2$, provided that $f(x)$, $f'(x)$, $f''(x)$ are continuous from x to $x+2h$. (Cf. Goursat, *loc. cit.*, **1** (4e éd., 1923), § 21.)

Then we have

$$\sum_r^s n\Delta^2 u_n = re^{-r^2t} - (r-1)e^{-(r+1)^2t} - (s+1)e^{-(s+1)^2t} + se^{-(s+2)^2t},$$

which differs from

$$r(e^{-r^2t} - e^{-(r+1)^2t}) - (s+1)(e^{-(s+1)^2t} - e^{-(s+2)^2t})$$

by at most unity.

But, when n is a positive integer and $t > 0$,

$$\begin{aligned} 0 < n(e^{-n^2t} - e^{-(n+1)^2t}) &= 2n(n+\theta)te^{-(n+\theta)^2t} \ldots (0<\theta<1) \\ &< 2(n+\theta)^2te^{-(n+\theta)^2t} \\ &< 2e^{-1}. \end{aligned}$$

Therefore, for the set of terms considered,

$$\sum_r^s |n\Delta^2 u_n| < 4e^{-1} + 1.$$

Then the argument above shows that the condition (α) is satisfied.

IV. *Let the sequence of Arithmetic Means for the series* $\sum_0^\infty a_n$ *converge to* σ. *Then the series* $\sum_0^\infty a_n x^n$ *will converge when* $0 < x < 1$, *and*

$$\lim_{x\to 1-0} \sum_0^\infty a_n x^n = \sigma.$$

This follows from the first part of (III) on putting $x = e^{-t}$.

V. *Let the terms of the series* $\sum_0^\infty a_n$ *be functions of* x, *and the sequence of Arithmetic Means for this series converge uniformly to the bounded function* $\sigma(x)$ *in* $a \leqq x \leqq b$. *Then* $\lim_{t\to+0} \sum_0^\infty a_n u_n$ *converges uniformly to* $\sigma(x)$ *in this interval, provided that* u_n *is a function of* t *satisfying the conditions* (α), (β) *and* (γ) *of Bromwich's Theorem, when* $t > 0$.

This follows at once by making slight changes in the argument of (II).

The theorems proved in this section will be found useful in the solution of problems in Applied Mathematics, when the differential equation, which corresponds to the problem, is solved by series. The solution has to satisfy certain initial and boundary conditions. What we really need is that, as we approach the boundary, or as the time tends to zero, our solution shall have the given value as its limit. What happens upon the boundaries, or at the instant $t = 0$, is not discussed. (See below § 123.)

74. Integration of Series. Infinite Integrals. Finite Interval. In the discussion of § 70 we dealt only with ordinary finite integrals. We shall now examine the question of term by term integration, both when the integrand has points of infinite discontinuity in the interval of integration, supposed finite, and when the integrand is bounded in any finite interval, but the interval of integration itself extends to infinity. In this section we shall deal with the first of these forms, and it will be sufficient to confine the discussion to the case when the infinity occurs at one end of the interval (a, b), say $x = b$.

I. *Let $u_1(x)$, $u_2(x)$, ... be continuous in (a, b) and the series $\sum_1^\infty u_n(x)$ converge uniformly to $f(x)$ in (a, b).*

*Also let $g(x)$ have an infinite discontinuity at $x=b$ and $\int_a^b g(x)dx$ be absolutely convergent.**

Then
$$\int_a^b f(x)g(x)dx = \sum_1^\infty \int_a^b u_n(x)g(x)dx.$$

From the uniform convergence of $\sum_1^\infty u_n(x)$ in (a, b), we know that its sum $f(x)$ is continuous in (a, b), and thus bounded and integrable.

Also $\int_a^b f(x)g(x)dx$ is absolutely convergent, since $\int_a^b g(x)dx$ is so (§ 61, VI).

Let
$$\int_a^b |g(x)|dx = A.$$

Then, having chosen the positive number ϵ, as small as we please, we may put
$$f(x) = s_n(x) + R_n(x),$$
where $|R_n(x)| < \dfrac{\epsilon}{A}$, when $n \geqq \nu$, the same ν serving for all values of x in (a, b).

But $\int_a^b f(x)\,g(x)dx$ and $\int_a^b s_n(x)\,g(x)dx$ both exist.

It follows that $\int_a^b R_n(x)\,g(x)dx$ also exists, and that
$$\int_a^b f(x)\,g(x)dx = \int_a^b s_n(x)\,g(x)dx + \int_a^b R_n(x)\,g(x)dx.$$

Thus we see that
$$\left|\int_a^b f(x)\,g(x)dx - \sum_1^n \int_a^b u_r(x)\,g(x)\,dx\right| = \left|\int_a^b R_n(x)\,g(x)\,dx\right|$$
$$< \frac{\epsilon}{A}\int_a^b |g(x)|\,dx$$
$$< \epsilon, \text{ when } n \geqq \nu,$$
which proves that the series $\sum_1^\infty \int_a^b u_n(x)\,g(x)\,dx$ is convergent and that its sum is $\int_a^b f(x)\,g(x)\,dx$.

Ex. This case is illustrated by
$$\int_0^1 \log x \log(1+x)\,dx = \sum_1^\infty (-1)^{n-1}\int_0^1 \frac{x^n}{n}\log x\,dx$$
$$= \sum_1^\infty \frac{(-1)^n}{n(n+1)^2}, \text{ since } \int_0^1 x^n \log x\,dx = -\frac{1}{(n+1)^2},$$
$$= \sum_1^\infty (-1)^n\left[\frac{1}{n} - \frac{1}{n+1} - \frac{1}{(n+1)^2}\right]$$
$$= 2 - 2\log 2 - \frac{1}{12}\pi^2, \text{ using the series for } \frac{1}{12}\pi^2.†$$

---

* It is clear that this proof also applies when $g(x)$ has a finite number of infinite discontinuities in (a, b) and $\int_a^b g(x)dx$ is absolutely convergent.

† Cf. Carslaw, *Plane Trigonometry* (2nd ed., 1915), 279.

Here the series for $\log(1+x)$ converges uniformly in $0 \leqq x \leqq 1$, and

$$\int_0^1 \log x\, dx$$

converges absolutely (as a matter of fact $\log x$ is always of the same sign in $0 < x \leqq 1$), while $|\log x| \to \infty$ as $x \to 0$.

On the other hand, we may still apply term by term integration in certain cases when the above conditions are not satisfied, as will be seen from the following theorems:

II. *Let $u_1(x)$, $u_2(x)$, ... be continuous and positive and the series $\sum_1^\infty u_n(x)$ converge uniformly to $f(x)$ in the arbitrary interval (a, α), where $a < \alpha < b$.*

Further, let $g(x)$ be positive, bounded and integrable in (a, α).

Then
$$\int_a^b f(x)g(x)\,dx = \sum_1^\infty \int_a^b u_n(x)g(x)\,dx,$$

provided that either the integral $\int_a^b f(x)g(x)dx$ or the series $\sum_1^\infty \int_a^b u_n(x)g(x)dx$ converges.

Let us suppose that $\int_a^b f(x)g(x)dx$ converges.

In other words, we are given that the repeated limit

$$\lim_{\xi \to 0} \int_a^{b-\xi} [\lim_{n\to\infty} \sum_1^n u_r(x)]\, g(x)dx \text{ exists.}$$

Since the functions $u_1(x)$, $u_2(x)$, ... are all positive, as well as $g(x)$, in (a, α), from the convergence of $\int_a^b f(x)g(x)dx$ there follows at once the convergence of

$$\int_a^b u_r(x)g(x)dx \quad (r=1, 2, \ldots).$$

Again, let $\qquad f(x) = u_1(x) + u_2(x) + \ldots + u_n(x) + R_n(x).$

Then $\int_a^b R_n(x)g(x)dx$ also converges, and for every positive integer n

$$\int_a^b f(x)g(x)dx - \sum_1^n \int_a^b u_r(x)g(x)dx = \int_a^b R_n(x)g(x)dx. \quad \text{..............(1)}$$

But from the convergence of $\int_a^b f(x)g(x)dx$ it follows that, when the arbitrary positive number ϵ has been chosen, as small as we please, there will be a positive number ξ such that

$$0 < \int_{b-\xi}^b f(x)g(x)dx < \tfrac{1}{2}\epsilon.$$

A fortiori, $\qquad 0 < \int_{b-\xi}^b R_n(x)g(x)dx < \tfrac{1}{2}\epsilon, \quad$ ....(2)

and this holds for all positive integers n.

Let the upper bound of $g(x)$ in $(a, b-\xi)$ be M.

The series $\sum_1^\infty u_n(x)$ converges uniformly in $(a, b-\xi)$.

Keeping the number ϵ we have chosen above, there will be a positive integer ν such that

$$0 < R_n(x) < \frac{\epsilon}{2M(b-a)}, \text{ when } n \geqq \nu,$$

the same ν serving for all values of x in $(a, b-\xi)$.

Thus
$$0 < \int_a^{b-\xi} R_n(x) g(x) dx < \frac{\epsilon}{2M(b-a)} \int_a^{b-\xi} g(x) dx$$
$$< \frac{\epsilon}{2(b-a)} \int_a^{b-\xi} dx$$
$$< \tfrac{1}{2}\epsilon, \text{ when } n \geqq \nu. \quad \text{.......................(3)}$$

Combining these results (2) and (3), we have

$$0 < \int_a^b R_n(x) g(x) dx < \epsilon, \text{ when } n \geqq \nu.$$

Then, from (1),

$$0 < \int_a^b f(x) g(x) dx - \sum_1^n \int_a^b u_r(x) g(x) dx < \epsilon, \text{ when } n \geqq \nu.$$

Therefore
$$\int_a^b f(x) g(x) dx = \sum_1^\infty \int_a^b u_r(x) g(x) dx.$$

The other alternative, stated in the enunciation, may be treated in the same way.

Bromwich has pointed out that in this case where the terms are all positive, as well as the multiplier $g(x)$, the argument is substantially the same as that employed in dealing with the convergence of a Double Series of positive terms, and the same remark applies to the corresponding theorem in § 75.*

Ex. 1. Show that
$$\int_0^1 \frac{\log x}{1-x} dx = \sum_0^\infty \int_0^1 x^n \log x \, dx \dagger$$
$$= -\sum_0^\infty \frac{1}{(n+1)^2}$$
$$= -\frac{\pi^2}{6}.$$

Ex. 2. Show that $\displaystyle\int_0^1 \log \frac{1+x}{1-x} \frac{dx}{x} = 2\sum_1^\infty \frac{1}{(2n-1)^2} = \frac{\pi^2}{4}.$

The condition imposed upon the terms $u_1(x), u_2(x), \ldots$ and $g(x)$, that they are positive, may be removed, and the following more general theorem stated :

III. *Let* $u_1(x), u_2(x), \ldots$ *be continuous and the series* $\sum_1^\infty |u_n(x)|$ *converge uniformly in the arbitrary interval* (a, α), *where* $a < \alpha < b$. *Also let* $\sum_1^\infty u_n(x) = f)x)$.

---

* Cf. Bromwich, *Infinite Series* (2nd ed., 1926), 496, and *Messenger of Math.*, **36** (1906), 1.

† The interval (0, 1) has to be broken up into two parts, $(0, a)$ and $(a, 1)$. In the first we use Theorem I and in the second Theorem II. Or we may apply § 72, VII.

Further, let $g(x)$ be bounded and integrable in (a, α).

Then $$\int_a^b f(x)g(x)dx = \sum_1^\infty \int_a^b u_n(x)g(x)dx,$$

provided that either the integral $\int_a^b \sum_1^\infty |u_n(x)| \, |g(x)| \, dx$ *or the series* $\sum^\infty \int_a^b |u_n(x)| \, |g(x)| \, dx$ *converges.*

This can be deduced from (II), using the identity:

$$u_n g = \{u_n + |u_n|\}\{g + |g|\} - \{u_n + |u_n|\}|g| - |u_n|\{g + |g|\} + |u_n| \, |g|,$$

since that theorem can be applied to each term on the right-hand side.

Ex. 1. Show that
$$\int_0^1 \frac{\log x}{1+x} dx = \sum_0^\infty (-1)^n \int_0^1 x^n \log x \, dx$$
$$= \sum_0^\infty \frac{(-1)^{n+1}}{(n+1)^2}$$
$$= -\frac{\pi^2}{12}.$$

Ex. 2. Show that
$$\int_0^1 \frac{x^p}{1+x} \log x \, dx = \sum_1^\infty (-1)^{n-1} \int_0^1 x^{n+p-1} \log x \, dx = \sum_1^\infty \frac{(-1)^n}{(n+p)^2},$$
when $p+1>0$.*

75. Integration of Series. Infinite Integrals. Interval Infinite.

For the second form of infinite integral we have results corresponding to the theorems proved in § 74.

I. *Let $u_1(x), u_2(x), \ldots$ be continuous and bounded in $x \geqq a$, and let the series $\sum_1^\infty u_n(x)$ converge uniformly to $f(x)$ in $x \geqq a$. Further, let $g(x)$ be bounded and integrable in the arbitrary interval (a, α), where $a < \alpha$, and α may be chosen as large as we please; and let $\int_a^\infty g(x)dx$ converge absolutely.*

Then $$\int_a^\infty f(x)g(x)dx = \sum_1^\infty \int_a^\infty u_n(x)g(x)dx.$$

The proof of this theorem follows exactly the same lines as (I) of § 74.

Ex. $$\int_1^\infty \frac{e^{-x}}{x} dx = \sum_1^\infty \int_1^\infty \frac{e^{-x}\,dx}{(x+n-1)(x+n)}.$$

This follows from the fact that the series
$$\frac{1}{x(x+1)} + \frac{1}{(x+1)(x+2)} + \ldots$$
converges uniformly to $1/x$ in $x \geqq 1$.

But it is often necessary to justify term by term integration when either

*If $p>0$, Theorem III can be used at once.

If $0>p>-1$, the interval has to be broken up into two parts $(0, a)$ and $(a, 1)$. In the first we use Theorem I and in the second Theorem III. Or we may apply § 72, VII.

$\int_a^\infty |g(x)|dx$ is divergent or $\sum_1^\infty u_n(x)$ can only be shown to converge uniformly in the arbitrary interval (a, α), where α can be taken as large as we please.

Many important cases are included in the following theorems, which correspond to (II) and (III) of § 74:

II. *Let $u_1(x)$, $u_2(x)$, ... be continuous and positive and the series $\sum_1^\infty u_n(x)$ converge uniformly to $f(x)$ in the arbitrary interval (a, α), where α may be taken as large as we please.*

Also let $g(x)$ be positive, bounded and integrable in (a, α).

Then
$$\int_a^\infty f(x)g(x)dx = \sum_1^\infty \int_a^\infty u_n(x)g(x)dx,$$
provided that either the integral $\int_a^\infty f(x)g(x)dx$ or the series $\sum_1^\infty \int_a^\infty u_n(x)g(x)dx$ converges.

Let us suppose that $\int_a^\infty f(x)g(x)dx$ converges.

In other words we are given that the repeated limit
$$\lim_{\alpha\to\infty} \int_a^\alpha [\lim_{n\to\infty} \sum_1^n u_r(x)]g(x)dx \text{ exists.}$$

Since the terms of the series $\sum_1^\infty u_r(x)$ are all positive, as well as $g(x)$, in $x \geqq a$, from the convergence of $\int_a^\infty f(x)g(x)dx$ there follows at once that of
$$\int_a^\infty u_r(x)g(x)dx \quad (r=1, 2, \ldots).$$

Again let $\qquad f(x) = u_1(x) + u_2(x) + \ldots + u_n(x) + R_n(x).$

Then $\int_a^\infty R_n(x)g(x)dx$ also converges, and for every positive integer n
$$\int_a^\infty f(x)g(x)dx - \sum_1^n \int_a^\infty u_r(x)g(x)dx = \int_a^\infty R_n(x)g(x)dx. \quad \text{.................(1)}$$

But from the convergence of $\int_a^\infty f(x)g(x)dx$, it follows that, when the arbitrary positive number ϵ has been chosen, as small as we please, there will be a positive number α such that
$$0 < \int_\alpha^\infty f(x)g(x)dx < \tfrac{1}{2}\epsilon.$$

A fortiori, $\qquad 0 < \int_\alpha^\infty R_n(x)g(x)dx < \tfrac{1}{2}\epsilon,$

and this holds for all positive integers n.

With this choice of α, let the upper bound of $g(x)$ in (a, α) be M.

The series $\sum_1^\infty u_n(x)$ converges uniformly in (a, α).

Keeping the number ν we have chosen above, there will be a positive integer ν such that
$$0 < R_n(x) < \frac{\epsilon}{2M(\alpha - a)}, \text{ when } n \geqq \nu,$$
the same ν serving for all values of x in (a, α).

Thus $$0 < \int_a^\alpha R_n(x)g(x)\,dx < \tfrac{1}{2}\epsilon, \text{ when } n \geqq \nu.$$

But $$0 < \int_\alpha^\infty R_n(x)g(x)\,dx < \tfrac{1}{2}\epsilon,$$

and this holds for all values of n.

Combining these two results, we have

$$0 < \int_a^\infty R_n(x)g(x)\,dx < \epsilon, \text{ when } n \geqq \nu,$$

and from (1),

$$0 < \int_a^\infty f(x)g(x)\,dx - \sum_1^n \int_a^\infty u_r(x)g(x)\,dx < \epsilon, \text{ when } n \geqq \nu.$$

Therefore $\int_a^\infty f(x)g(x)\,dx = \sum_1^\infty \int_a^\infty u_r(x)g(x)\,dx.$

The other alternative can be treated in the same way.

Ex. 1. $$\int_0^\infty e^{-ax}\cosh bx\,dx = \sum_0^\infty \frac{b^{2n}}{2n!}\int_0^\infty e^{-ax}x^{2n}\,dx \text{ if } 0 < |b| < a.$$

Ex. 2. $$\int_0^\infty e^{-a^2x^2}\cosh bx\,dx = \sum_0^\infty \frac{b^{2n}}{2n!}\int_0^\infty e^{-a^2x^2}x^{2n}\,dx.$$

Further, the condition imposed upon $u_1(x)$, $u_2(x)$, ... and $g(x)$, that they shall be positive, may be removed, leading to the theorem:

III. *Let $u_1(x)$, $u_2(x)$, ... be continuous and the series $\sum_1^\infty |u_n(x)|$ converge uniformly in the arbitrary interval (a, α), where α may be taken as large as we please. Also let $\sum_1^\infty u_n(x) = f(x)$.*

Further, let $g(x)$ be bounded and integrable in (a, α).

Then $$\int_a^\infty f(x)g(x)\,dx = \sum_1^\infty \int_a^\infty u_n(x)g(x)\,dx,$$

provided that either the integral $\int_a^\infty \sum_1^\infty |u_n(x)|\;|g(x)|dx$ or the series

$$\sum_1^\infty \int_a^\infty |u_n(x)|\,|g(x)|\,dx \textit{ converges.}$$

This is deduced, as before, from the identity

$$u_n g = \{u_n + |u_n|\}\{g + |g|\} - \{u_n + |u_n|\}|g| - |u_n|\{g + |g|\} + |u_n|\;|g|,$$

since the Theorem II can be applied to each term in the right-hand side.

Ex. 1. Show that $\int_0^\infty e^{-ax}\cos bx\,dx = \sum_0^\infty \frac{(-1)^n b^{2n}}{2n!}\int_0^\infty e^{-ax}x^{2n}\,dx$, if $0 < |b| < a$.

Ex. 2. Show that $\int_0^\infty e^{-a^2x^2}\cos bx\,dx = \sum_0^\infty \frac{(-1)^n b^{2n}}{2n!}\int_0^\infty e^{-a^2x^2}x^{2n}\,dx$.

76. Certain cases to which the theorems of § 75 do not apply are covered by the following test:

Let $u_1(x)$, $u_2(x)$, ... be continuous in $x \geqq a$, and let the series $\sum_1^\infty u_n(x)$ converge uniformly to $f(x)$ in the arbitrary interval (a, α), where α may be taken as large as we please.

Further, let the integrals

$$\int_a^\infty u_1(x)\,dx, \quad \int_a^\infty u_2(x)\,dx, \quad etc.,$$

converge, and the series of integrals

$$\int_a^x u_1(x)\,dx + \int_a^x u_2(x)\,dx + \ldots$$

converge uniformly in $x \geqq a$.

Then the series of integrals

$$\int_a^\infty u_1(x)\,dx + \int_a^\infty u_2(x)\,dx + \ldots$$

converges, and the integral $\int_a^\infty f(x)\,dx$ *converges.*

Also
$$\int_a^\infty f(x)\,dx = \int_a^\infty u_1(x)\,dx + \int_a^\infty u_2(x)\,dx + \ldots.$$

Let
$$s_n(x) = \int_a^x u_1(x)\,dx + \int_a^x u_2(x)\,dx + \ldots + \int_a^x u_n(x)\,dx.$$

Then we know that $\lim_{n\to\infty} s_n(x)$ exists in $x \geqq a$, and we denote it by $F(x)$.

Also we know that $\lim_{x\to\infty} s_n(x)$ exists, and we denote it by $G(n)$.

We shall now show that $\lim_{x\to\infty} F(x)$ and $\lim_{n\to\infty} G(n)$ both exist, and that the two limits are equal.

From this result our theorem as to term by term integration will follow at once.

I. *To prove* $\lim_{x\to\infty} F(x)$ *exists.*

Since $\lim_{n\to\infty} s_n(x)$ converges uniformly to $F(x)$ in $x \geqq a$, with the usual notation, we have
$$|F(x) - s_n(x)| < \tfrac{1}{3}\epsilon, \text{ when } n \geqq \nu,$$
the same ν serving for every x greater than or equal to a.

Choose some value of n in this range.

Then we have
$$\lim_{x\to\infty} s_n(x) = G(n).$$

Therefore we can choose X so that
$$|s_n(x'') - s_n(x')| < \tfrac{1}{3}\epsilon, \text{ when } x'' > x' \geqq X > a.$$

But
$$\begin{aligned} &| F(x'') - F(x') | \\ &\leqq | F(x'') - s_n(x'') | + | s_n(x'') - s_n(x') | + | s_n(x') - F(x') | \\ &< \tfrac{1}{3}\epsilon + \tfrac{1}{3}\epsilon + \tfrac{1}{3}\epsilon \\ &< \epsilon, \text{ when } x'' > x' \geqq X > a. \end{aligned}$$

Thus $F(x)$ has a limit as $x \to \infty$.

II. *To prove* $\lim_{n\to\infty} G(n)$ *exists.*

Since $\lim_{n\to\infty} s_n(x)$ converges uniformly to $F(x)$ in $x \geqq a$,
$$| s_{n''}(x) - s_{n'}(x) | < \tfrac{1}{3}\epsilon, \text{ when } n'' > n' \geqq \nu,$$
the same ν serving for every x greater than or equal to a.

But
$$\lim_{x\to\infty} s_n(x) = G(n).$$

Therefore we can choose X_1, X_2 such that
$$|G(n') - s_{n'}(x)| < \tfrac{1}{3}\epsilon, \text{ when } x \geqq X_1 > a,$$
$$|G(n'') - s_{n''}(x)| < \tfrac{1}{3}\epsilon, \text{ when } x \geqq X_2 > a.$$

Then, taking a value of x not less than X_1 or X_2,
$$\begin{aligned} &|G(n'') - G(n')| \\ &\leqq |G(n'') - s_{n''}(x)| + |s_{n''}(x) - s_{n'}(x)| + |s_{n'}(x) - G(n')| \\ &< \tfrac{1}{3}\epsilon + \tfrac{1}{3}\epsilon + \tfrac{1}{3}\epsilon \\ &< \epsilon, \text{ when } n'' > n' \geqq \nu. \end{aligned}$$

Therefore $\lim_{n\to\infty} G(n)$ exists.

III. *To prove* $\lim_{x\to\infty} F(x) = \lim_{n\to\infty} G(n)$.

Since $\lim_{n\to\infty} G(n)$ exists, we can choose ν_1 so that
$$\left|\sum_{n+1}^{\infty}\int_a^{\infty} u_r(x)\,dx\right| < \tfrac{1}{3}\epsilon, \text{ when } n \geqq \nu_1.$$

Again $\sum_1^{\infty}\int_a^{\infty} u_r(x)\,dx$ converges uniformly to $F(x)$ in $x \geqq a$.

Therefore we can choose ν_2 so that
$$\left|\sum_{n+1}^{\infty}\int_a^{x} u_r(x)\,dx\right| < \tfrac{1}{3}\epsilon, \text{ when } n \geqq \nu_2,$$
the same ν_2 serving for every x greater than or equal to a.

Choose ν not less than ν_1 or ν_2.

From the convergence of the integral $\int_a^{\infty}\sum_1^{\nu} u_r(x)\,dx$, we can choose X so that
$$\left|\int_x^{\infty}\sum_1^{\nu} u_r(x)\,dx\right| < \tfrac{1}{3}\epsilon, \text{ when } x \geqq X > a.$$

But
$$\begin{aligned} &\left|\sum_1^{\infty}\int_a^{\infty} u_r(x)\,dx - F(x)\right| \\ &= \left|\int_x^{\infty}\sum_1^{\nu} u_r(x)\,dx + \sum_{\nu+1}^{\infty}\int_a^{\infty} u_r(x)\,dx - \sum_{\nu+1}^{\infty}\int_a^{x} u_r(x)\,dx\right| \\ &\leqq \left|\int_x^{\infty}\sum_1^{\nu} u_r(x)\,dx\right| + \left|\sum_{\nu+1}^{\infty}\int_a^{\infty} u_r(x)\,dx\right| + \left|\sum_{\nu+1}^{\infty}\int_a^{x} u_r(x)\,dx\right| \\ &< \tfrac{1}{3}\epsilon \qquad\qquad + \tfrac{1}{3}\epsilon \qquad\qquad + \tfrac{1}{3}\epsilon, \\ &< \epsilon, \text{ when } x \geqq X > a. \end{aligned}$$

Therefore
$$\lim_{x\to\infty} F(x) = \sum_1^{\infty}\int_a^{\infty} u_r(x)\,dx = \lim_{n\to\infty} G(n).$$

IV. But we are given that the series
$$u_1(x) + u_2(x) + \ldots$$
converges uniformly to $f(x)$ in any arbitrary interval (a, α).

Therefore we have
$$\int_a^x f(x)\,dx = \int_a^x u_1(x)\,dx + \int_a^x u_2(x)\,dx + \ldots \text{ in } x \geqq a.$$

Thus, with the above notation,

$$F(x)=\int_a^x f(x)dx.$$

It follows from I that $\int_a^\infty f(x)dx$ converges, and from III that

$$\int_a^\infty f(x)dx=\int_a^\infty u_1(x)dx+\int_a^\infty u_2(x)dx+\dots.$$

REFERENCES.

Bromwich, *loc. cit.* (2nd ed., 1926), Ch. VII-VIII and App. III.

De la Vallée Poussin, *loc. cit.* 1 (3e éd., 1923), Ch. XI.

Dini, *Lezioni di Analisi Infinitesimale,* 2 (Pisa, 1909), 1a Parte, Cap. VIII.

Goursat, *loc. cit.*, 1 (4e éd., 1923), Ch. II.

Hobson, *Theory of Functions of a Real Variable,* 1 (3rd ed., 1927), Ch. VI; 2 (2nd ed., 1926), Ch. III, V.

Knopp, *loc. cit.*, English translation, Ch. V, XI.

Kowalewski, *loc. cit.*, Kap. VII, XV.

Osgood, *loc. cit.*, 1 (4 Aufl., 1923), Kap. III.

Pierpont, *loc. cit.*, 2 (1912), Ch. V.

Stolz, *loc. cit.*, Bd. I, Absch. X.

And

Brunel, *loc. cit., Enc. d. math. Wiss.*, Bd. II, Tl. I (Leipzig, 1899).

Montel-Rosenthal, *loc. cit., Enc. d. math. Wiss.*, Bd. II, Tl. III, 2 (Leipzig, 1923).

Pringsheim, "Grundlagen der allgemeinen Funktionenlehre," *Enc. d. math. Wiss.*, Bd. II, Tl. I (Leipzig, 1899).

EXAMPLES ON CHAPTER V.

UNIFORM CONVERGENCE.

1. Let

$$\phi_n(x)=\frac{\sin\frac{1}{n}\sin\frac{2}{n}}{\sin^2\frac{1}{n}+x\cos^2\frac{1}{n}}.$$

Show that

$$\lim_{n\to\infty}\phi_n(x)=0,\ \text{when } x\gtrless 0,$$

and

$$\lim_{n\to\infty}\phi_n(x)=2,\ \text{when } x=0.$$

Also show that $\phi_n(x)$ converges uniformly to zero, when $x\geqq x_0$, where x_0 is any positive number.

2. If $\phi(x)$ is continuous in the interval $0\leqq x\leqq 1$, show that $\lim_{n\to\infty} x^n\phi(x)$ exists in that interval.

Also show that $x^n\phi(x)$ converges uniformly to its limit in $0\leqq x\leqq x_0<1$, and that it converges uniformly in the interval $0\leqq x\leqq 1$, only if $\phi(1)=0$.

3. Examine the convergence of the series

$$\sum_1^\infty \frac{x}{(nx+1)[(n+1)x+1][(n+2)x+1]}, \quad x \geqq 0,$$

and by its means illustrate the effect of non-uniform convergence upon the continuity of a function of x represented by an infinite series. (Cf. Example on page 144.)

4. Show that the series $\sum_1^\infty \operatorname{cosech}^2 nx$ is uniformly convergent in any interval $x \geqq x_0$, where $x_0 > 0$.

5. Prove that the series

$$x^2 + \frac{x^2}{(1+x^2)} + \frac{x^2}{(1+x^2)^2} + \cdots$$

is convergent for all values of x, but is not uniformly convergent in an interval including the origin.

6. Prove that the series

$$\text{(i)} \sum_1^\infty \frac{1}{n^3+n^4x^2}, \qquad \text{(ii)} \sum_1^\infty \frac{x^{\frac{3}{2}}}{1+n^2x^2},$$

are uniformly convergent for all values of x.

7. If $u_n(x) = \dfrac{x}{1+n^2x^2}$, show that $\sum_1^\infty u_n(x)$ is uniformly convergent in any interval $x \geqq x_0$, where $x_0 > 0$.

Also show that, if m is any positive integer, $\sum_1^\infty u_n\left(\dfrac{1}{m}\right) > \frac{1}{2}$, and deduce that the convergence is not uniform in any interval including the origin.

8. Find for what values of x the series $\sum_1^\infty u_n$ converges where

$$u_n = \frac{x^n - x^{-n-1}}{(x^n + x^{-n})(x^{n+1} + x^{-n-1})}$$

$$= \frac{1}{(x-1)(x^n+x^{-n})} - \frac{1}{(x-1)(x^{n+1}+x^{-(n+1)})}.$$

Find also whether the series is

(i) uniformly convergent through an interval including $+1$;

(ii) continuous when x passes through the value $+1$.

9. Discuss the uniformity or non-uniformity of the convergence of the series whose general term is

$$u_n = \frac{1-(1+x)^n}{1+(1+x)^n} - \frac{1-(1+x)^{n-1}}{1+(1+x)^{n-1}}.$$

10. Let $$a_0 + a_1 + \cdots$$

be an absolutely convergent series of constant terms, and let

$$f_0(x), \ f_1(x), \ \ldots$$

be a set of functions each continuous in the interval $\alpha \leqq x \leqq \beta$, and each comprised between certain fixed limits,

$$A \leqq f_r(x) \leqq B, \qquad r = 0, 1, \ldots,$$

where A, B are constants.

Show that the series

$$a_0 f_0(x) + a_1 f_1(x) + \dots$$

represents a continuous function of x in the interval $\alpha \leqq x \leqq \beta$.

11. Show that the function defined by the series

$$\sum_1^\infty \frac{x}{n(1+nx^2)}$$

is finite and continuous for all values of x. Examine whether the series is uniformly convergent for all such values.

12. Show that if $$u_0(x) + u_1(x) + \dots$$
is a series of functions each continuous and having no roots in the interval $a \leqq x \leqq b$, and if

$$\left| \frac{u_{n+1}(x)}{u_n(x)} \right| \leqq \gamma < 1, \text{ when } n \geqq \nu,$$

where γ, ν do not depend on x, then the given series is uniformly convergent in this interval.

Apply this test to the series

$$1 + xa + x(x-1)\frac{a^2}{2!} + x(x-1)(x-2)\frac{a^3}{3!} + \dots,$$

where $0 < a < 1$.

13. Using the inequality $\dfrac{\sin x}{x} > \dfrac{2}{\pi}$, when $0 < x < \frac{1}{2}\pi$,

show that if $$s_n(x) = \sum_1^n \frac{\sin^3 rx}{r}, \text{ then } s_n\left(\frac{\pi}{2n}\right) > \tfrac{1}{3}.$$

Also show that the series $\sum\limits_1^\infty \dfrac{\sin^3 nx}{n^k}$ converges uniformly in $0 \leqq x \leqq \frac{1}{2}\pi$, when $k > 1$; and that if, $0 \leqq k \leqq 1$, it converges, but not uniformly, in this interval of x.

14. Let the series $\sum\limits_1^\infty u_n(x)$ converge uniformly to $f(x)$ in the open interval $a < x < b$, and for every positive integer let $\lim\limits_{x \to c} u_n(x) = l_n$, where c is a point of the open interval. Then the series $\sum\limits_1^\infty l_n$ converges, and $\lim\limits_{x \to c} f(x) = \sum\limits_1^\infty l_n$.

15. Let the series $\sum\limits_1^\infty u_n(x)$ converge uniformly to $f(x)$ in the open interval $a < x < b$, and for every positive integer let $\lim\limits_{x \to a+0} u_n(x) = a_n$ and $\lim\limits_{x \to b-0} u_n(x) = b_n$. Then the series $\sum\limits_1^\infty u_n$ converges, and $\lim\limits_{x \to a+0} f(x) = \sum\limits_1^\infty a_n$, with a similar result for $\lim\limits_{x \to b-0} f(x)$.

16. Let the series $\sum\limits_1^\infty u_n(x)$ converge uniformly to $f(x)$ in the infinite interval $x \geqq a$, and, for every positive integer, let $\lim\limits_{x \to \infty} u_n(x) = l_n$. Then $\sum\limits_1^\infty l_n$ converges, and $\lim\limits_{x \to \infty} f(x) = \sum\limits_1^\infty l_n$.

THE POWER SERIES.

17. From the equation $\sin^{-1}x=\int_0^x \frac{dx}{\sqrt{(1-x^2)}}$,
show that the series for $\sin^{-1}x$ is

$$x+\frac{1}{2}\frac{x^3}{3}+\frac{1\cdot 3}{2\cdot 4}\frac{x^5}{5}+\dots \text{ when } |x|<1.$$

Prove that the expansion is also valid when $|x|=1$.*

18. From the equation $\tan^{-1}x=\int_0^x \frac{dx}{1+x^2}$,
obtain Gregory's Series,

$$\tan^{-1}x=x-\tfrac{1}{3}x^3+\tfrac{1}{5}x^5-\dots .$$

Within what range of x does this hold?

19. Show that we may substitute the series for $\sin^{-1}x$ and $\tan^{-1}x$ in the integrals

$$\int_0^x \frac{\sin^{-1}x}{x}dx \text{ and } \int_0^x \frac{\tan^{-1}x}{x}dx,$$

and integrate term by term, when $|x|<1$.

Also show that

$$\int_0^1 \frac{\sin^{-1}x}{x}dx=\sum_0^\infty \frac{1\cdot 3\dots 2n-1}{2\cdot 4\dots 2n}\frac{1}{(2n+1)^2},\quad \int_0^1 \frac{\tan^{-1}x}{x}dx=\sum_0^\infty (-1)^n\frac{1}{(2n+1)^2},$$

and, from the integration by parts of $\frac{\sin^{-1}x}{x}$ and $\frac{\tan^{-1}x}{x}$, prove that these series also represent

$$\int_0^1 \frac{|\log x|}{\sqrt{(1-x^2)}}dx \text{ and } \int_0^1 \frac{|\log x|}{1+x^2}dx.$$

20. If $|x|<1$, prove that

$$\int_0^x \tan^{-1}x\,dx=\frac{x^2}{1\cdot 2}-\frac{x^4}{3\cdot 4}+\frac{x^6}{5\cdot 6}-\dots .$$

Show that the result also holds for $x=1$, and deduce that

$$1-\tfrac{1}{2}-\tfrac{1}{3}+\tfrac{1}{4}+\tfrac{1}{5}-\dots=0{\cdot}43882\dots .$$

21. If $|x|<1$, prove that

$$\int_0^x \log(1+x)\,dx=\frac{x^2}{1\cdot 2}-\frac{x^3}{2\cdot 3}+\frac{x^4}{3\cdot 4}-\dots .$$

Does the result hold for $x=\pm 1$?

22. Prove that

$$\int_0^x \log(1-x)\frac{dx}{x}=-\sum_1^\infty \frac{x^n}{n^2},\quad 0<x<1.$$

$$\int_0^x \log(1+x)\frac{dx}{x}=\sum_1^\infty (-1)^{n-1}\frac{x^n}{n^2},\quad 0<x<1.$$

$$\int_0^x \log\left(\frac{1+x}{1-x}\right)\frac{dx}{x}=2\sum_1^\infty \frac{x^{2n-1}}{(2n-1)^2},\quad 0<x<1.$$

---

*When $x=1$, $\lim_{n\to\infty}\frac{u_{n+1}}{u_n}=1$, but Raabe's test (cf. Bromwich, *loc. cit.*, p. 35, or Goursat, *loc. cit.*, p. 404) shows that the series converges for this value of x.

Express the integrals

$$\int_0^1 \log(1-x)\frac{dx}{x},\quad \int_0^1 \log(1+x)\frac{dx}{x} \quad\text{and}\quad \int_0^1 \log\left(\frac{1+x}{1-x}\right)\frac{dx}{x}$$

as infinite series.

23. Show that

$$1-\tfrac{1}{5}-\tfrac{1}{7}+\tfrac{1}{11}+\tfrac{1}{13}-\tfrac{1}{17}-\tfrac{1}{19}+\ldots$$
$$=\int_0^1 \frac{1-x^4}{1+x^6}\,dx$$
$$=\frac{1}{\sqrt{3}}\log(2+\sqrt{3}).$$

24. Prove that, when $0 \leqq x \leqq 1$, $x|\log x| \leqq e^{-1}$.

Hence show

(i) that
$$1+a(x\log x)+\frac{a^2}{2!}(x\log x)^2+\ldots$$
converges uniformly in the interval $0 \leqq x \leqq 1$, and

(ii) that
$$\int_0^1 x^{ax}\,dx=\sum_0^\infty (-1)^n \frac{a^n}{(n+1)^{n+1}}.$$

25. Prove that, when $a>1$, the series

$$x^{a-1}(1-x+x^2-\ldots)$$

converges uniformly to $\dfrac{x^{a-1}}{1+x}$ in the interval $0 \leqq x \leqq x_0$, where $x_0<1$.

Hence show that

$$x_0^{-a}\int_0^{x_0}\frac{x^{a-1}}{1+x}\,dx=\sum_0^\infty (-1)^n\frac{x_0^{\,n}}{n+a},$$

and deduce that

$$\int_0^1 \frac{x^{a-1}}{1+x}\,dx=\sum_0^\infty \frac{(-1)^n}{n+a},\quad \text{when } a>1. \quad \text{(See also Ex. 36 below.)}$$

INTEGRATION AND DIFFERENTIATION OF SERIES.

26. Prove that the series $\sum_1^\infty (\alpha e^{-n\alpha x}-\beta e^{-n\beta x})$ is uniformly convergent in (c, γ), where α, β, γ and c are positive and $c<\gamma$.

Verify that

$$\int_c^\gamma \sum_1^\infty (\alpha e^{-n\alpha x}-\beta e^{-n\beta x})\,dx=\sum_1^\infty \int_c^\gamma (\alpha e^{-n\alpha x}-\beta e^{-n\beta x})\,dx.$$

Is
$$\int_0^\gamma \sum_1^\infty (\alpha e^{-n\alpha x}-\beta e^{-n\beta x})\,dx=\sum_1^\infty \int_0^\gamma (\alpha e^{-n\alpha x}-\beta e^{-n\beta x})\,dx\,?$$

27. If it be given that for values of x between 0 and π,

$$\pi\cosh ax=2\sinh a\pi\left\{\frac{1}{2a}-\frac{a\cos x}{1+a^2}+\frac{a\cos 2x}{2^2+a^2}-\ldots\right\},$$

prove rigorously that

$$\pi\sinh ax=2\sinh a\pi\left\{\frac{\sin x}{1+a^2}-\frac{2\sin 2x}{2^2+a^2}+\frac{3\sin 3x}{3^2+a^2}-\ldots\right\}.$$

28. Show that if $f(x)=\sum_1^\infty \frac{1}{n^3+n^4x^2}$, then it has a differential coefficient equal to $-2x\sum_1^\infty \frac{1}{n^2(1+nx^2)^2}$ for all values of x.

29. When a stands for a positive number, then the series

$$\sum_1^\infty \frac{1}{r!}\frac{a^{-r}}{a^{-2r}+x^2}, \quad \sum_1^\infty \frac{(-1)^r}{r!}\frac{a^{-r}}{a^{-2r}+x^2},$$

are uniformly convergent for all values of x; and, if their sums are $f(x)$ and $F(x)$ respectively,

$$f'(x)=\sum_0^\infty \frac{1}{r!}\frac{d}{dx}\left(\frac{a^{-r}}{a^{-2r}+x^2}\right),$$

$$F'(x)=\sum_0^\infty \frac{1}{r!}\frac{d}{dx}\left(\frac{(-1)^r a^{-r}}{a^{-2r}+x^2}\right).$$

30. Find all the values of x for which the series

$$e^x \sin x + e^{2x}\sin 2x + \dots$$

converges. Does it converge uniformly for these values? For what values of x can the series be differentiated term by term?

31. Let
$$\begin{cases} u_n(x)=x^2\left(x^{\frac{1}{2n-1}}-x^{\frac{1}{2n-3}}\right)\sin\frac{1}{x} & \text{for } x\gtrless 0,\\ u_n(0)=0, \end{cases}$$

for any positive integer greater than unity; and

$$\begin{cases} u_1(x)=x^3\sin\frac{1}{x} & \text{for } x\gtrless 0,\\ u_1(0)=0. \end{cases}$$

Show that $\sum_1^\infty u_n(x)$ converges for all values of x to $f(x)$, where

$$f(x)=x^2\sin\frac{1}{x} \text{ for } x\gtrless 0 \text{ and } f(0)=0.$$

Also show that $f'(x)$ is discontinuous at $x=0$; that $\sum_1^\infty u_n'(x)$ is not uniformly convergent in any interval including the origin; and that $f'(x)=\sum_1^\infty u_n'(x)$ for all values of x.

32.* Show that the series

$$2xe^{-x^2}=\sum_1^\infty 2x\left(\frac{1}{n^2}e^{-\frac{x^2}{n^2}}-\frac{1}{(n+1)^2}e^{-\frac{x^2}{(n+1)^2}}\right)$$

can be integrated term by term between any two finite limits. Can the function defined by the series be integrated between the limits 0 and ∞? If so, is the value of this integral given by integrating the series term by term between these limits?

---

*Ex. 32–38 depend upon the theorems of §§ 74–76.

33. If each of the terms of the series

$$u_1(x)+u_2(x)+\dots$$

is a continuous function of x in $x \geqq a > 0$ and if the series

$$x^\kappa u_1(x)+x^\kappa u_2(x)+\dots \quad (\kappa>1)$$

satisfies the M-test (§ 67. 1), then the original series may be integrated term by term from a to ∞.

34. Show that the series

$$\frac{1}{(1+x)^2}+\frac{1}{(2+x)^2}+\dots \quad (x \geqq 0)$$

can be integrated term by term between any two positive finite limits. Can this series be integrated term by term between the limits 0 and ∞? Show that the function defined by the series cannot be integrated between these limits.

35. Show that the function defined by the series

$$\frac{1}{(1+x)^3}+\frac{1}{(2+x)^3}+\dots$$

can be integrated from 0 to ∞, and that its value is given by the term by term integration of the series.

36. Prove that

$$\int_0^1 \frac{x^{a-1}}{1+x}\,dx=\frac{1}{a}-\frac{1}{a+1}+\frac{1}{a+2}-\dots \quad (a>0).$$

Explain the nature of the difficulties involved in your proof, and justify the process you have used.

37. By expansion in powers of a, prove that, if $|a|<1$,

$$\int_0^\infty e^{-x}(1-e^{-ax})\frac{dx}{x}=\log(1+a),$$

$$\int_0^{\frac{1}{2}\pi} \tan^{-1}(a\sin x)\frac{dx}{\sin x}=\tfrac{1}{2}\pi\sinh^{-1}a,$$

examining carefully the legitimacy of term by term integration in each case.

38. Assuming that $$J_0(bx)=\sum_0^\infty \frac{(-1)^n\left(\frac{bx}{2}\right)^{2n}}{(n!)^2},$$

show that $$\int_0^\infty e^{-ax}J_0(bx)\,dx=\frac{1}{\sqrt{(a^2+b^2)}},$$

when $a>0$.

NOTE

A valuable collection of *Examples in Infinite Series with Solutions*, by Francis and Littlewood (Cambridge, 1928) will be found useful by the student who makes a special study of Infinite Series and Integrals.

CHAPTER VI

DEFINITE INTEGRALS CONTAINING AN ARBITRARY PARAMETER

77. Continuity of the Integral. Finite Interval. In the ordinary definite integral $\int_a^{a'} \phi(x, y)dx$ let a, a' be constants. Then the integral will be a function of y.*

The properties of such integrals will be found to correspond very closely to those of infinite series whose terms are functions of a single variable. Indeed this chapter will follow almost the same lines as the preceding one, in which such infinite series were treated.

I. *If $\phi(x, y)$ is a continuous function of (x, y) in the region*

$$a \leqq x \leqq a', \quad b \leqq y \leqq b',$$

then $\int_a^{a'} \phi(x, y)dx$ is a continuous function of y in the interval (b, b').

Since $\phi(x, y)$ is a continuous function of (x, y)†, as defined in § 37, it is also a continuous function of x and a continuous function of y.

Thus $\phi(x, y)$ is integrable with regard to x.

Let
$$f(y) = \int_a^{a'} \phi(x, y)\, dx.$$

---

*As already remarked in § 62, it is understood that before proceeding to the limit involved in the integration, the value of y, for which the integral is required, is to be inserted in the integrand.

† When a function of two variables x, y is continuous with respect to the two variables as defined in § 37, we speak of it as a continuous function of (x, y).

It will be noticed that we do not use the full consequences of this continuity in the following argument.

We know that, since $\phi(x, y)$ is a continuous function of (x, y) in the given region, to any positive number ϵ, chosen as small as we please, there corresponds a positive number η such that

$$|\phi(x, y+\Delta y)-\phi(x, y)|<\epsilon, \text{ when } |\Delta y| \leqq \eta,$$

the same η serving for all values of x in (a, a').*

Let Δy satisfy this condition, and write

$$f(y+\Delta y)=\int_a^{a'} \phi(x, y+\Delta y)dx.$$

Then $$f(y+\Delta y)-f(y)=\int_a^{a'} [\phi(x, y+\Delta y)-\phi(x, y)]\,dx.$$

Therefore

$$|f(y+\Delta y)-f(y)| \leqq \int_a^{a'} |\phi(x, y+\Delta y)-\phi(x, y)|dx$$
$$<(a'-a)\epsilon, \text{ when } |\Delta y| \leqq \eta.$$

Thus $f(y)$ is continuous in the interval (b, b').

II. *If* $\phi(x, y)$ *is a continuous function of* (x, y) *in* $a \leqq x \leqq a'$, $b \leqq y \leqq b'$, *and* $\psi(x)$ *is bounded and integrable in* (a, a'), *then* $\int_a^{a'} \phi(x, y)\psi(x)dx$ *is a continuous function of* y *in* (b, b').

Let $$f(y)=\int_a^{a'} \phi(x, y)\,\psi(x)\,dx.$$

The integral exists, since the product of two integrable functions is integrable.

Also, with the same notation as in (I),

$$f(y+\Delta y)-f(y)=\int_a^{a'} [\phi(x, y+\Delta y)-\phi(x, y)]\,\psi(x)\,dx.$$

Let M be the upper bound of $|\psi(x)|$ in (a, a').

Then $\quad |f(y+\Delta y)-f(y)|<M(a'-a)\epsilon$, when $|\Delta y| \leqq \eta$.

Thus $f(y)$ is continuous in (b, b').

78. Differentiation of the Integral.

I. *Let* $f(y)=\int_a^{a'} \phi(x, y)\,dx$, *where* $\phi(x, y)$ *is a continuous function of* (x, y) *in* $a \leqq x \leqq a'$, $b \leqq y \leqq b'$, *and* $\dfrac{\partial \phi}{\partial y}$ *exists and satisfies the same condition.*

Then $f'(y)$ *exists and is equal to* $\int_a^{a'} \dfrac{\partial \phi}{\partial y}\,dx$.

---

* This follows from the theorem on the uniform continuity of a continuous function (cf. § 37, p. 87).

Since $\frac{\partial \phi}{\partial y}$ is a continuous function of (x, y) in the given region, to any positive number ϵ, chosen as small as we please, there corresponds a positive number η, such that, with the usual notation,

$$\left| \frac{\partial \phi(x, y+\Delta y)}{\partial y} - \frac{\partial \phi(x, y)}{\partial y} \right| < \epsilon, \text{ when } |\Delta y| \leqq \eta,$$

the same η serving for all values of x in (a, a').

Let Δy satisfy this condition.

Then

$$\frac{f(y+\Delta y)-f(y)}{\Delta y} = \int_a^{a'} \frac{\phi(x, y+\Delta y) - \phi(x, y)}{\Delta y} dx$$

$$= \int_a^{a'} \frac{\partial \phi(x, y+\theta\Delta y)}{\partial y} dx, \quad \text{where } 0<\theta<1,$$

$$= \int_a^{a'} \frac{\partial \phi(x, y)}{\partial y} dx + \int_a^{a'} \left[\frac{\partial \phi(x, y+\theta\Delta y)}{\partial y} - \frac{\partial \phi(x, y)}{\partial y} \right] dx.$$

Thus we have

$$\left| \frac{f(y+\Delta y)-f(y)}{\Delta y} - \int_a^{a'} \frac{\partial \phi}{\partial y} dx \right| = \left| \int_a^{a'} \left[\frac{\partial \phi(x, y+\theta\Delta y)}{\partial y} - \frac{\partial \phi(x, y)}{\partial y} \right] dx \right|$$

$$< (a'-a)\epsilon, \text{ when } |\Delta y| \leqq \eta.$$

And this establishes that $\lim_{\Delta y \to 0} \left\{ \frac{f(y+\Delta y)-f(y)}{\Delta y} \right\}$ exists and is equal to $\int_a^{a'} \frac{\partial \phi}{\partial y} dx$ at any point in (b, b').

II. *Let* $f(y) = \int_a^{a'} \phi(x, y)\psi(x) dx$, *where* $\phi(x, y)$ *and* $\frac{\partial \phi}{\partial y}$ *are as in* (I), *and* $\psi(x)$ *is bounded and integrable in* (a, a').

Then $f'(y)$ *exists and is equal to* $\int_a^{a'} \frac{\partial \phi}{\partial y} \psi(x) dx$.

Let the upper bound of $|\psi(x)|$ in (a, a') be M.

Then we find, as above, that

$$\left| \frac{f(y+\Delta y)-f(y)}{\Delta y} - \int_a^{a'} \frac{\partial \phi}{\partial y} \psi(x) dx \right| < M(a'-a)\epsilon, \text{ when } |\Delta y| \leqq \eta.$$

And the result follows.

The theorems of this section show that if we have to differentiate the integral $\int_a^{a'} F(x, y) dx$, where $F(x, y)$ is of the form $\phi(x, y)$ or $\phi(x, y)\psi(x)$, and these functions satisfy the conditions named above, we may put the symbol of differentiation under the integral sign. In other words, we may reverse the order of the two limiting operations involved without affecting the result.

It will be noticed that so far we are dealing with ordinary integrals. The interval of integration is finite, and the function has no points of infinite discontinuity in the interval.

79. Integration of the Integral.

Let $f(y)=\int_a^{a'} \phi(x, y)\psi(x)\,dx$, *where* $\phi(x, y)$ *is a continuous function of* (x, y) *in* $a \leqq x \leqq a'$, $b \leqq y \leqq b'$, *and* $\psi(x)$ *is bounded and integrable in* (a, a').

Then
$$\int_{y_0}^{y} f(y)\,dy=\int_a^{a'} dx \int_{y_0}^{y} \phi(x, y)\,\psi(x)\,dy,$$
where y_0, y *are any two points in* (b, b').

Let
$$\Phi(x, y)=\int_b^y \phi(x, y)\,dy.$$

Then we know that
$$\frac{\partial \Phi}{\partial y}=\phi(x, y) \quad [\S\ 49],$$
and it is easy to show that $\Phi(x, y)$ is a continuous function of (x, y) in the region $a \leqq x \leqq a'$, $b \leqq y \leqq b'$.

Now let
$$g(y)=\int_a^{a'} \Phi(x, y)\,\psi(x)\,dx.$$

From § 78 we know that
$$g'(y)=\int_a^{a'} \frac{\partial \Phi}{\partial y}\,\psi(x)\,dx$$
$$=\int_a^{a'} \phi(x, y)\,\psi(x)\,dx.$$

Also $g'(y)$ is continuous in the interval (b, b') by § 77.

Therefore
$$\int_{y_0}^{y} g'(y)\,dy=\int_{y_0}^{y} dy \int_a^{a'} \phi(x, y)\,\psi(x)\,dx,$$
where y_0, y are any two points in (b, b').

Thus
$$\int_{y_0}^{y} dy \int_a^{a'} \phi(x, y)\,\psi(x)\,dx$$
$$=g(y)-g(y_0)$$
$$=\int_a^{a'} [\Phi(x, y)-\Phi(x, y_0)]\,\psi(x)\,dx$$
$$=\int_a^{a'}\left[\int_b^y \phi(x, y)\,dy-\int_b^{y_0} \phi(x, y)\,dy\right]\psi(x)\,dx$$
$$=\int_a^{a'} dx \int_{y_0}^{y} \phi(x, y)\,\psi(x)\,dy.$$

Thus we have shown that we may invert the order of integration with respect to x and y in the repeated integral

$$\int_a^{a'} dx \int_{y_0}^{y} F(x, y)\,dy,$$

when the integrand satisfies the above conditions ; and in particular, since we may put $\psi(x)=1$, when $F(x, y)$ is a continuous function of (x, y) in the region with which the integral deals.

80. In the preceding sections of this chapter the intervals (a, a') and (b, b') have been supposed finite, and the integrand bounded in $a \leqq x \leqq a'$, $b \leqq y \leqq b'$. The argument employed does not apply to infinite integrals.

For example, the infinite integral

$$f(y)=\int_0^{\infty} ye^{-xy}\,dx$$

converges when $y \geqq 0$, but it is discontinuous at $y=0$, since $f(y)=1$ when $y>0$, and $f(0)=0$.

Similarly $$\int_0^{\infty} \sin \pi y\, e^{-x \sin^2 \pi y}\,dx$$

converges for all values of y, but it is discontinuous for every positive and negative integral value of y, as well as for $y=0$.

Under what conditions then, it may be asked, will the infinite integrals

$$\int_a^{\infty} F(x, y)\,dx \quad \text{and} \quad \int_a^{a'} F(x, y)\,dx,$$

if convergent when $b \leqq y \leqq b'$, define continuous functions of y in (b, b') ? And when can we differentiate and integrate under the sign of integration ?

In the case of infinite series, we have met with the same questions and partly answered them [cf. §§ 68, 70, 71]. We proceed to discuss them for both types of infinite integral. The discussion requires the definition of the form of convergence of infinite integrals which corresponds to uniform convergence in infinite series.

81. Uniform Convergence of Infinite Integrals. We deal first with the convergent infinite integral

$$\int_a^{\infty} F(x, y)\,dx,$$

where $F(x, y)$ is bounded in the region $a \leqq x \leqq a'$, $b \leqq y \leqq b'$, the number a' being arbitrary.

I. *The integral* $\int_a^\infty F(x, y)dx$ *is said to converge* **uniformly** *to its value* $f(y)$ *in the interval* (b, b'), *if, any positive number* ϵ *having been chosen, as small as we please, there is a positive number* X *such that*

$$\left| f(y) - \int_a^x F(x, y)dx \right| < \epsilon, \text{ when } x \geqq X,$$

the same X serving for every y in (b, b′).

And, just as in the case of infinite series, we have a useful test for uniform convergence, corresponding to the general principle of convergence (§ 15):

II. *A necessary and sufficient condition for the uniform convergence of the integral* $\int_a^\infty F(x, y)dx$ *in the interval* (b, b') *is that, if any positive number* ϵ *has been chosen, as small as we please, there shall be a positive number* X *such that*

$$\left| \int_{x'}^{x''} F(x, y)dx \right| < \epsilon, \text{ when } x'' > x' \geqq X,$$

the same X serving for every y in (b, b′).

The proof that (II) forms a necessary and sufficient condition for the uniform convergence of the integral, as defined in (I), follows exactly the same lines as the proof in § 66 for the corresponding theorems in infinite series.

Further, it will be seen that *if* $\int_a^\infty F(x, y)dx$ *converges uniformly in* (b, b'), *to the arbitrary positive number* ϵ *there corresponds a positive number* X *such that*

$$\left| \int_x^\infty F(x, y)\, dx \right| < \epsilon, \text{ when } x \geqq X,$$

the same X serving for every y in (b, b′).

The definition and theorem given above correspond exactly to those for the series

$$u_1(x) + u_2(x) + \dots,$$

uniformly convergent in $a \leqq x \leqq b$; namely,

$$|R_n(x)| < \epsilon, \text{ when } n \geqq \nu,$$

and $\quad |{}_pR_n(x)| < \epsilon$, when $n \geqq \nu$, for every positive integer p,

the same ν serving for every value of x in the interval (a, b).

82. Uniform Convergence of Infinite Integrals (*continued*).

We now consider the convergent infinite integral

$$\int_a^{a'} F(x, y)\, dx,$$

where the interval (a, a') is finite, but the integrand is not bounded in the region $a \leqq x \leqq a'$, $b \leqq y \leqq b'$. This case is more complex than the preceding, since the points of infinite discontinuity can be distributed in more or less complicated fashion over the given region. We shall confine ourselves in our definition, and in the theorems which follow, to the simplest case, which is also the most important, where the integrand $F(x, y)$ has points of infinite discontinuity only on certain lines

$$x = a_1, a_2, \ldots a_n \quad (a \leqq a_1 < a_2 \ldots < a_n \leqq a'),$$

and is bounded in the given region, except in the neighbourhood of these lines.

This condition can be realised in two different ways. The infinities may be at isolated points, or they may be distributed right along the lines.

$$\textit{E.g.}\ \text{(i)} \int_0^a \frac{dx}{\sqrt{(x+y)}}, \text{ when } 0 \leqq y \leqq b.$$

$$\text{(ii)} \int_0^1 x^{y-1} e^{-x} dx, \text{ when } 0 \leqq y < 1.$$

In the first of these integrals there is a single infinity in the given region, at the origin ; in the second, there are infinities right along the line $x=0$ from the origin up to but not including $y=1$.

In the definitions and theorems which follow there is no need for any distinction between the two cases.

Consider, first of all, the convergent integral

$$\int_a^{a'} F(x, y)\, dx,$$

where $F(x, y)$ has points of infinite discontinuity on $x=a'$, and is bounded in $a \leqq x \leqq a' - \xi$, $b \leqq y \leqq b'$, where $a < a' - \xi < a'$.

For this integral we have the following definition of uniform convergence :

I. *The integral* $\int_a^{a'} F(x, y)\, dx$ *is said to converge uniformly to its value* $f(y)$ *in the interval* (b, b'), *if, any positive number* ϵ *having been chosen, as small as we please, there is a positive number* η *such that*

$$|f(y)-\int_a^{a'-\xi} F(x, y)\,dx\,|<\epsilon, \text{ when } 0<\xi \leqq \eta,$$

the same η serving for every y in (b, b′).

And, from this definition, the following test for uniform convergence can be established as before :

II. *A necessary and sufficient condition for the uniform convergence of the integral* $\int_a^{a'} F(x, y)\,dx$ *in the interval* (b, b') *is that, if any positive number* ϵ *has been chosen, as small as we please, there shall be a positive number* η *such that*

$$\left|\int_{a'-\xi'}^{a'-\xi''} F(x, y)\,dx\right|<\epsilon, \text{ when } 0<\xi''<\xi' \leqq \eta,$$

the same η serving for every y in (b, b′).

Also we see that *if this infinite integral is uniformly convergent in* (b, b'), *to the arbitrary positive number* ϵ *there will correspond a positive number* η *such that*

$$\left|\int_{a'-\xi}^{a'} F(x, y)\,dx\right|<\epsilon, \text{ when } 0<\xi \leqq \eta,$$

the same η serving for every y in (b, b′).

The definition, and the above condition, require obvious modifications when the points of infinite discontinuity lie on $x=a$, instead of $x=a'$.

And when they lie on lines $x=a_1, a_2, \ldots a_n$ in the given region, by the definition of the integral it can be broken up into several others in which $F(x, y)$ has points of infinite discontinuity only at one of the limits.

In this case the integral is said to converge uniformly in (b, b'), when each of these integrals converges uniformly in that interval.

And, as before, if the integrand $F(x, y)$ has points of infinite discontinuity on $x=a_1, a_2, \ldots a_n$, and we are dealing with the integral $\int_a^{\infty} F(x, y)\,dx$, this integral must be broken up into several integrals of the preceding type, followed by an integral of the form discussed in § 81.

The integral is now said to be uniformly convergent in (b, b') when the integrals into which it has been divided are each uniformly convergent in this interval.

83. Tests for Uniform Convergence. The simplest test for the uniform convergence of the integral $\int_a^\infty F(x, y)\,dx$, taking the first type of infinite integral, corresponds to Weierstrass's M-test for the uniform convergence of infinite series (§ 67. 1).

I. *Let $F(x, y)$ be bounded in $a \leqq x \leqq a'$, $b \leqq y \leqq b'$ and integrable in (a, a'), where a' is arbitrary, for every y in (b, b'). Then the integral $\int_a^\infty F(x, y)\,dx$ will converge uniformly in (b, b'), if there is a function $\mu(x)$, independent of y, such that*

(i) $\mu(x) \geqq 0$, *when* $x \geqq a$;

(ii) $|F(x, y)| \leqq \mu(x)$, *when* $x \geqq a$ *and* $b \leqq y \leqq b'$;

and (iii) $\int_a^\infty \mu(x)\,dx$ *exists.*

For, by (i) and (ii), when $x'' > x' \geqq a$ and $b \leqq y \leqq b'$,

$$\left|\int_{x'}^{x''} F(x, y)\,dx\right| \leqq \int_{x'}^{x''} \mu(x)\,dx,$$

and, from (iii), there is a positive number X such that

$$\int_{x'}^{x''} \mu(x)\,dx < \epsilon, \text{ when } x'' > x' \geqq X.$$

These conditions will be satisfied if $x^n F(x, y)$ is bounded when $x \geqq a$, and $b \leqq y' \leqq b'$ for some constant n greater than 1.

Corollary. *Let $F(x, y) = \phi(x, y)\,\psi(x)$, where $\phi(x, y)$ is bounded in $x \geqq a$ and $b \leqq y \leqq b'$, and integrable in the interval (a, a'), where a' is arbitrary, for every y in (b, b'). Also let $\int_a^\infty \psi(x)\,dx$ be absolutely convergent. Then $\int_a^\infty F(x, y)\,dx$ is uniformly convergent in (b, b').*

Ex. 1. $\int_1^\infty \frac{dx}{x^{1+y}}$, $\int_0^\infty e^{-xy}\,dx$ converge uniformly in $y \geqq y_0 > 0$.

Ex. 2. $\int_0^\infty e^{-\frac{x}{y}}\,dx$ converges uniformly in $0 < y \leqq Y$, where Y is an arbitrary positive number.

Ex. 3. $\int_1^\infty \frac{\cos xy}{x^{1+n}}\,dx$, $\int_1^\infty \frac{\sin xy}{x^{1+n}}\,dx$, $\int_0^\infty \frac{\cos xy}{1+x^2}\,dx$, $\int_0^\infty \frac{\sin xy}{1+x^2}\,dx$ converge uniformly for all values of y, where $n > 0$.

II. *Let $\phi(x, y)$ be bounded in $x \geqq a$, $b \leqq y \leqq b'$, and a monotonic function of x for every y in (b, b'). Also let $\psi(x)$ be bounded and* *not change sign more than a finite number of times in the arbitrary interval* (a, a'),* *and let* $\int_a^\infty \psi(x)\,dx$ *exist.*

Then $\int_a^\infty \phi(x, y)\,\psi(x)\,dx$ *converges uniformly* in (b, b').

This follows immediately from the Second Theorem of Mean Value, which gives, subject to the conditions named above,

$$\int_{x'}^{x''} \phi(x, y)\,\psi(x)\,dx = \phi(x', y)\int_{x'}^{\xi} \psi(x)\,dx + \phi(x'', y)\int_{\xi}^{x''} \psi(x)\,dx,$$

where ξ satisfies $a < x' \leqq \xi \leqq x''$.

But $\phi(x, y)$ is bounded in $x \geqq a$ and $b \leqq y \leqq b'$, and $\int_a^\infty \psi(x)\,dx$ converges.

Thus it follows from the relation

$$\left|\int_{x'}^{x''} \phi(x, y)\,\psi(x)\,dx\right| \leqq \left|\phi(x', y)\right| \left|\int_{x'}^{\xi} \psi(x)\,dx\right| + \left|\phi(x'', y)\right| \left|\int_{\xi}^{x''} \psi(x)\,dx\right|$$

that $\int_a^\infty \phi(x, y)\,\psi(x)\,dx$ converges uniformly in (b, b').

It is evident that $\psi(x)$ in this theorem may be replaced by $\psi(x, y)$, if $\int_a^\infty \psi(x, y)\,dx$ converges uniformly in (b, b').

Ex. $\int_0^\infty e^{-xy}\frac{\sin x}{x}\,dx$, $\int_a^\infty e^{-xy}\frac{\cos x}{x}\,dx$ $(a > 0)$ converge uniformly in $y \geqq 0$.

III. *Let* $\phi(x, y)$ *be a monotonic function of* x *for each* y *in* (b, b'), *and tend uniformly to zero as* x *increases,* y *being kept constant. Also let* $\psi(x)$ *be bounded and integrable in the arbitrary interval* (a, a'), *and not change sign more than a finite number of times in such an interval.** *Further, let* $\int_a^x \psi(x)\,dx$ *be bounded in* $x \geqq a$, *without converging as* $x \to \infty$.

Then $\int_a^\infty \phi(x, y)\,\psi(x)\,dx$ *is uniformly convergent in* (b, b').

This follows at once from the Second Theorem of Mean Value,

* This condition is borrowed from the enunciation in the Second Theorem of Mean Value, as proved in § 50.1. If the more general proof is taken, a corresponding extension of (II) and (III) follows.

as in (II). Also it will be seen that $\psi(x)$ may be replaced by $\psi(x, y)$, if $\int_a^x \psi(x, y)\,dx$ is bounded in $x \geqq a$ and $b \leqq y \leqq b'$.*

Ex. 1. $\int_0^\infty e^{-xy} \sin x\,dx$, $\int_0^\infty e^{-xy} \cos x\,dx$ converge uniformly in $y \geqq y_0 > 0$.

Ex. 2. $\int_0^\infty \frac{\sin xy}{x}\,dx$ and $\int_0^\infty \frac{x \sin xy}{1+x^2}\,dx$ both converge uniformly in $y \geqq y_0 > 0$ and $y \leqq -y_0 < 0$.

It can be left to the reader to enunciate and prove similar theorems for the second type of infinite integral $\int_a^{a'} F(x, y)\,dx$. The most useful test for uniform convergence in this case is that corresponding to (I) above.

Ex. 1. $\int_0^1 x^{y-1}\,dx$, $\int_0^1 x^{y-1} e^{-x}\,dx$ converge uniformly in $1 > y \geqq y_0 > 0$.

Ex. 2. $\int_0^1 \frac{\sin x}{x^{1+y}}\,dx$ converges uniformly in $0 < y \leqq y_0 < 1$.

84. Continuity of the Integral $\int_a^\infty$ F(x, y) dx. We shall now consider, to begin with, the infinite integral $\int_a^\infty F(x, y)\,dx$, where $F(x, y)$ is bounded in the region $a \leqq x \leqq a'$, $b \leqq y \leqq b'$, a' being arbitrary. Later we shall return to the other form of infinite integral in which the region contains points of infinite discontinuity.

Let $f(y) = \int_a^\infty F(x, y)\,dx$, *where* $F(x, y)$ *either is a continuous function of* (x, y) *in* $a \leqq x \leqq a'$, $b \leqq y \leqq b'$, a' *being arbitrary, or is of the form* $\phi(x, y)\psi(x)$, *where* $\phi(x, y)$ *is continuous as above, and* $\psi(x)$ *is bounded and integrable in the arbitrary interval* (a, a').

Also let $\int_a^\infty F(x, y)\,dx$ *converge uniformly in* (b, b').

Then $f(y)$ *is a continuous function of* y *in* (b, b').

Let the positive number ϵ be chosen, as small as we please.

Then to $\frac{1}{3}\epsilon$ there corresponds a positive number X such that

$$\left| \int_x^\infty F(x, y)\,dx \right| < \tfrac{1}{3}\epsilon, \text{ when } x \geqq X,$$

the same X serving for every y in (b, b').

---

*In Examples 1 and 2 of § 88 illustrations of this theorem will be found.

But we have proved in § 77 that, under the given conditions, $\int_a^X F(x, y)\,dx$ is continuous in y in (b, b').

Therefore, for some positive number η,

$$\left|\int_a^X F(x, y+\Delta y)\,dx - \int_a^X F(x, y)\,dx\right| < \tfrac{1}{3}\epsilon, \text{ when } |\Delta y| \leqq \eta.$$

Also $$f(y) = \int_a^X F(x, y)\,dx + \int_X^\infty F(x, y)\,dx.$$

Thus $$f(y+\Delta y) - f(y) = \left[\int_a^X F(x, y+\Delta y)\,dx - \int_a^X F(x, y)\,dx\right] + \int_X^\infty F(x, y+\Delta y)\,dx - \int_X^\infty F(x, y)\,dx.$$

Also we have $$\left|\int_X^\infty F(x, y)\,dx\right| < \tfrac{1}{3}\epsilon,$$

and $$\left|\int_X^\infty F(x, y+\Delta y)\,dx\right| < \tfrac{1}{3}\epsilon.$$

Therefore, finally,

$$|f(y+\Delta y) - f(y)| < \tfrac{1}{3}\epsilon + \tfrac{1}{3}\epsilon + \tfrac{1}{3}\epsilon$$
$$< \epsilon, \text{ when } |\Delta y| \leqq \eta.$$

Thus $f(y)$ is continuous in (b, b').

85. Integration of the Integral $\int_a^\infty F(x, y)\,dx$.

Let $F(x, y)$ satisfy the same conditions as in § 84.

Then $$\int_{y_0}^y dy \int_a^\infty F(x, y)\,dx = \int_a^\infty dx \int_{y_0}^y F(x, y)\,dy,$$

where y_0, y are any two points in (b, b').

Let $$f(y) = \int_a^\infty F(x, y)\,dx.$$

Then we have shown that under the given conditions $f(y)$ is continuous, and therefore integrable, in (b, b').

Also from § 79, for any arbitrary interval (a, x), where x can be taken as large as we please,

$$\int_a^x dx \int_{y_0}^y F(x, y)\,dy = \int_{y_0}^y dy \int_a^x F(x, y)\,dx,$$

y_0, y being any two points in (b, b').

Therefore

$$\int_a^\infty dx \int_{y_0}^{y} F(x, y)\, dy = \lim_{x\to\infty} \int_{y_0}^{y} dy \int_a^x F(x, y)\, dx,$$

provided we can show that the limit on the right-hand exists.

But

$$\int_{y_0}^{y} dy \int_a^x F(x, y)\, dx = \int_{y_0}^{y} dy \int_a^\infty F(x, y)\, dx - \int_{y_0}^{y} dy \int_x^\infty F(x, y)\, dx.$$

Thus we have only to show that

$$\lim_{x\to\infty} \int_{y_0}^{y} dy \int_x^\infty F(x, y)\, dx = 0.$$

Of course we cannot reverse the order of these limiting processes and write this as

$$\int_{y_0}^{y} dy \lim_{x\to\infty} \int_x^\infty F(x, y)\, dx,$$

for we have not shown that this inversion would not alter the result.

But we are given that $\int_a^\infty F(x, y)\, dx$ is uniformly convergent in (b, b').

Let the positive number ϵ be chosen, as small as we please. Then take $\epsilon/(b'-b)$. To this number there corresponds a positive number X such that

$$\left| \int_x^\infty F(x, y)\, dx \right| < \frac{\epsilon}{b'-b}, \text{ when } x \geqq X,$$

the same X serving for every y in (b, b').

It follows that

$$\left| \int_{y_0}^{y} dy \int_x^\infty F(x, y) dx \right| < \epsilon, \text{ when } x \geqq X,$$

if y_0, y lie in (b, b').

In other words,

$$\lim_{x\to\infty} \left[\int_{y_0}^{y} dy \int_x^\infty F(x, y)\, dx \right] = 0.$$

And from the preceding remarks this establishes our result.

86. Differentiation of the Integral $\int_a^\infty F(x, y)\, dx$.

Let $F(x, y)$ either be a continuous function of (x, y) in the region $a \leqq x \leqq a'$, $b \leqq y \leqq b'$, a' being arbitrary, or be of the form $\phi(x, y)\, \psi(x)$, where $\phi(x, y)$ is continuous as above, and $\psi(x)$ is bounded *and integrable in the arbitrary interval (a, a'). Also let $F(x, y)$ have a partial differential coefficient $\frac{\partial F}{\partial y}$ which satisfies the same conditions.*

Then, if the integral $\int_a^\infty F(x, y)\,dx$ converges to $f(y)$, and the integral $\int_a^\infty \frac{\partial F}{\partial y}\,dx$ converges uniformly in (b, b'), $f(y)$ has a differential coefficient at every point in (b, b'), and

$$f'(y)=\int_a^\infty \frac{\partial F}{\partial y}\,dx.$$

We know from § 84 that, on the assumption named above, $\int_a^\infty \frac{\partial F}{\partial y}\,dx$ is a continuous function of y in (b, b').

Let
$$g(y)=\int_a^\infty \frac{\partial F}{\partial y}\,dx.$$

Then, by § 85,
$$\int_{y_0}^{y_1} g(y)dy=\int_a^\infty dx\int_{y_0}^{y_1}\frac{\partial F}{\partial y}\,dy,$$
where y_0, y_1 are any two points in (b, b').

Let
$$y_0=y \quad\text{and}\quad y_1=y+\Delta y.$$

Then
$$\int_y^{y+\Delta y} g(y)dy=\int_a^\infty [F(x, y+\Delta y)-F(x, y)]\,dx.$$

Therefore
$$g(\xi)\Delta y=f(y+\Delta y)-f(y),$$
where
$$y \leqq \xi \leqq y+\Delta y \quad\text{and}\quad f(y)=\int_a^\infty F(x, y)\,dx.$$

Thus
$$g(\xi)=\frac{f(y+\Delta y)-f(y)}{\Delta y}.$$

But $\lim_{\Delta y\to 0} g(\xi)=g(y)$, since $g(x)$ is continuous.

It follows that $f(y)$ is differentiable, and that
$$f'(y)=\int_a^\infty \frac{\partial F}{\partial y}\,dx,$$
where y is any point in (b, b').

87. Properties of the Infinite Integral $\int_a^{a'} F(x, y)\,dx$. The results of §§ 84-86 can be readily extended to the second type of infinite integral. It will be sufficient to state the theorems without proof. The steps in the argument are in each case parallel to those in the preceding discussion. As before, the region with which we deal is

$$a \leqq x \leqq a', \quad b \leqq y \leqq b'.$$

I. Continuity of $\int_a^{a'} F(x, y)\,dx$.

Let $f(y) = \int_a^{a'} F(x, y)\,dx$, *where* $F(x, y)$ *has points of infinite discontinuity on certain lines (e.g.,* $x = a_1, a_2, \ldots a_n$*) between* $x = a$ *and* $x = a'$*, and is either a continuous function of* (x, y)*, or the product of a continuous function* $\phi(x, y)$ *and a bounded and integrable function* $\psi(x)$*, except in the neighbourhood of the said lines.*

Then, if $\int_a^{a'} F(x, y)\,dx$ *is uniformly convergent in* (b, b')*,* $f(y)$ *is a continuous function of* y *in* (b, b').

II. Integration of the Integral $\int_a^{a'} F(x, y)\,dx$.

Let $F(x, y)$ *satisfy the same conditions as in (I).*

Then
$$\int_{y_0}^{y} dy \int_a^{a'} F(x, y)dx = \int_a^{a'} dx \int_{y_0}^{y} F(x, y)\,dy,$$
where y_0, y *are any two points in* (b, b').

III. Differentiation of the Integral $\int_a^{a'} F(x, y)\,dx$.

Let $f(y) = \int_a^{a'} F(x, y)\,dx$, *where* $F(x, y)$ *has points of infinite discontinuity on certain lines (e.g.,* $x = a_1, a_2, \ldots a_n$*) between* $x = a$ *and* $x = a'$*, and is either a continuous function of* (x, y)*, or the product of a continuous function* $\phi(x, y)$ *and a bounded and integrable function* $\psi(x)$*, except in the neighbourhood of the said lines.*

Also let $F(x, y)$ *have a partial differential coefficient* $\frac{\partial F}{\partial y}$*, which satisfies the same conditions.*

Further, let $\int_a^{a'} F(x, y)\,dx$ *converge, and* $\int_a^{a'} \frac{\partial F}{\partial y}\,dx$ *converge uniformly in* (b, b').

Then $f'(y)$ *exists and is equal to* $\int_a^{a'} \frac{\partial F}{\partial y}\,dx$ *in* (b, b').

88. Applications of the preceding Theorems.

Ex. 1. *To prove*
$$\int_0^{\infty} \frac{\sin x}{x}\,dx = \frac{\pi}{2}.*$$

(i) Let
$$F(\alpha) = \int_0^{\infty} e^{-\alpha x} \frac{\sin x}{x}\,dx \quad (\alpha \geqq 0).$$

---

*For other proofs, see Ex. 11 and Ex. 12 on pp. 135-6, and Ex. 10 on p. 213.

This integral converges uniformly when $a \geqq 0$, (Cf. § 83, III.) for e^{-ax}/x is a monotonic function of x when $x > 0$.

Thus, by the Second Theorem of Mean Value,

$$\int_{x'}^{x''} \frac{e^{-ax}}{x} \sin x\, dx = \frac{e^{-ax'}}{x'} \int_{x'}^{\xi} \sin x\, dx + \frac{e^{-ax''}}{x''} \int_{\xi}^{x''} \sin x\, dx,$$

where $0 < x' \leqq \xi \leqq x''$.

Therefore

$$\left| \int_{x'}^{x''} \frac{e^{-ax}}{x} \sin x\, dx \right| \leqq \frac{e^{-ax'}}{x'} \left| \int_{x'}^{\xi} \sin x\, dx \right| + \frac{e^{-ax''}}{x''} \left| \int_{\xi}^{x''} \sin x\, dx \right|$$

$$< 4\frac{e^{-ax'}}{x'}, \quad \text{since} \left| \int_{p}^{q} \sin x\, dx \right| \leqq 2 \text{ for all values of } p \text{ and } q,$$

$$< \frac{4}{x'}, \text{ since } a \geqq 0.$$

It follows that $\left| \int_{x'}^{x''} e^{-ax} \frac{\sin x}{x} dx \right| < \epsilon$ when $x'' > x' \geqq X$,

provided that $X > 4/\epsilon$, and this holds for every a greater than or equal to 0.

This establishes the uniform continuity of the integral, and it follows from § 84 that $F(a)$ is continuous in $a \geqq 0$.

Thus
$$F(0) = \lim_{a \to 0} \int_0^{\infty} e^{-ax} \frac{\sin x}{x} dx,$$

i.e.
$$\int_0^{\infty} \frac{\sin x}{x} dx = \lim_{a \to 0} \int_0^{\infty} e^{-ax} \frac{\sin x}{x} dx.$$

(ii) Again, the integral $\int_0^{\infty} e^{-ax} \sin x\, dx$

is uniformly convergent in $a \geqq a_0 > 0$.

This follows as above, and is again an example of § 83, III.

Thus, by § 86, when $a > 0$,

$$F'(a) = \int_0^{\infty} \frac{\partial}{\partial a}\left(e^{-ax} \frac{\sin x}{x}\right) dx$$

$$= -\int_0^{\infty} e^{-ax} \sin x\, dx.$$

But
$$\frac{d}{dx} e^{-ax}(\cos x + a \sin x) = -(a^2+1) e^{-ax} \sin x.$$

Therefore
$$\int_0^{\infty} e^{-ax} \sin x\, dx = \frac{1}{a^2+1}.$$

Thus
$$F'(a) = -\frac{1}{a^2+1}.$$

And
$$F(a) = -\tan^{-1} a + \frac{\pi}{2},$$

since $\lim_{a \to \infty} F(a) = 0$.*

---

* If a formal proof of this is required, we might proceed as follows:
Let the arbitrary positive ϵ be chosen, as small as we please.

It follows from (i) that $\int_0^\infty \frac{\sin x}{x}\,dx = \frac{\pi}{2}$.

Ex. 2. *To prove*

$$\int_0^\infty \frac{\cos ax}{1+x^2}\,dx = \frac{\pi}{2}e^{-a} \quad \textit{and} \quad \int_0^\infty \frac{\sin ax}{x(1+x^2)}\,dx = \frac{\pi}{2}(1-e^{-a}) \quad (a \geqq 0).$$

(i) Let
$$f(a) = \int_0^\infty \frac{\cos ax}{1+x^2}\,dx.$$

The integral is uniformly convergent for every a, so that, by § 84, $f(a)$ is continuous for all values of a, and we can integrate under the sign of integration (§ 85).

(ii) Let
$$\phi(a) = \int_0^a f(a)\,da.$$

Then $\phi(a)$ is a continuous function of a for all values of a (§ 49).

Also
$$\phi(a) = \int_0^\infty dx \int_0^a \frac{\cos ax}{1+x^2}\,da$$
$$= \int_0^\infty \frac{\sin ax}{x(1+x^2)}\,dx.$$

(iii) Again, we know that $f'(a)$ will be equal to $-\int_0^\infty \frac{x}{1+x^2}\sin ax\,dx$, if a lies within an interval of uniform convergence of this integral.

But $\frac{x}{1+x^2}$ is monotonic when $x \geqq 1$, and $\lim_{x\to\infty} \frac{x}{1+x^2} = 0$.

Since $\int_0^\infty e^{-\alpha x}\frac{\sin x}{x}\,dx$ converges uniformly in $\alpha \geqq 0$, there is a positive number X such that
$$\left|\int_X^\infty e^{-\alpha x}\frac{\sin x}{x}\,dx\right| < \tfrac{1}{3}\epsilon,$$
and this number X is independent of α.

Also we can choose x_0, independent of α, so that
$$\left|\int_0^{x_0} e^{-\alpha x}\frac{\sin x}{x}\,dx\right| < \tfrac{1}{3}\epsilon.$$

But
$$\int_{x_0}^X e^{-\alpha x}\frac{\sin x}{x}\,dx = e^{-\alpha x_0}\int_{x_0}^\xi \frac{\sin x}{x}\,dx + e^{-\alpha X}\int_\xi^X \frac{\sin x}{x}\,dx,$$
where $x_0 \leqq \xi \leqq X$.

And we know that
$$\left|\int_p^q \frac{\sin x}{x}\,dx\right| < \pi \text{ for } q > p > 0 \cdot \text{(cf. p. 222)}.$$

Therefore
$$\left|\int_{x_0}^X e^{-\alpha x}\frac{\sin x}{x}\,dx\right| < 2\pi e^{-\alpha x_0}.$$

Thus we can choose A so large that
$$\left|\int_{x_0}^X e^{-\alpha x}\frac{\sin x}{x}\,dx\right| < \tfrac{1}{3}\epsilon, \text{ when } \alpha \geqq A.$$

It follows that $\left|\int_0^\infty e^{-\alpha x}\frac{\sin x}{x}\,dx\right| < \tfrac{1}{3}\epsilon + \tfrac{1}{3}\epsilon + \tfrac{1}{3}\epsilon$, when $\alpha \geqq A$.

Thus we have, when $x'' > x' \geqq 1$,

$$\int_{x'}^{x''} \frac{x}{1+x^2} \sin ax\, dx = \frac{x'}{1+x'^2}\int_{x'}^{\xi} \sin ax\, dx + \frac{x''}{1+x''^2}\int_{\xi}^{x''} \sin ax\, dx,$$

where $x' \leqq \xi \leqq x''$.

It follows that

$$\int_{x'}^{x''} \frac{x}{1+x^2} \sin ax\, dx = \frac{x'}{a(1+x'^2)}\int_{ax'}^{a\xi} \sin x\, dx + \frac{x''}{a(1+x''^2)}\int_{a\xi}^{ax''} \sin x\, dx.$$

Therefore $$\left|\int_{x'}^{x''} \frac{x}{1+x^2} \sin ax\, dx\right| \leqq \frac{4x'}{a(1+x'^2)}.$$

Thus $\int_0^\infty \frac{x}{1+x^2} \sin ax\, dx$ is uniformly convergent when $a \geqq a_0 > 0$, and

$$f'(a) = -\int_0^\infty \frac{x}{1+x^2} \sin ax\, dx.$$

(iv) Now $$\phi(a) = \int_0^a f(a)\, da.$$

Therefore $$\phi'(a) = f(a)$$

and $$\phi''(a) = f'(a) = -\int_0^\infty \frac{x}{1+x^2} \sin ax\, dx.$$

Thus
$$\begin{aligned}\phi''(a) &= -\int_0^\infty \left[1 - \frac{1}{1+x^2}\right] \frac{\sin ax}{x}\, dx\\ &= -\int_0^\infty \frac{\sin ax}{x}\, dx + \int_0^\infty \frac{\sin ax}{x(1+x^2)}\, dx\\ &= -\frac{\pi}{2} + \phi(a).\end{aligned}$$

This result has been established on the understanding that $a > 0$.

(v) From (iv) we have

$$\phi(a) = Ae^a + Be^{-a} + \frac{\pi}{2}, \text{ when } a > 0.$$

But $\phi(a)$ is continuous in $a \geqq 0$, and $\phi(0) = 0$.

Therefore $$\lim_{a \to 0} \phi(a) = 0,$$

and $$A + B + \frac{\pi}{2} = 0.$$

Also $\phi'(a)$ is continuous in $a \gtreqless 0$. and $\phi'(0) = \frac{\pi}{2}$.

Therefore $$A - B - \frac{\pi}{2} = 0.$$

It follows that $$A = 0 \quad \text{and} \quad B = -\frac{\pi}{2}.$$

Thus $$\phi(a) = \frac{\pi}{2}(1 - e^{-a}), \text{ when } a > 0.$$

And $$f(a) = \phi'(a) = \frac{\pi}{2} e^{-a}, \text{ when } a > 0.$$

Both these results obviously hold for $a = 0$ as well.

Ex. 3. *The Gamma Function* $\Gamma(n)=\int_0^\infty e^{-x}x^{n-1}\,dx$, $n>0$, *and its derivatives.*

(i) *To prove* $\Gamma(n)$ *is uniformly convergent when* $N \geqq n \geqq n_0 > 0$, *however large* N *may be and however near zero* n_0 *may be.*

When $n \geqq 1$, the integral $\int_0^\infty e^{-x}x^{n-1}\,dx$ has to be examined for convergence only at the upper limit. When $0<n<1$, the integrand becomes infinite at $x=0$. In this case we break up the integral into

$$\int_0^1 e^{-x}x^{n-1}\,dx+\int_1^\infty e^{-x}x^{n-1}\,dx.$$

Take first the integral $\int_0^1 e^{-x}x^{n-1}\,dx.$

When $0<x<1$, $\quad x^{n-1} \leqq x^{n_0-1}$, $\quad$ if $n \geqq n_0 > 0$.

Therefore $\quad e^{-x}x^{n-1} \leqq e^{-x}x^{n_0-1}$, if $n \geqq n_0 > 0$.

It follows from the theorem which corresponds to § 83, I, that $\int_0^1 e^{-x}x^{n-1}\,dx$ converges uniformly when $n \geqq n_0 > 0$.

Again consider $\quad \int_1^\infty e^{-x}x^{n-1}\,dx$, $n>0$.

When $\quad x>1$, $x^{n-1} \leqq x^{N-1}$, $\quad$ if $0<n \leqq N$.

Therefore $\quad e^{-x}x^{n-1} \leqq e^{-x}x^{N-1}$, if $0<n \leqq N$.

It follows as above (from § 83, I) that $\int_1^\infty e^{-x}x^{n-1}\,dx$ converges uniformly for $0<n \leqq N$.

Combining these two results, we see that $\int_0^\infty e^{-x}x^{n-1}\,dx$ converges uniformly when $N \geqq n \geqq n_0 > 0$, however large N may be and however near zero n_0 may be.

(ii) *To prove* $\Gamma'(n)=\int_0^\infty e^{-x}x^{n-1}\log x\,dx$, $\quad n>0$.

We know that $\quad \lim_{x\to 0}\,(x^r \log x)=0$, when $r>0$,

so that the integrand has an infinity at $x=0$ for positive values of n only when $0<n \leqq 1$.

But when $0<x<1$, $\quad x^{n-1} \leqq x^{n_0-1}$, $\quad$ if $n \geqq n_0 > 0$.

Therefore $\quad e^{-x}x^{n-1}|\log x| \leqq x^{n_0-1}|\log x|$, if $n \geqq n_0 > 0$.

And we have seen [Ex. 7, p. 133] that $\int_0^1 x^{n_0-1}\log x\,dx$ converges when $0<n_0 \leqq 1$.

It follows as above that $\int_0^1 e^{-x}x^{n-1}\log x\,dx$ is uniformly convergent when $n \geqq n_0 > 0$.

Also for $\int_1^\infty e^{-x}x^{n-1}\log x\,dx$, we proceed as follows:

When $\quad x>1$, $\quad x^{n-1} \leqq x^{N-1}$, if $0<n \leqq N$.

Therefore $\quad e^{-x}x^{n-1}\log x \leqq e^{-x}x^{N-1}\log x$

$$< e^{-x}x^{N}, \quad \text{since } \frac{\log x}{x}<1 \text{ when } x>1.$$

But $\int_1^\infty e^{-x}x^{N}\,dx$ is convergent.

Therefore $\int_1^\infty e^{-x}x^{n-1}\log x\,dx$ is uniformly convergent when $0<n\leqq N$.

Combining these two results, we see that $\int_0^\infty e^{-x}x^{n-1}\log x\,dx$ is uniformly convergent when $N\geqq n\geqq n_0>0$, however large N may be and however near zero n_0 may be.

We are thus able to state, relying on §§ 86, 87, that

$$\Gamma'(n)=\int_0^\infty e^{-x}x^{n-1}\log x\,dx \text{ for } n>0.$$

It can be shown in the same way that the successive derivatives of $\Gamma(n)$ can be obtained by differentiating under the integral sign.

Ex. 4. (i) *To prove* $\int_0^\pi \log(1-2y\cos x+y^2)\,dx$ *is uniformly convergent for any interval of* y *(e.g.* $b\leqq y\leqq b'$*); and* (ii) *to deduce that* $\int_0^{\frac{\pi}{2}}\log\sin x\,dx=-\frac{1}{2}\pi\log 2$.

(i) Since $1-2y\cos x+y^2=(y-\cos x)^2+\sin^2 x$, this expression is positive for all values of x, y, unless when $x=m\pi$ and $y=(-1)^m$, $m=0,\ \pm1,\ \pm2$, etc., and for these values it is zero.

It follows that the integrand becomes infinite at $x=0$ and $x=\pi$; in the one case when $y=1$, and in the other when $y=-1$.

We consider first the infinity at $x=0$.

As the integrand is bounded in any strip $0\leqq x\leqq X$, where $X<\pi$, for any interval of y which does not include $y=1$, we have only to examine the integral

$$\int_0^x \log(1-2y\cos x+y^2)\,dx$$

in the neighbourhood of $y=1$.

Put $y=1+h$, where $|h|\leqq a$ and a is some positive number less than unity, to be fixed more definitely later.

Since
$$\int_0^x \log(1-2y\cos x+y^2)\,dx$$
$$=\int_0^x \log\left(1-\cos x+\frac{h^2}{2(1+h)}\right)dx+x\log 2(1+h),$$

it is clear that we need only discuss the convergence of the integral

$$\int_0^x \log\left(1-\cos x+\frac{h^2}{2(1+h)}\right)dx.$$

Take a value of $a(0<a<1)$ such that $\dfrac{a^2}{2(1-a)}<1$.

Then
$$1>\frac{a^2}{2(1-a)}\geqq\frac{h^2}{2(1+h)}\geqq 0,\quad \text{since } |h|\leqq a.$$

Now let
$$\beta=\cos^{-1}\frac{a^2}{2(1-a)}.$$

It will be seen that, when $|h|\leqq a$,

$$0<1-\cos x\leqq 1-\cos x+\frac{h^2}{2(1+h)}\leqq 1,$$

provided that $0<x\leqq\beta$.

Therefore, under the same conditions,

$$\left|\log\left(1-\cos x+\frac{h^2}{2(1+h)}\right)\right| \leqq \left|\log(1-\cos x)\right|.$$

But the μ-test shows that the integral

$$\int_0^\beta \log(1-\cos x)\,dx$$

converges.

It follows that $$\int_0^\beta \log\left(1-\cos x+\frac{h^2}{2(1+h)}\right)dx$$

converges uniformly for $|h| \leqq a$. [Cf. § 83, I.]

And therefore $$\int_0^\beta \log(1-2y\cos x+y^2)\,dx$$

converges uniformly for any interval (b, b') of y.

The infinity at $x=\pi$ can be treated in the same way, and the uniform convergence of the integral

$$\int_0^\pi \log(1-2y\cos x+y^2)\,dx$$

is thus established for any interval (b, b') of y.

(ii) Let $$f(y)=\int_0^\pi \log(1-2y\cos x+y^2)\,dx.$$

We know from § 70. 1 that

$$f(y)=0, \quad \text{when } |y|<1,$$

and $$f(y)=\pi\log y^2, \quad \text{when } |y|>1.$$

But we have just seen that the integral converges uniformly for any finite interval of y.

It follows, from § 87, I, that

$$f(1)=\lim_{y\to 1} f(y)=0$$

and $$f(-1)=\lim_{y\to -1} f(y)=0.$$

But $$f(1)=\int_0^\pi \log 2(1-\cos x)dx$$

$$=2\pi\log 2+2\int_0^\pi \log\sin\frac{x}{2}\,dx$$

$$=2\pi\log 2+4\int_0^{\frac{\pi}{2}} \log\sin x\,dx.$$

Thus $$\int_0^{\frac{\pi}{2}} \log\sin x\,dx=-\tfrac{1}{2}\pi\log 2.*$$

From $f(-1)=0$, we find in the same way that

$$\int_0^{\frac{\pi}{2}} \log\cos x\,dx=-\tfrac{1}{2}\pi\log 2,$$

a result which, of course, could have been deduced from the preceding.

---

*This integral was obtained otherwise in Ex. 4, p. 131

89. The Repeated Integral $\int_a^\infty dx \int_b^\infty f(x, y)\,dy$. It is not easy to determine general conditions under which the equation

$$\int_a^\infty dx \int_b^\infty f(x, y)dy = \int_b^\infty dy \int_a^\infty f(x, y)dx$$

is satisfied.

The problem is closely analogous to that of term by term integration of an infinite series between infinite limits. We shall discuss only a case somewhat similar to that in infinite series given in § 76.

Let $f(x, y)$ be a continuous function of (x, y) in $x \geqq a$, $y \geqq b$, and let the integrals

$$\text{(i)} \int_a^\infty f(x, y)dx, \qquad \text{(ii)} \int_b^\infty f(x, y)dy,$$

respectively, converge uniformly in the arbitrary intervals

$$b \leqq y \leqq b', \quad a \leqq x \leqq a'.$$

Also let the integral $\quad \text{(iii)} \int_a^\infty dx \int_b^y f(x, y)dy$

converge uniformly in $y \geqq b$.

Then the integrals

$$\int_a^\infty dx \int_b^\infty f(x, y)dy \quad \text{and} \quad \int_b^\infty dy \int_a^\infty f(x, y)dx$$

exist and are equal.

Since we are given that $\int_a^\infty f(x, y)dx$ converges uniformly in the arbitrary interval $b \leqq y \leqq b'$, we know from § 85 that

$$\int_b^y dy \int_a^\infty f(x, y)dx = \int_a^\infty dx \int_b^y f(x, y)dy, \text{ when } y > b.$$

It follows that

$$\int_b^\infty dy \int_a^\infty f(x, y)dx = \lim_{y \to \infty} \int_a^\infty dx \int_b^y f(x, y)dy,$$

provided that the limit on the right-hand side exists.

To prove the existence of this limit, it is sufficient to show that to the arbitrary positive number ϵ there corresponds a positive number Y such that

$$\left| \int_a^\infty dx \int_{y'}^{y''} f(x, y)dy \right| < \epsilon, \text{ when } y'' > y' \geqq Y.$$

But from the uniform convergence of

$$\int_a^\infty dx \int_b^y f(x, y)dy$$

in $y \geqq b$, we can choose the positive number X such that

$$\left| \int_x^\infty dx \int_b^y f(x, y)dy \right| < \tfrac{1}{3}\epsilon, \text{ when } x \geqq X, \quad \ldots\ldots\ldots\ldots\ldots\ldots(1)$$

the same X serving for every y greater than or equal to b.

Also we are given that $\quad \int_b^\infty f(x, y)dy$

is uniformly convergent in the arbitrary interval (a, a').

Therefore we can choose the positive number Y so that

$$\left|\int_{y'}^{y''} f(x, y)dy\right| < \frac{\epsilon}{3(X-a)}, \text{ when } y'' > y' \geqq Y,$$

the same Y serving for every x in (a, X).

Thus we have $\left|\int_a^X dx \int_{y'}^{y''} f(x, y)dy\right| < \frac{1}{3}\epsilon$, when $y'' > y' \geqq Y$. ....(2)

But it is clear that

$$\int_a^\infty dx \int_{y'}^{y''} f(x, y)dy = \int_a^X dx \int_{y'}^{y''} f(x, y)dy + \int_X^\infty dx \int_b^{y''} f(x, y)dy - \int_X^\infty dx \int_b^{y'} f(x, y)dy.$$

Therefore from (1) and (2) we have

$$\left|\int_a^\infty dx \int_{y'}^{y''} f(x, y)dy\right| < \tfrac{1}{3}\epsilon + \tfrac{1}{3}\epsilon + \tfrac{1}{3}\epsilon$$

$$< \epsilon, \text{ when } y'' > y' \geqq Y.$$

We have thus shown that

$$\int_b^\infty dy \int_a^\infty f(x, y)dx = \lim_{y\to\infty} \int_a^\infty dx \int_b^y f(x, y)dy. \quad \text{...................(3)}$$

It remains to prove that

$$\lim_{y\to\infty} \int_a^\infty dx \int_b^y f(x, y)dy = \int_a^\infty dx \int_b^\infty f(x, y)dy.$$

Let the limit on the left-hand side be l.

Then ϵ being any positive number, as small as we please, there is a positive number Y_1 such that

$$\left|l - \int_a^\infty dx \int_b^y f(x, y)dy\right| < \tfrac{1}{3}\epsilon, \text{ when } y \geqq Y_1. \quad \text{..................(4)}$$

Also, from the uniform convergence of

$$\int_a^\infty dx \int_b^y f(x, y)dy, \text{ when } y \geqq b,$$

we know that there is a positive number X such that

$$\left|\int_a^\infty dx \int_a^y f(x, y)\,dy - \int_a^{X'} dx \int_b^y f(x, y)\,dy\right| < \tfrac{1}{3}\epsilon, \text{ when } X' \geqq X, \quad \text{.......(5)}$$

the same X serving for every y greater than or equal to b.

Choose a number X' such that $X' \geqq X > a$.

Then, from the uniform convergence of $\int_b^\infty f(x, y)dy$ in any arbitrary interval, we know that there is a positive number Y_2 such that

$$\left|\int_b^y f(x, y)\,dy - \int_b^\infty f(x, y)\,dy\right| < \frac{\epsilon}{3(X'-a)}, \text{ when } y \geqq Y_2,$$

the same Y_2 serving for every x in (a, X').

Thus $\left|\int_a^{X'} dx \int_b^y f(x, y)\,dy - \int_a^{X'} dx \int_b^\infty f(x, y)\,dy\right| < \frac{1}{3}\epsilon$, when $y \geqq Y_2$. ....(6)

Now take a number Y greater than Y_1 and Y_2.

Equations (4), (5) and (6) hold for this number Y.

But

$$\left| l - \int_b^{X'} dx \int_b^{\infty} f(x, y)\,dy \right|$$

$$\leqq \left| l - \int_a^{\infty} dx \int_b^{Y} f(x, y)\,dy \right| + \left| \int_a^{\infty} dx \int_b^{Y} f(x,y)\,dy - \int_a^{X'} dx \int_b^{Y} f(x, y)\,dy \right|$$

$$+ \left| \int_a^{X'} dx \int_b^{Y} f(x, y)\,dy - \int_a^{X'} dx \int_b^{\infty} f(x, y)\,dy \right|$$

$< \frac{1}{3}\epsilon + \frac{1}{3}\epsilon + \frac{1}{3}\epsilon$, from (4), (5) and (6),

$< \epsilon$.

This result holds for every number X' greater than or equal to X.

Thus we have shown that

$$l = \lim_{x \to \infty} \int_a^{x} dx \int_b^{\infty} f(x, y)\,dy = \int_a^{\infty} dx \int_b^{\infty} f(x, y)\,dy.$$

Also, from (3), we have

$$\int_a^{\infty} dx \int_b^{\infty} f(x, y)\,dy = \int_b^{\infty} dy \int_a^{\infty} f(x, y)\,dx$$

under the conditions stated in the theorem.

It must be noticed that the conditions we have taken are *sufficient*, but not *necessary*. For a more complete discussion of the conditions under which the integrals

$$\int_a^{\infty} dx \int_b^{\infty} f(x, y)\,dy, \quad \int_b^{\infty} dy \int_a^{\infty} f(x, y)\,dx,$$

when they both exist, are equal, reference should be made to the works of de la Vallée Poussin,* to whom the above treatment is due. A valuable discussion of the whole subject is also given in Pierpont's *Theory of Functions of a Real Variable.* The question is treated in Hobson's *Theory of Functions of a Real Variable*, but from a more difficult standpoint.

REFERENCES.

BROMWICH, *loc. cit.*, App. III.

DE LA VALLÉE POUSSIN, *loc. cit.*, 2 (4e éd., 1922), Ch. I.

DINI, *Lezioni di Analisi Infinitesimale*, 2 (Pisa, 1909), 1a Parte, Cap. IX, X.

GOURSAT, *loc. cit.*, 1 (4e éd., 1923), Ch. IV and V.

HOBSON, *loc. cit.*, 1 (3rd ed., 1927), Ch. VI; 2 (2nd ed., 1926), Ch. V.

OSGOOD, *loc. cit.*, 1 (4 Aufl., 1923), Kap. III.

PIERPONT, *loc. cit.*, 1 (1905), Ch. XIII-XV.

STOLZ, *loc. cit.*, Bd. I, Absch. X.

BRUNEL, *loc. cit.*, *Enc. d. math. Wiss.*, Bd. II, Tl. I (Leipzig, 1899).

And

MONTEL-ROSENTHAL, *loc. cit.*, *Enc. d. math. Wiss.*, Bd. II, Tl. III, 2 (Leipzig, 1923).

---

* His investigations are contained in three memoirs; the first in *Ann. soc. scient. Bruxelles*, **16** (1892); the second in *Journal de math.*, (4), **8** (1892); and the third in *Journal de math.*, (5), **5** (1899).

See also Bromwich, *Proc. London Math. Soc.*, (2), **1** (1903), 176.

EXAMPLES ON CHAPTER VI.

1. Prove that $\int_0^\infty e^{-ax^2}dx$ is uniformly convergent in $a \geqq a_0 > 0$, and that $\int_b^\infty \frac{dx}{x^2+a^2}$ is uniformly convergent in $a \geqq 0$, when $b > 0$.

2. Prove that $\int_0^\infty e^{-xy}\frac{\sin xy}{y}\,dx$ is uniformly convergent in $y \geqq y_0 > 0$, and that $\int_0^\infty e^{-ax}\frac{\sin xy}{y}\,dx$ is uniformly convergent in $y \geqq 0$, when $a > 0$.

3. Prove that $\int_0^\infty e^{-ax}x^{n-1}\cos x\,dx$ is uniformly convergent in the interval $a \geqq a_0 > 0$, when $n \geqq 1$, and in the interval $a \geqq 0$, when $0 < n < 1$.

4. Prove that $\int_0^\infty e^{-ax}x^{n-1}\sin x\,dx$ is uniformly convergent in the interval $a \geqq a_0 > 0$, when $n \geqq 0$, and in the interval $a \geqq 0$, when $-1 < n < 1$.

5. Using the fact that $\int_0^\infty \frac{\sin xy}{x}\,dx$ is uniformly convergent in

$$y \geqq y_0 > 0 \quad \text{and} \quad y \leqq -y_0 < 0,$$

show that $\int_0^\infty \frac{\sin xy \cos ax}{x}\,dx$ is uniformly convergent in any interval of y which does not include $y = \pm a$.

6. Show that (i) $\int_0^\infty (1-e^{-xy})\frac{\sin x}{x}\,dx$,

(ii) $\int_0^\infty e^{-xy}\frac{(\cos ax - \cos bx)}{x}\,dx, \quad b \neq a$,

are uniformly convergent for $y \geqq 0$.

7. Discuss the uniform convergence of the integrals:

(i) $\int_0^1 \frac{\tan^{-1} xy}{x\sqrt{(1-x^2)}}\,dx$. (ii) $\int_0^1 \frac{dx}{(1+x^2y^2)\sqrt{(1-x^2)}}$. (iii) $\int_0^1 x^y\,dx$.

(iv) $\int_0^1 \frac{x^y - 1}{\log x}\,dx$. (v) $\int_0^1 x^{y-1}(\log x)^n\,dx, \quad n > 0$.

8. Show that differentiation under the integral sign is allowable in the following integrals, and hence obtain the results that are given opposite each:

(i) $\int_0^\infty e^{-ax^2}\,dx = \frac{1}{2}\sqrt{\frac{\pi}{a}}$; $\quad \int_0^\infty x^{2n}e^{-ax^2}\,dx = \frac{\sqrt{\pi}}{2}\cdot\frac{1\,.\,3\ldots(2n-1)}{2^n a^{n+\frac{1}{2}}}$.

(ii) $\int_0^\infty \frac{dx}{x^2+a} = \frac{\pi}{2\sqrt{a}}, \ a > 0$; $\quad \int_0^\infty \frac{dx}{(x^2+a)^{n+1}} = \frac{\pi}{2}\cdot\frac{1\,.\,3\ldots(2n-1)}{2^n n!\,a^{n+\frac{1}{2}}}$.

(iii) $\int_0^1 x^n\,dx = \frac{1}{n+1}, \ n > -1$; $\quad \int_0^1 x^n(-\log x)^m\,dx = \frac{m!}{(n+1)^{m+1}}$.

(iv) $\int_0^\infty \frac{x^n}{1+x^2}\,dx = \frac{\pi}{2\cos\frac{n\pi}{2}}, \ 0 < n < 1$; $\quad \int_0^\infty \frac{x^n \log x}{1+x^2}\,dx = \frac{\pi^2}{4}\sec\frac{n\pi}{2}\tan\frac{n\pi}{2}$.

9. Establish the right to integrate under the integral sign in the following integrals:

(i) $\int_0^\infty e^{-ax}\,dx$; interval $a \geqq a_0 > 0$.

(ii) $\int_0^\infty e^{-ax}\cos bx\,dx$; interval $a \geqq a_0 > 0$, or any interval of b.

(iii) $\int_0^\infty e^{-ax}\sin bx\,dx$; interval $a \geqq a_0 > 0$, or any interval of b.

(iv) $\int_0^1 x^a\,dx$; interval $a \geqq a_0 > -1$.

10. Assuming that $\int_0^\infty e^{-ax}\sin bx\,dx = \dfrac{b}{a^2+b^2}$, $a > 0$, show that

$$\int_0^\infty \frac{e^{-fx}-e^{-gx}}{x}\sin bx\,dx = \tan^{-1}\frac{g}{b} - \tan^{-1}\frac{f}{b},\quad g > f > 0.$$

Deduce that (i) $\displaystyle\int_0^\infty \frac{1-e^{-gx}}{x}\sin bx\,dx = \tan^{-1}\frac{g}{b}$.

(ii) $\displaystyle\int_0^\infty \frac{\sin bx}{x}\,dx = \tfrac{1}{2}\pi,\ b > 0.$

11. Show that the integrals

$$\int_0^\infty e^{-ax}\cos bx\,dx = \frac{a}{a^2+b^2},\quad \int_0^\infty e^{-ax}\sin bx\,dx = \frac{b}{a^2+b^2},\quad a > 0,$$

can be differentiated under the integral sign, either with regard to a or b, and hence obtain the values of

$$\int_0^\infty xe^{-ax}\cos bx\,dx,\quad \int_0^\infty xe^{-ax}\sin bx\,dx,$$

$$\int_0^\infty x^2e^{-ax}\cos bx\,dx,\quad \int_0^\infty x^2e^{-ax}\sin bx\,dx.$$

12. Let $$f(y) = \int_0^\infty \frac{1-\cos xy}{x}e^{-x}\,dx.$$

Show that $f'(y) = \int_0^\infty \sin xy\, e^{-x}\,dx$ for all values of y, and deduce that

$$f(y) = \tfrac{1}{2}\log(1+y^2).$$

13. Let $$f(y) = \int_0^\infty e^{-x^2}\cos 2xy\,dx.$$

Show that $f'(y) = -2\int_0^\infty xe^{-x^2}\sin 2xy\,dx$ for all values of y.

On integrating by parts, it will be found that

$$f'(y) + 2y\,f(y) = 0.$$

From this result, show that $f(y) = \frac{1}{2}\sqrt{\pi}e^{-y^2}$, assuming that $\Gamma(\frac{1}{2}) = \sqrt{\pi}$.

Also show that $$\int_0^\infty dx\int_0^y e^{-x^2}\cos 2xy\,dy = \int_0^y f(y)\,dy,$$

and deduce that $$\int_0^\infty e^{-x^2}\frac{\sin 2xy}{x}\,dx = \sqrt{\pi}\int_0^y e^{-y^2}\,dy.$$

14. Let $\quad U=\int_0^\infty e^{-ax}x^{n-1}\cos bx\,dx, \quad V=\int_0^\infty e^{-ax}x^{n-1}\sin bx\,dx,$

where $a>0,\ n>0.$

Make the following substitutions :

$$a=r\cos\theta,\quad b=r\sin\theta,\quad \text{where}\ -\tfrac{1}{2}\pi<\theta<\tfrac{1}{2}\pi,$$
$$rx=y,\quad Ur^n=u,\quad Vr^n=v.$$

Then show that $\qquad \dfrac{du}{d\theta}=-nv,\quad \dfrac{dv}{d\theta}=nu.$

From these it follows that $\qquad \dfrac{d^2u}{d\theta^2}+n^2u=0.$

Deduce that $\qquad u=\Gamma(n)\cos n\theta,\quad v=\Gamma(n)\sin n\theta.$

Thus $\qquad U=\Gamma(n)\dfrac{\cos n\theta}{r^n},\quad V=\Gamma(n)\dfrac{\sin n\theta}{r^n}.$

Also show that, if $0<n<1$, $\lim\limits_{a\to 0} U=\int_0^\infty x^{n-1}\cos bx\,dx,$

$$\lim_{a\to 0} V=\int_0^\infty x^{n-1}\sin bx\,dx.$$

And deduce that

(i) $\displaystyle\int_0^\infty \frac{\cos bx}{x^{1-n}}\,dx=\frac{\Gamma(n)\cos\frac{n\pi}{2}}{b^n}$, (ii) $\displaystyle\int_0^\infty \frac{\sin bx}{x^{1-n}}\,dx=\frac{\Gamma(n)\sin\frac{n\pi}{2}}{b^n}$,

(iii) $\displaystyle\int_0^\infty \frac{\cos x}{\sqrt{x}}\,dx=\sqrt{\frac{\pi}{2}}=\int_0^\infty \frac{\sin x}{\sqrt{x}}\,dx.$

[Compare Gibson, *Treatise on the Calculus* (2nd ed., 1906), 471.]

15. Prove that

$$\int_0^\infty dx\int_b^\infty e^{-xy}\sin x\,dy=\int_b^\infty dy\int_0^\infty e^{-xy}\ \sin x\,dx,$$

where b is any positive number.

CHAPTER VII

FOURIER'S SERIES

90. Trigonometrical Series and Fourier's Series. We have already discussed some of the properties of infinite series whose terms are functions of x, confining our attention chiefly to those whose terms are continuous functions.

The trigonometrical series,

$$a_0 + (a_1 \cos x + b_1 \sin x) + (a_2 \cos 2x + b_2 \sin 2x) + \dots, \quad \dots\dots(1)$$

where a_0, a_1, b_1, etc., are constants, is a special type of such series.

Let $f(x)$ be given in the interval $(-\pi, \pi)$. If bounded, let it be integrable in this interval; if unbounded, let the infinite integral $\int_{-\pi}^{\pi} f(x)\,dx$ be absolutely convergent. Then

$$\int_{-\pi}^{\pi} f(x') \cos nx'\,dx' \quad \text{and} \quad \int_{-\pi}^{\pi} f(x') \sin nx'\,dx'$$

exist for all values of n. (§ 61, VI.)

The trigonometrical series (1) is said to be a *Fourier's Series*, when the coefficients a_0, a_1, b_1, etc., are given by

$$\left.\begin{aligned} a_0 &= \frac{1}{2\pi}\int_{-\pi}^{\pi} f(x')\,dx' \text{ and, when } n \geqq 1, \\ a_n &= \frac{1}{\pi}\int_{-\pi}^{\pi} f(x') \cos nx'\,dx', \quad b_n = \frac{1}{\pi}\int_{-\pi}^{\pi} f(x') \sin nx'\,dx'. \end{aligned}\right\} \quad \dots\dots(2)$$

These coefficients are called Fourier's Coefficients or Fourier's Constants for the integrable function $f(x)$; and the Fourier's Series is said to correspond to the function.

This nomenclature is used quite independently of any assumption as to the convergence of the series (1) when Fourier's Constants are substituted for a_0, a_1, b_1, etc.

The most important thing about Fourier's Series is that, when $f(x)$ satisfies very general conditions in the interval $(-\pi, \pi)$, the sum of this series is equal to $f(x)$, or in special cases to

$$\tfrac{1}{2}[f(x+0)+f(x-0)],$$

when x lies in this interval.

If we assume that the arbitrary function $f(x)$, given in the interval $(-\pi, \pi)$, can be expanded in a trigonometrical series of the form (1), and that the series may be integrated term by term after multiplying both sides by $\cos nx$ or $\sin nx$, we obtain these values for the coefficients.

For, multiply both sides of the equation

$$f(x)=a_0+(a_1 \cos x+b_1 \sin x)+(a_2 \cos 2x+b_2 \sin 2x)+\dots,$$
$$-\pi \leqq x \leqq \pi, \quad \dots\dots\dots(3)$$

by $\cos nx$, and integrate from $-\pi$ to π.

Then
$$\int_{-\pi}^{\pi} f(x) \cos nx\, dx=\pi a_n,$$

since
$$\int_{-\pi}^{\pi} \cos mx \cos nx\, dx=\int_{-\pi}^{\pi} \sin mx \cos nx\, dx=0,$$

when m, n are different integers, and

$$\int_{-\pi}^{\pi} \cos^2 nx\, dx=\pi.$$

Thus we have

$$a_n=\frac{1}{\pi}\int_{-\pi}^{\pi} f(x') \cos nx'\, dx', \text{ when } n \geqq 1.$$

And similarly,

$$b_n=\frac{1}{\pi}\int_{-\pi}^{\pi} f(x') \sin nx'\, dx',$$

$$a_0=\frac{1}{2\pi}\int_{-\pi}^{\pi} f(x')\, dx'.$$

Inserting these values in the series (3), the result may be written

$$f(x)=\frac{1}{2\pi}\int_{-\pi}^{\pi} f(x')\, dx'+\frac{1}{\pi}\sum_{1}^{\infty}\int_{-\pi}^{\pi} f(x') \cos n(x'-x)\, dx',$$
$$-\pi \leqq x \leqq \pi. \quad \dots\dots\dots(4)$$

This is the *Fourier's Series* for $f(x)$.

If the arbitrary function, given in $(-\pi, \pi)$, is an even function—

in other words, if $f(x)=f(-x)$, when $0<x<\pi$—the Fourier's Series becomes the Cosine Series :

$$f(x)=\frac{1}{\pi}\int_0^\pi f(x')\,dx'+\frac{2}{\pi}\sum_1^\infty \cos nx\int_0^\pi f(x')\cos nx'\,dx',$$
$$0\leqq x\leqq \pi. \quad \text{.........(5)}$$

Again, if it is an odd function—*i.e.* if $f(x)=-f(-x)$, when $0<x<\pi$—the Fourier's Series becomes the Sine Series :

$$f(x)=\frac{2}{\pi}\sum_1^\infty \sin nx\int_0^\pi f(x')\sin nx'\,dx',\quad 0\leqq x\leqq \pi. \quad \text{......(6)}$$

The expansions in (5) and (6) could have been obtained in the same way as the expansion in (3) by assuming a series in cosines only, or a series in sines only, and multiplying by $\cos nx$ or $\sin nx$, as the case may be, integrating now from 0 to π.

Further, if we take the interval $(-l, l)$ instead of $(-\pi, \pi)$, we find the following expansions, corresponding to (4), (5) and (6) :

$$f(x)=\frac{1}{2l}\int_{-l}^l f(x')\,dx'+\frac{1}{l}\sum_1^\infty\int_{-l}^l f(x')\cos\frac{n\pi}{l}(x'-x)\,dx',$$
$$-l\leqq x\leqq l. \quad \text{......(7)}$$

$$f(x)=\frac{1}{l}\int_0^l f(x')\,dx'+\frac{2}{l}\sum_1^\infty \cos\frac{n\pi}{l}x\int_0^l f(x')\cos\frac{n\pi}{l}x'\,dx',$$
$$0\leqq x\leqq l. \quad \text{......(8)}$$

$$f(x)=\frac{2}{l}\sum_1^\infty \sin\frac{n\pi}{l}x\int_0^l f(x')\sin\frac{n\pi}{l}x'\,dx',\quad 0\leqq x\leqq l. \quad \text{......(9)}$$

However, this method does not give a rigorous proof of these very important expansions for the following reasons :

(i) We have assumed the possibility of the expansion of the function in the series.

(ii) We have integrated the series term by term.

This would have been allowable if the convergence of the series were uniform, since multiplying right through by $\cos nx$ or $\sin nx$ does not affect the uniformity ; but this property has not been proved, and indeed is not generally applicable to the whole interval in these expansions.

(iii) The discussion does not give us any information as to the behaviour of the series at points of discontinuity, if such arise, nor does it give any suggestion as to the conditions to which $f(x)$ must be subject if the expansion is to hold.

Another method of obtaining the coefficients, due to Lagrange,* may be illustrated by the case of the Sine Series.

Consider the curve

$$y = a_1 \sin x + a_2 \sin 2x + \dots + a_{n-1} \sin (n-1)x.$$

We can obtain the values of the coefficients

$$a_1, \quad a_2, \dots a_{n-1},$$

so that this curve shall pass through the points of the curve

$$y = f(x),$$

at which the abscissae are

$$\frac{\pi}{n}, \quad \frac{2\pi}{n}, \dots (n-1)\frac{\pi}{n}.$$

In this way we find $a_1, a_2, \dots a_{n-1}$ as functions of n. Proceeding to the limit as $n \to \infty$, we have the values of the coefficients in the infinite series

$$f(x) = a_1 \sin x + a_2 \sin 2x + \dots .$$

But this passage from a finite number of equations to an infinite number requires more complete examination before the results can be accepted.

The most satisfactory method of discussing the possibility of expressing an arbitrary function $f(x)$, given in the interval $(-\pi, \pi)$, by the corresponding Fourier's Series, is to take the series

$$a_0 + (a_1 \cos x + b_1 \sin x) + (a_2 \cos 2x + b_2 \cos 2x) + \dots ,$$

where the constants have the values given in (2), and sum the terms up to $(a_n \cos nx + b_n \sin nx)$; then to find the limit of this sum, if it has a limit, as $n \to \infty$.

In this way we shall show that, when $f(x)$ satisfies very general conditions, the Fourier's Series for $f(x)$ converges to $f(x)$ at every point in $(-\pi, \pi)$, where $f(x)$ is continuous; that it converges to $\frac{1}{2}[f(x+0) + f(x-0)]$ at every point of ordinary discontinuity; also that it converges to $\frac{1}{2}[f(-\pi+0) + f(\pi-0)]$ at $x = \pm\pi$, when these limits exist.

Since the series is periodic in x with period 2π, when the sum is known in $(-\pi, \pi)$, it is also known for every value of x.

*Lagrange, *Œuvres*, 1 (Paris, 1867), 553: Byerly, *Fourier's Series*, etc. (1893), 30.

If it is more convenient to take the interval in which $f(x)$ is defined as $(0, 2\pi)$, the values of the coefficients in the corresponding expansion would be

$$a_0=\frac{1}{2\pi}\int_0^{2\pi} f(x')\,dx',$$

$$a_n=\frac{1}{\pi}\int_0^{2\pi} f(x')\cos nx'\,dx', \quad b_n=\frac{1}{\pi}\int_0^{2\pi} f(x')\sin nx'\,dx', \quad n \geqq 1.$$

It need hardly be added that the function $f(x)$ can have different analytical expressions in different parts of the given interval. And in particular we can obtain any number of such expansions which will hold in the interval $(0, \pi)$, since we can give $f(x)$ any value we please, subject to the general conditions we shall establish, in the interval $(-\pi, 0)$.

The following discussion of the possibility of the expansion of an arbitrary function in the corresponding Fourier's Series depends upon a modified form of the integrals by means of which Dirichlet* gave the first rigorous proof that, for a large class of functions, the Fourier's Series converges to $f(x)$. With the help of the Second Theorem of Mean Value the sum of the series can be deduced at once from these integrals, which we shall call Dirichlet's Integrals.

91. Dirichlet's Integrals (First Form).

$$\lim_{\mu\to\infty}\int_0^a f(x)\frac{\sin \mu x}{x}\,dx=\frac{\pi}{2}f(+0), \quad \lim_{\mu\to\infty}\int_a^b f(x)\frac{\sin \mu x}{x}\,dx=0,$$

where $0<a<b$.

When we apply the Second Theorem of Mean Value to the integral

$$\int_{b'}^{c'}\frac{\sin x}{x}\,dx, \quad 0<b'<c',$$

we see that $\displaystyle\int_{b'}^{c'}\frac{\sin x}{x}\,dx=\frac{1}{b'}\int_{b'}^{\xi}\sin x\,dx+\frac{1}{c'}\int_{\xi}^{c'}\sin x\,dx,$

where $b' \leqq \xi \leqq c'$.

Thus
$$\left|\int_{b'}^{c'}\frac{\sin x}{x}\,dx\right| \leqq 2\left(\frac{1}{b'}+\frac{1}{c'}\right)$$
$$<\frac{4}{b'}.$$

Journal für Math., 4 (1829), 157, and Dove's *Repertorium der Physik*, 1 (1837), 152.

It follows that the integral

$$\int_0^\infty \frac{\sin x}{x}\,dx$$

is convergent. Its value has been found in § 88 to be $\frac{1}{2}\pi$.*

The integrals $\displaystyle\int_0^a \frac{\sin \mu x}{x}\,dx$ and $\displaystyle\int_a^b \frac{\sin \mu x}{x}\,dx$

can be transformed, by putting $\mu x = x'$, into

$$\int_0^{\mu a} \frac{\sin x}{x}\,dx, \quad \int_{\mu a}^{\mu b} \frac{\sin x}{x}\,dx,$$

respectively.

It follows that

$$\lim_{\mu\to\infty} \int_0^a \frac{\sin \mu x}{x}\,dx = \int_0^\infty \frac{\sin x}{x}\,dx = \tfrac{1}{2}\pi, \quad 0<a,$$

and
$$\lim_{\mu\to\infty} \int_a^b \frac{\sin \mu x}{x}\,dx = \lim_{\mu\to\infty} \int_{\mu a}^{\mu b} \frac{\sin x}{x}\,dx = 0, \quad 0<a<b.$$

These results are special cases of the theorem that, *when* $f(x)$ *satisfies certain conditions, given below,*

$$\lim_{\mu\to\infty} \int_0^a f(x)\frac{\sin \mu x}{x}\,dx = \frac{\pi}{2} f(+0), \quad 0<a,$$

$$\lim_{\mu\to\infty} \int_a^b f(x)\frac{\sin \mu x}{x}\,dx = 0, \quad 0<a<b.$$

In the discussion of this theorem we shall, first of all, assume that $f(x)$ satisfies the conditions we have imposed upon $\phi(x)$ in our notation for the Second Theorem of Mean Value ; viz., it is to be bounded and monotonic (and therefore integrable) in the interval with which we are concerned.

It is clear that $\dfrac{\sin \mu x}{x}$ satisfies the conditions imposed upon $\psi(x)$ in the theorem as proved in § 50. 1. It is bounded and integrable, and does not change sign more than a finite number of times in the interval.

We shall remove some of the restrictions placed upon $f(x)$ later.

I. Consider the integral

$$\int_a^b f(x)\frac{\sin \mu x}{x}\,dx, \quad 0<a<b.$$

---

*See also the footnote on p. 202.

From the Second Theorem of Mean Value

$$\int_a^b f(x)\frac{\sin \mu x}{x}\,dx = f(a+0)\int_a^\xi \frac{\sin \mu x}{x}\,dx + f(b-0)\int_\xi^b \frac{\sin \mu x}{x}\,dx,$$

where ξ is some definite value of x in $a \leqq x \leqq b$.

Since $f(x)$ is monotonic in $a \leqq x \leqq b$, the limits $f(a+0)$ and $f(b-0)$ exist.

And we have seen that the limits of the integrals on the right-hand are zero when $\mu \to \infty$.

It follows that, under the conditions named above,

$$\lim_{\mu\to\infty}\int_a^b f(x)\frac{\sin \mu x}{x}\,dx = 0, \text{ when } 0 < a < b.$$

II. Consider the integral

$$\int_0^a f(x)\frac{\sin \mu x}{x}\,dx, \quad 0 < a.$$

Put $f(x) = \phi(x) + f(+0)$. The limit $f(+0)$ exists, since $f(x)$ is monotonic in $0 \leqq x \leqq a$.

Then $\phi(x)$ is monotonic and $\phi(+0) = 0$.

Also

$$\int_0^a f(x)\,\frac{\sin \mu x}{x}\,dx = f(+0)\int_0^a \frac{\sin \mu x}{x}\,dx + \int_0^a \phi(x)\,\frac{\sin \mu x}{x}\,dx.$$

As $\mu \to \infty$ the first integral on the right-hand has the limit $\frac{1}{2}\pi$. We shall now show that the second integral has the limit zero.

To prove this, it is sufficient to show that, to the arbitrary positive number ϵ, there corresponds a positive number ν such that

$$\left|\int_0^a \phi(x)\,\frac{\sin \mu x}{x}\,dx\right| < \epsilon \text{ when } \mu \geqq \nu.$$

Let us break up the interval $(0, a)$ into two parts, $(0, \alpha)$ and (α, a), where α is chosen so that

$$|\phi(\alpha - 0)| < \epsilon/2\pi.$$

We can do this, since we are given that $\phi(+0) = 0$, and thus there is a positive number α such that

$$|\phi(x)| < \epsilon/2\pi, \text{ when } 0 < x \leqq \alpha.$$

Then, by the Second Theorem of Mean Value,

$$\int_0^\alpha \phi(x)\,\frac{\sin \mu x}{x}\,dx = \phi(\alpha - 0)\int_\xi^\alpha \frac{\sin \mu x}{x}\,dx,$$

since $\phi(+0) = 0$, ξ being some definite value of x in $0 \leqq x \leqq \alpha$.

But, in the curve

$$y = \frac{\sin x}{x}, \quad x \geqq 0,$$

the successive waves have the same breadth and diminishing amplitude, and the area between 0 and π is greater than that between π and 2π in absolute value: that between π and 2π is greater than that between 2π and 3π, and so on; since $|\sin x|$ goes through the same set of values in each case, and $1/x$ diminishes as x increases.

Thus
$$\int_0^x \frac{\sin x}{x}\,dx \leqq \int_0^\pi \frac{\sin x}{x}\,dx < \pi,$$
whatever positive value x may have.

Also
$$\int_p^q \frac{\sin x}{x}\,dx = \int_0^q \frac{\sin x}{x}\,dx - \int_0^p \frac{\sin x}{x}\,dx,$$
and each of the integrals on the right-hand is positive and less than π for $0 < p < q$.

Therefore
$$\left|\int_p^q \frac{\sin x}{x}\,dx\right| < \pi, \text{ when } 0 \leqq p < q.$$

It follows that
$$\left|\int_0^\alpha \phi(x)\frac{\sin \mu x}{x}\,dx\right| < \tfrac{1}{2}\frac{\epsilon}{\pi} \times \pi$$
$$< \tfrac{1}{2}\epsilon,$$
and this is independent of μ.

But we have seen in (I) that
$$\lim_{\mu\to\infty} \int_\alpha^a \phi(x)\frac{\sin \mu x}{x}\,dx = 0, \quad 0 < \alpha < a.$$
Therefore, to the arbitrary positive number $\frac{1}{2}\epsilon$, there corresponds a positive number ν such that
$$\left|\int_\alpha^a \phi(x)\frac{\sin \mu x}{x}\,dx\right| < \tfrac{1}{2}\epsilon, \text{ when } \mu \geqq \nu.$$

Also
$$\left|\int_0^a \phi(x)\frac{\sin \mu x}{x}\,dx\right| \leqq \left|\int_0^\alpha \phi(x)\frac{\sin \mu x}{x}\,dx\right| + \left|\int_\alpha^a \phi(x)\frac{\sin \mu x}{x}\,dx\right|.$$

Therefore
$$\left|\int_0^a \phi(x)\frac{\sin \mu x}{x}\,dx\right| < \tfrac{1}{2}\epsilon + \tfrac{1}{2}\epsilon$$
$$< \epsilon, \text{ when } \mu \geqq \nu.$$

Thus
$$\lim_{\mu\to\infty} \int_0^a \phi(x)\frac{\sin \mu x}{x}\,dx = 0.$$

And, finally, under the conditions named above,

$$\lim_{\mu\to\infty}\int_0^a f(x)\frac{\sin\mu x}{x}\,dx=\frac{\pi}{2}f(+0).$$

92. In the preceding section we have assumed that $f(x)$ is bounded and monotonic in the intervals $(0, a)$ and (a, b). We shall now show that these restrictions may be somewhat relaxed.

I. *Dirichlet's Integrals still hold when $f(x)$ is bounded, and the interval of integration can be broken up into a finite number of open partial intervals, in each of which $f(x)$ is monotonic.**

This follows at once from the fact that under these conditions we may write

$$f(x)=F(x)-G(x),$$

where $F(x)$ and $G(x)$ are positive, bounded, and monotonic increasing in the interval with which we are concerned [cf. § 36. 1 or § 36. 2].

This result can be obtained, as follows, without the use of the theorems of § 36. 1 or § 36. 2.

Let the interval $(0, a)$ be broken up into the n open intervals,

$$(0, a_1),\quad (a_1, a_2),\ \ldots,\ (a_{n-1}, a),$$

in each of which $f(x)$ is bounded and monotonic.

Then, writing $a_0=0$ and $a_n=a$, we have

$$\int_0^a f(x)\frac{\sin\mu x}{x}\,dx=\sum_{r=1}^{n}\int_{a_{r-1}}^{a_r} f(x)\frac{\sin\mu x}{x}\,dx.$$

The first integral in this sum has the limit $\frac{\pi}{2}f(+0)$, and the others have the limit zero when $\mu\to\infty$.

It follows that, under the given conditions,

$$\lim_{\mu\to\infty}\int_0^a f(x)\frac{\sin\mu x}{x}\,dx=\frac{\pi}{2}f(+0),\quad 0<a.$$

The proof that, under the same conditions,

$$\lim_{\mu\to\infty}\int_a^b f(x)\frac{\sin\mu x}{x}\,dx=0,\quad 0<a<b,$$

is practically contained in the above.

It will be seen that we have used the condition that the number of partial intervals is finite, as we have relied upon the theorem that the limit of a sum is equal to the sum of the limits.

II. *The integrals still hold for certain cases where a finite number of points of infinite discontinuity of $f(x)$ (as defined in § 33) occur in the interval of integration.*

---

*They also hold when $f(x)$ is of bounded variation (§ 36. 2) in the interval, since if $f(x)$ is of bounded variation, it can be replaced by the difference of two positive, bounded and monotonic increasing functions.

*We shall suppose that, when arbitrarily small neighbourhoods of these points of infinite discontinuity are excluded, the remainder of the interval of integration can be broken up into a finite number of open partial intervals, in each of which $f(x)$ is bounded and monotonic.**

Further, we shall assume that the infinite integral $\int f(x)\,dx$ *is absolutely convergent in the interval of integration, and that* $x=0$ *is not a point of infinite discontinuity.*

We may take first the case when an infinite discontinuity occurs at the upper limit b of the integral $\int_a^b f(x)\frac{\sin\mu x}{x}\,dx$, and only there.

Since we are given that $\int_a^b f(x)\,dx$ is absolutely convergent, we know that $\int_a^b f(x)\frac{\sin\mu x}{x}\,dx$ also converges, for $\left|\frac{\sin\mu x}{x}\right| \leqq \frac{1}{a}$ in (a, b). And this convergence is uniform.

To the arbitrary positive number ϵ there corresponds a positive number η, which we take less than $(b-a)$, such that

$$\left|\int_{b-\xi}^b f(x)\frac{\sin\mu x}{x}\,dx\right| < \tfrac{1}{2}\epsilon, \text{ when } 0<\xi\leqq\eta, \quad \text{.........(1)}$$

and the same η serves for all values of μ.

But

$$\int_a^b f(x)\frac{\sin\mu x}{x}\,dx = \int_a^{b-\eta} f(x)\frac{\sin\mu x}{x}\,dx + \int_{b-\eta}^b f(x)\frac{\sin\mu x}{x}\,dx. \quad (2)$$

And, by (I) above,

$$\lim_{\mu\to\infty}\int_a^{b-\eta} f(x)\frac{\sin\mu x}{x}\,dx = 0.$$

It follows that there is a positive number ν such that

$$\left|\int_a^{b-\eta} f(x)\frac{\sin\mu x}{x}\,dx\right| < \tfrac{1}{2}\epsilon, \text{ when } \mu\geqq\nu. \quad \text{..............(3)}$$

From (1), (2) and (3), we have at once

$$\left|\int_a^b f(x)\frac{\sin\mu x}{x}\,dx\right| < \tfrac{1}{2}\epsilon + \tfrac{1}{2}\epsilon$$

$$< \epsilon, \text{ when } \mu\geqq\nu.$$

Thus we have shown that, with the conditions described above,

$$\lim_{\mu\to\infty}\int_a^b f(x)\frac{\sin\mu x}{x}\,dx = 0, \quad 0<a<b.$$

---

*The integrals still hold when the function is of bounded variation in the remainder of the interval of integration.

A similar argument applies to the case when an infinite discontinuity occurs at the lower limit a of the integral, and only there.

When there is an infinite discontinuity at a and at b, and only there, the result follows from these two, since

$$\int_a^b f(x)\frac{\sin\mu x}{x}\,dx=\int_a^c f(x)\frac{\sin\mu x}{x}\,dx+\int_c^b f(x)\frac{\sin\mu x}{x}\,dx,\quad a<c<b.$$

When an infinite discontinuity occurs between a and b we proceed in the same way; and, as we have assumed that the number of points of infinite discontinuity is finite, we can break up the given interval into a definite number of partial intervals, to which we can apply the results just obtained.

Thus, *under the conditions stated above in (II)*,

$$\lim_{\mu\to\infty}\int_a^b f(x)\frac{\sin\mu x}{x}\,dx=0,\ \text{when}\ 0<a<b.$$

Further, we have assumed that $x=0$ is not a point of infinite discontinuity of $f(x)$. Thus the interval $(0, a)$ can be broken up into two intervals, $(0, \alpha)$ and (α, a), where $f(x)$ is bounded in $(0, \alpha)$, and satisfies the conditions given in (I) of this section in $(0, \alpha)$.

It follows that

$$\lim_{\mu\to\infty}\int_0^\alpha f(x)\frac{\sin\mu x}{x}\,dx=\frac{\pi}{2}f(+0),$$

and we have just shown that

$$\lim_{\mu\to\infty}\int_\alpha^a f(x)\frac{\sin\mu x}{x}\,dx=0.$$

Therefore, *under the conditions stated above in (II)*,

$$\lim_{\mu\to\infty}\int_0^a f(x)\frac{\sin\mu x}{x}\,dx=\frac{\pi}{2}f(+0).$$

93. Dirichlet's Conditions. The results which we have obtained in §§ 91, 92 can be conveniently expressed in terms of what we shall call Dirichlet's Conditions.*

* If the functions of bounded variation of § 36. 2 are included in the class of functions available for discussion, $f(x)$ may be said to satisfy Dirichlet's Conditions (i) when it is of bounded variation in the whole interval; or (ii) when it has a finite number of points of infinite discontinuity in the interval, and it is of bounded variation in the remainder of that interval, when the arbitrarily small neighbourhoods of these points have been excluded; provided that the infinite integral $\int_a^b f(x)\,dx$ be absolutely convergent.

A function $f(x)$ will be said to satisfy Dirichlet's Conditions* in an interval (a, b), in which it is defined, when it is subject to one of the two following conditions:

(i) *$f(x)$ is bounded in (a, b), and the interval can be broken up into a finite number of open partial intervals, in each of which $f(x)$ is monotonic;*

(ii) *$f(x)$ has a finite number of points of infinite discontinuity in the interval. When arbitrarily small neighbourhoods of these points are excluded, $f(x)$ is bounded in the remainder of the interval, and this can be broken up into a finite number of open partial intervals, in each of which $f(x)$ is monotonic. Further, the infinite integral $\int_a^b f(x)\,dx$ is to be absolutely convergent.*

We may now say that:

When $f(x)$ satisfies Dirichlet's Conditions in the intervals $(0, a)$ and (a, b) respectively, where $0<a<b$, and $f(+0)$ exists, then

$$\lim_{\mu\to\infty}\int_0^a f(x)\frac{\sin\mu x}{x}\,dx=\frac{\pi}{2}f(+0),$$

and
$$\lim_{\mu\to\infty}\int_a^b f(x)\frac{\sin\mu x}{x}\,dx=0.$$

It follows from the properties of monotonic functions (cf. § 34) that except at the points, if any, where $f(x)$ becomes infinite, or oscillates infinitely, a function which satisfies Dirichlet's Conditions, as defined above, can only have *ordinary* discontinuities.† But we have not assumed ‡ that the function $f(x)$ shall have only a finite number of ordinary discontinuities. A bounded function which is monotonic in an open interval can have an infinite number of ordinary discontinuities in that interval [cf. § 34].

Perhaps it should be added that the conditions which Dirichlet

*But see footnote on p. 225.

†The conditions in the text can be further extended so as to include a finite number of points of oscillatory discontinuity in the neighbourhood of which the function is bounded [*e.g.* $\sin 1/(x-c)$ at $x=c$], or of continuity, with an infinite number of maxima and minima in their neighbourhood [*e.g.* $(x-c)\sin 1/(x-c)$ at $x=c$].

This generalisation would also apply to the sections in which Dirichlet's Conditions are employed.

‡The same remark applies to the case when $f(x)$ is a function of bounded variation.

himself imposed upon the function $f(x)$ in a given interval (a, b) were not so general as those to which we have given the name *Dirichlet's Conditions.* He contemplated at first only bounded functions, continuous, except at a finite number of ordinary discontinuities, and with only a finite number of maxima and minima. Later he extended his results to the case in which there are a finite number of points of infinite discontinuity in the interval, provided that the infinite integral $\int_a^b f(x)\,dx$ is absolutely convergent.

94. Dirichlet's Integrals (Second Form).

$$\lim_{\mu\to\infty}\int_0^a f(x)\frac{\sin\mu x}{\sin x}\,dx=\frac{\pi}{2}f(+0),\quad \lim_{\mu\to\infty}\int_a^b f(x)\frac{\sin\mu x}{\sin x}\,dx=0,$$

where $0<a<b<\pi$.

In the discussion of Fourier's Series the integrals which we shall meet are slightly different from Dirichlet's Integrals, the properties of which we have just established.

The second type of integral—and this is the one which Dirichlet himself used in his classical treatment of Fourier's Series—is

$$\int_0^a f(x)\frac{\sin\mu x}{\sin x}\,dx,\quad \int_a^b f(x)\frac{\sin\mu x}{\sin x}\,dx,$$

where $0<a<b<\pi$.

We shall now prove that :

When $f(x)$ satisfies Dirichlet's Conditions (as defined in § 93) in the intervals $(0, a)$ and (a, b) respectively, where $0<a<b<\pi$, and $f(+0)$ exists, then

$$\lim_{\mu\to\infty}\int_0^a f(x)\frac{\sin\mu x}{\sin x}\,dx=\frac{\pi}{2}f(+0),$$

and

$$\lim_{\mu\to\infty}\int_a^b f(x)\frac{\sin\mu x}{\sin x}\,dx=0.$$

Let us suppose that $f(x)$ satisfies the first* of the two conditions given in § 93 as Dirichlet's Conditions :

$f(x)$ is bounded, and the intervals $(0, a)$ and (a, b) can be broken up into a finite number of open partial intervals, in each of which $f(x)$ is monotonic.

---

*Or, alternatively, that $f(x)$ is of bounded variation in the intervals $(0, a)$ and (a, b).

Then, by § 36. 1 or § 36. 2 we can write

$$f(x)=F(x)-G(x),$$

where $F(x)$, $G(x)$ are positive, bounded and monotonic increasing in the interval with which we are concerned.

Thus $$f(x)\frac{\sin \mu x}{\sin x}=\left[F(x)\frac{x}{\sin x}-G(x)\frac{x}{\sin x}\right]\frac{\sin \mu x}{x}.$$

But $x/\sin x$ is bounded, positive and monotonic increasing in $(0, a)$ or (a, b), when $0<a<b<\pi$.*

Therefore $F(x)\dfrac{x}{\sin x}$ and $G(x)\dfrac{x}{\sin x}$ are both bounded, positive and monotonic increasing in the interval $(0, a)$ or (a, b), as the case may be, provided that $0<a<b<\pi$.

It follows from § 91 that

$$\lim_{\mu\to\infty}\int_0^a f(x)\frac{\sin \mu x}{\sin x}dx=\frac{\pi}{2}(F(+0)-G(+0))$$

$$=\frac{\pi}{2}f(+0),$$

and $$\lim_{\mu\to\infty}\int_a^b f(x)\frac{\sin \mu x}{\sin x}dx=0, \text{ when } 0<a<b<\pi.$$

Next, let $f(x)$ satisfy the second † of the conditions given in § 93, and let $f(+0)$ exist.

We can prove, just as in § 92, II, that

$$\lim_{\mu\to\infty}\int_a^b f(x)\frac{\sin \mu x}{\sin x}dx=0, \text{ when } 0<a<b<\pi.$$

For we are given that $\int_a^b f(x)\,dx$ is absolutely convergent, and we know that $x/\sin x$ is bounded and integrable in (a, b).

It follows that $$\int_a^b f(x)\frac{x}{\sin x}dx$$

is absolutely convergent; and the preceding proof [§ 92, II] applies to the neighbourhood of the point, or points, of infinite discontinuity, when we write $f(x)\dfrac{x}{\sin x}$ in place of $f(x)$.

---

* We assign to $x/\sin x$ the value 1 at $x=0$.

† Or, alternatively, that $f(x)$ is of bounded variation in the remainder of the interval, when the arbitrarily small neighbourhoods of the points of infinite discontinuity are excluded.

Also, for the case $\lim_{\mu\to\infty}\int_0^a f(x)\frac{\sin\mu x}{\sin x}\,dx,$

we need only, as before, break up the interval $(0, a)$ into $(0, \alpha)$ and (α, a), where $f(x)$ is bounded in $(0, \alpha)$ and from the results we have already obtained in this section the limit is found as stated.

If it is desired to obtain the second form of Dirichlet's Integrals for the cases stated below without the use of the theorems of § 36. 1 or § 36. 2 the reader may proceed as follows :

(i) Let $f(x)$ be positive, bounded and monotonic increasing in $(0, a)$ and (a, b).

Then $\frac{x}{\sin x}$ is so also, and $\phi(x)=f(x)\frac{x}{\sin x}$ is so also.

But, by § 91,

$$\lim_{\mu\to\infty}\int_0^a \phi(x)\frac{\sin\mu x}{x}\,dx=\frac{\pi}{2}\phi(+0)=\frac{\pi}{2}f(+0).$$

Therefore

$$\lim_{\mu\to\infty}\int_0^a f(x)\frac{\sin\mu x}{\sin x}\,dx=\frac{\pi}{2}f(+0).$$

Also

$$\int_a^b f(x)\frac{\sin\mu x}{\sin x}\,dx=\int_0^b f(x)\frac{\sin\mu x}{\sin x}\,dx-\int_0^a f(x)\frac{\sin\mu x}{\sin x}\,dx.$$

Therefore $$\lim_{\mu\to\infty}\int_a^b f(x)\frac{\sin\mu x}{\sin x}\,dx=0.$$

(ii) Let $f(x)$ be positive, bounded and monotonic decreasing.

Then for some value of c the function $c-f(x)$ is positive, bounded and monotonic increasing.

Also

$$\int_0^a [c-f(x)]\frac{\sin\mu x}{\sin x}\,dx=c\int_0^a\frac{\sin\mu x}{\sin x}\,dx-\int_0^a f(x)\frac{\sin\mu x}{\sin x}\,dx.$$

Using (i), the result follows.

(iii) If $f(x)$ is bounded and monotonic increasing, but not positive all the time, by adding a constant we can make it positive, and proceed as in (ii) ; and a similar remark applies to the case of the monotonic decreasing function.

(iv) When $f(x)$ is bounded and the interval can be broken up into a finite number of open partial intervals in which it is monotonic, the result follows from (i)-(iii).

(v) And if $f(x)$ has a finite number of points of infinite discontinuity, as stated in the second of Dirichlet's Conditions, so far as these points are concerned the proof is similar to that given above.

95. Proof of the Convergence of Fourier's Series. In the opening sections of this chapter we have given the usual elementary, but quite incomplete, argument, by means of which the coefficients in the expansion

$$f(x)=a_0+(a_1 \cos x+b_1 \sin x)+(a_2 \cos 2x+b_2 \sin 2x)+\dots$$
$$-\pi \leqq x \leqq \pi$$

are obtained.

We now return to this question, which we approach in quite a different way.

We take the Fourier's Series

$$a_0+(a_1 \cos x+b_1 \sin x)+(a_2 \cos 2x+b_2 \sin 2x)+\dots,$$

$$\left.\begin{aligned}&\text{where } a_0=\frac{1}{2\pi}\int_{-\pi}^{\pi} f(x')\,dx' \text{ and, when } n\geqq 1,\\ &a_n=\frac{1}{\pi}\int_{-\pi}^{\pi} f(x')\cos nx'\,dx',\quad b_n=\frac{1}{\pi}\int_{-\pi}^{\pi} f(x')\sin nx'\,dx'.\end{aligned}\right\}$$

We find the sum of the terms of this series up to $\cos nx$ and $\sin nx$, and we then examine whether this sum has a limit as $n\to\infty$.

We shall prove that, *when $f(x)$ is given in the interval $(-\pi, \pi)$, and satisfies Dirichlet's Conditions in that interval,* this sum has a limit as $n\to\infty$. It is equal to $f(x)$ at any point in $-\pi<x<\pi$, where $f(x)$ is continuous ; and to*

$$\tfrac{1}{2}[f(x+0)+f(x-0)],$$

when there is an ordinary discontinuity at the point ; and to $\frac{1}{2}[f(-\pi+0)+f(\pi-0)]$ at $x=\pm\pi$, when the limits $f(\pi-0)$ and $f(-\pi+0)$ exist.

Let

$$s_n(x)=a_0+(a_1 \cos x+b_1 \sin x)+\dots+(a_n \cos nx+b_n \sin nx),$$

where a_0, a_1, b_1, etc., have the values given above.

Then we find, without difficulty, that

$$\begin{aligned}s_n(x)&=\frac{1}{2\pi}\int_{-\pi}^{\pi} f(x')[1+2\cos(x'-x)+\dots+2\cos n(x'-x)]\,dx'\\ &=\frac{1}{2\pi}\int_{-\pi}^{\pi} f(x')\frac{\sin\frac{1}{2}(2n+1)(x'-x)}{\sin\frac{1}{2}(x'-x)}\,dx'\end{aligned}$$

---

* In this and later sections, when reference is made to Dirichlet's Conditions, it will be understood that these can be modified, if desired, by the introduction of the functions of bounded variation as explained in the footnote on p. 225.

$$=\frac{1}{2\pi}\int_{-\pi}^{x} f(x')\frac{\sin\frac{1}{2}(2n+1)(x'-x)}{\sin\frac{1}{2}(x'-x)}\,dx'$$

$$+\frac{1}{2\pi}\int_{\pi}^{x} f(x')\frac{\sin\frac{1}{2}(2n+1)(x'-x)}{\sin\frac{1}{2}(x'-x)}\,dx'.$$

Thus

$$s_n(x)=\frac{1}{\pi}\int_0^{\frac{1}{2}\pi+\frac{1}{2}x} f(x-2a)\frac{\sin(2n+1)a}{\sin a}\,da$$

$$+\frac{1}{\pi}\int_0^{\frac{1}{2}\pi-\frac{1}{2}x} f(x+2a)\frac{\sin(2n+1)a}{\sin a}\,da, \quad \text{............(1)}$$

on changing the variable by the substitutions $x'-x=\mp 2a$.

If $-\pi<x<\pi$ and $f(x)$ satisfies Dirichlet's Conditions in the interval $(-\pi, \pi)$, $f(x\mp 2a)$ considered as functions of a in the integrals of (1) satisfy Dirichlet's Conditions in the intervals $(0, \frac{1}{2}\pi+\frac{1}{2}x)$ and $(0, \frac{1}{2}\pi-\frac{1}{2}x)$ respectively, and these functions of a have limits as $a\to 0$, provided that at the point x with which we are concerned $f(x+0)$ and $f(x-0)$ exist.

It follows from § 94 that, when x lies between $-\pi$ and π and $f(x-0)$ and $f(x+0)$ both exist,

$$\lim_{n\to\infty} s_n(x)=\frac{1}{\pi}\left[\frac{\pi}{2}f(x-0)+\frac{\pi}{2}f(x+0)\right]$$

$$=\tfrac{1}{2}[f(x-0)+f(x+0)],$$

giving the value $f(x)$ at a point where $f(x)$ is continuous.

We have yet to examine the cases $x=\pm\pi$.

In finding the sum of the series for $x=\pi$, we must insert this value for x in $s_n(x)$ before proceeding to the limit.

Thus $$s_n(\pi)=\frac{1}{\pi}\int_0^{\pi} f(\pi-2a)\frac{\sin(2n+1)a}{\sin a}\,da,$$

since the second integral in (1) is zero.

It follows that

$$s_n(\pi)=\frac{1}{\pi}\int_0^{\pi-\xi} f(\pi-2a)\frac{\sin(2n+1)a}{\sin a}\,da$$

$$+\frac{1}{\pi}\int_{\pi-\xi}^{\pi} f(\pi-2a)\frac{\sin(2n+1)a}{\sin a}\,da$$

$$=\frac{1}{\pi}\int_0^{\pi-\xi} f(\pi-2a)\frac{\sin(2n+1)a}{\sin a}\,da$$

$$+\frac{1}{\pi}\int_0^{\xi} f(-\pi+2a)\frac{\sin(2n+1)a}{\sin a}\,da,$$

where ξ is any number between 0 and π.

We can apply the theorem of § 94 to these integrals, if $f(x)$ satisfies Dirichlet's Conditions in $(-\pi, \pi)$, and the limits $f(\pi-0)$, $f(-\pi+0)$ exist.

Thus we have $\lim\limits_{n\to\infty} s_n(\pi) = \tfrac{1}{2}[f(-\pi+0)+f(\pi-0)]$.

A similar discussion gives the same value for the sum at $x=-\pi$, which is otherwise obvious since the series has a period 2π.

Thus we have shown that *when the arbitrary function $f(x)$ satisfies Dirichlet's Conditions in the interval $(-\pi, \pi)$, and*

$$\left.\begin{aligned} a_0 &= \frac{1}{2\pi}\int_{-\pi}^{\pi} f(x')\,dx' \text{ and, when } n \geqq 1, \\ a_n &= \frac{1}{\pi}\int_{-\pi}^{\pi} f(x')\cos nx'\,dx', \quad b_n = \frac{1}{\pi}\int_{-\pi}^{\pi} f(x')\sin nx'\,dx', \end{aligned}\right\}$$

the Fourier's Series

$$a_0 + (a_1\cos x + b_1\sin x) + (a_2\cos 2x + b_2\sin 2x) + \dots$$

converges to

$$\tfrac{1}{2}[f(x+0)+f(x-0)]$$

at every point in $-\pi<x<\pi$ where $f(x+0)$ and $f(x-0)$ exist; and at $x=\pm\pi$ it converges to

$$\tfrac{1}{2}[f(-\pi+0)+f(\pi-0)],$$

*when $f(-\pi+0)$ and $f(\pi-0)$ exist.**

There is, of course, no reason why the arbitrary function should be defined by the same analytical expression in all the interval [cf. Ex. 2 below].

Also it should be noticed that if we first sum the series, and then let x approach a point of ordinary discontinuity x_0, we would obtain $f(x_0+0)$ or $f(x_0-0)$, according to the side from which we approach the point. On the other hand, if we insert the value x_0 in the terms of the series and then sum the series, we obtain $\tfrac{1}{2}[f(x_0+0)+f(x_0-0)]$.

We have already pointed out more than once that when we speak of the sum of the series for any value of x, it is understood that we first insert this value of x in the terms of the series, then find the sum of n terms, and finally obtain the limit of this sum.

Ex. 1. Find a series of sines and cosines of multiples of x which will represent $\dfrac{\pi}{2\sinh\pi}e^x$ in the interval $-\pi<x<\pi$.

---

*If the reader refers to § 101, he will see that, if $f(x)$ is defined outside the interval $(-\pi, \pi)$ by the equation $f(x+2\pi)=f(x)$, we can replace (1) by

$$s_n(x) = \frac{1}{\pi}\int_0^{\frac{1}{2}\pi} f(x-2\alpha)\frac{\sin(2n+1)\alpha}{\sin\alpha}\,d\alpha + \frac{1}{\pi}\int_0^{\frac{1}{2}\pi} f(x+2\alpha)\frac{\sin(2n+1)\alpha}{\sin\alpha}\,d\alpha.$$

In this form we can apply the result of § 94 at once to every point in the closed interval $(-\pi, \pi)$, except points of infinite discontinuity.

What is the sum of the series for $x = \pm\pi$?

Here $\quad f(x) = \dfrac{\pi}{2\sinh\pi} e^x \quad$ and $\quad a_n = \dfrac{1}{2\sinh\pi}\int_{-\pi}^{\pi} e^x \cos nx\, dx, \quad n \geqq 1.$

Integrating by parts,

$$(1+n^2)\int_{-\pi}^{\pi} e^x \cos nx\, dx = (e^{\pi} - e^{-\pi}) \cos n\pi.$$

Therefore $$a_n = \frac{(-1)^n}{1+n^2}.$$

Also we find $\quad a_0 = \frac{1}{2}.$

Similarly, $$b_n = \frac{1}{2\sinh\pi}\int_{-\pi}^{\pi} e^x \sin nx\, dx$$
$$= \frac{n}{1+n^2}(-1)^{n-1}.$$

Therefore

$$\frac{\pi}{2\sinh\pi} e^x = \frac{1}{2} + \left(-\frac{1}{1+1^2}\cos x + \frac{1}{1+1^2}\sin x\right)$$
$$+ \left(\frac{1}{1+2^2}\cos 2x - \frac{2}{1+2^2}\sin 2x\right) + \ldots,$$

when $-\pi < x < \pi$.

When $x = \pm\pi$, the sum of the series is $\frac{1}{2}\pi \coth\pi$, since

$$f(-\pi+0) + f(\pi-0) = \pi\coth\pi.$$

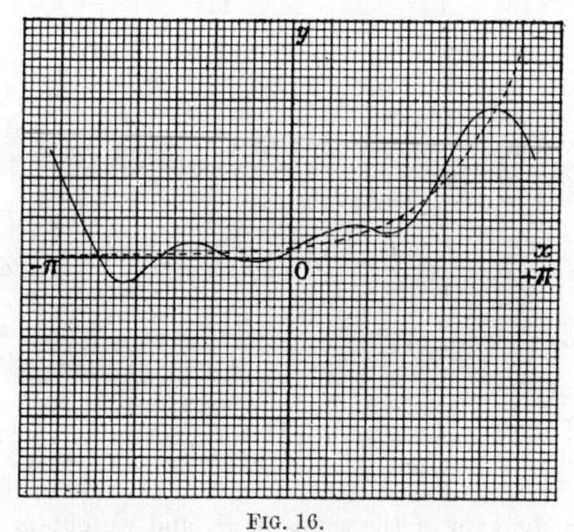

FIG. 16.

In Fig. 16, the curves

$$\left.\begin{aligned} y &= \frac{\pi}{2\sinh\pi} e^x, \\ y &= \tfrac{1}{2} + \left(-\frac{1}{1+1^2}\cos x + \frac{1}{1+1^2}\sin x\right) + (\ldots) + \left(-\frac{1}{1+3^2}\cos 3x + \frac{3}{1+3^2}\sin 3x\right) \end{aligned}\right\}$$

are drawn for the interval $(-\pi, \pi)$.

It will be noticed that the expansion we have obtained converges very slowly, and that more terms would have to be taken to bring the approximation curves $[y = s_n(x)]$ near the curve of the given function in $-\pi < x < \pi$.

At $x = \pm\pi$, the sum of the series is discontinuous. The behaviour of the approximation curves at a point of discontinuity of the sum is examined in Chapter IX.

Ex. 2. Find a series of sines and cosines of multiples of x which will represent $f(x)$ in the interval $-\pi < x < \pi$, when

$$\left.\begin{aligned} f(x) &= 0, \quad -\pi < x \leqq 0, \\ f(x) &= \tfrac{1}{4}\pi x, \quad 0 < x < \pi. \end{aligned}\right\}$$

Here

$$a_0 = \frac{1}{2\pi}\int_0^\pi \tfrac{1}{4}\pi x\, dx = \frac{\pi^2}{16},$$

$$a_n = \frac{1}{\pi}\int_0^\pi \tfrac{1}{4}\pi x \cos nx\, dx = \frac{1}{4n^2}(\cos n\pi - 1),$$

$$b_n = \frac{1}{\pi}\int_0^\pi \tfrac{1}{4}\pi x \sin nx\, dx = -\frac{\pi}{4n}\cos n\pi.$$

Therefore

$$f(x) = \frac{\pi^2}{16} + \frac{1}{2}\left[-\cos x + \frac{\pi}{2}\sin x\right] - \frac{\pi}{8}\sin 2x + \dots,$$

when $-\pi < x < \pi$.

When $x = \pm\pi$ the sum of the series is $\frac{1}{8}\pi^2$, and we obtain the well-known result,

$$\frac{\pi^2}{8} = 1 + \frac{1}{3^2} + \frac{1}{5^2} + \dots.$$

Ex. 3. Find a series of sines and cosines of multiples of x which will represent $x + x^2$ in the interval $-\pi < x < \pi$.

Here

$$a_0 = \frac{1}{2\pi}\int_{-\pi}^\pi (x + x^2)\, dx = \frac{1}{\pi}\int_0^\pi x^2 dx = \frac{\pi^2}{3},$$

$$a_n = \frac{1}{\pi}\int_{-\pi}^\pi (x + x^2)\cos nx\, dx = \frac{2}{\pi}\int_0^\pi x^2 \cos nx\, dx,$$

and, after integration by parts, we find that

$$a_n = \frac{4}{n^2}\cos n\pi.$$

Also,

$$b_n = \frac{1}{\pi}\int_{-\pi}^\pi (x + x^2)\sin nx\, dx = \frac{2}{\pi}\int_0^\pi x \sin nx\, dx,$$

which reduces to $b_n = (-1)^{n-1}\dfrac{2}{n}$.

Therefore

$$x + x^2 = \frac{\pi^2}{3} + 4\left(-\cos x + \frac{1}{2}\sin x\right) - 4\left(\frac{1}{2^2}\cos 2x + \frac{1}{4}\sin 2x\right) + \dots,$$

when $-\pi < x < \pi$.

When $x = \pm\pi$ the sum of the series is π^2, and we obtain the well-known result that

$$\frac{\pi^2}{6} = 1 + \frac{1}{2^2} + \frac{1}{3^2} + \dots.$$

96. The Cosine Series. Let $f(x)$ be given in the interval $(0, \pi)$, and satisfy Dirichlet's Conditions in that interval. Define $f(x)$ in $-\pi \leqq x < 0$ by the equation $f(-x) = f(x)$. The function thus defined for $(-\pi, \pi)$ satisfies Dirichlet's Conditions in this interval, and we can apply to it the results of § 95.

But it is clear that in this case

$$a_0=\frac{1}{2\pi}\int_{-\pi}^{\pi} f(x')\,dx' \qquad \text{leads to } a_0=\frac{1}{\pi}\int_0^{\pi} f(x')\,dx',$$

$$a_n=\frac{1}{\pi}\int_{-\pi}^{\pi} f(x')\cos nx'dx' \text{ leads to } a_n=\frac{2}{\pi}\int_0^{\pi} f(x')\cos nx'dx',$$

and $$b_n=\frac{1}{\pi}\int_{-\pi}^{\pi} f(x')\sin nx'\,dx' \text{ leads to } b_n=0.$$

Thus the sine terms disappear from the Fourier's Series for this function.

Also, from the way in which $f(x)$ was defined in $-\pi \leqq x < 0$, we have

$$\tfrac{1}{2}[f(+0)+f(-0)]=f(+0),$$

and

$$\tfrac{1}{2}[f(-\pi+0)+f(\pi-0)]=f(\pi-0),$$

provided the limits $f(+0)$ and $f(\pi-0)$ exist.

In this case the sum of the series for $x=0$ is $f(+0)$, and for $x=\pi$ it is $f(\pi-0)$.

It follows that, *when $f(x)$ is an arbitrary function satisfying Dirichlet's Conditions* in the interval $(0, \pi)$, the sum of the Cosine Series*

$$\frac{1}{\pi}\int_0^{\pi} f(x')\,dx' + \frac{2}{\pi}\sum_1^{\infty} \cos nx\int_0^{\pi} f(x')\cos nx'\,dx'$$

is equal to

$$\tfrac{1}{2}[f(x+0)+f(x-0)]$$

at every point between 0 *and* π *where* $f(x+0)$ *and* $f(x-0)$ *exist; and, when* $f(+0)$ *and* $f(\pi-0)$ *exist, the sum is* $f(+0)$ *at* $x=0$ *and* $f(\pi-0)$ *at* $x=\pi$.

Thus, when $f(x)$ is continuous and satisfies Dirichlet's Conditions in the interval $(0, \pi)$, the Cosine Series represents it in this *closed* interval.

Ex. 1. Find a series of cosines of multiples of x which will represent x in the interval $(0, \pi)$.

Here $$a_0=\frac{1}{\pi}\int_0^{\pi} x\,dx=\tfrac{1}{2}\pi,$$

and $$a_n=\frac{2}{\pi}\int_0^{\pi} x\cos nx\,dx=\frac{2}{n^2\pi}(\cos n\pi-1).$$

Therefore $$x=\frac{\pi}{2}-\frac{4}{\pi}\left[\cos x+\frac{1}{3^2}\cos 3x+\ldots\right], \quad 0\leqq x\leqq \pi.$$

Since the sum of the series is zero at $x=0$, we have again

$$\frac{\pi^2}{8}=1+\frac{1}{3^2}+\frac{1}{5^2}+\ldots.$$

---

*See footnote, p. 230.

In Fig. 17, the lines
$$\left.\begin{aligned} y&=x, & 0&\leqq x\leqq \pi,\\ y&=-x, & -\pi&\leqq x\leqq 0,\end{aligned}\right\}$$
and the approximation curve
$$y=\frac{\pi}{2}-\frac{4}{\pi}\left(\cos x+\frac{1}{3^2}\cos 3x+\frac{1}{5^2}\cos 5x\right), \quad -\pi\leqq x\leqq \pi,$$
are drawn.

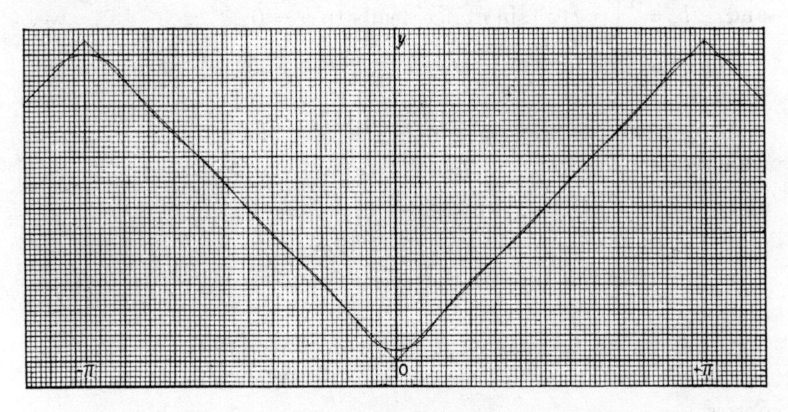

FIG. 17.

It will be seen how closely this approximation curve approaches the lines $y=\pm x$ in the whole interval.

Since the Fourier's Series has a period 2π, this series for unrestricted values

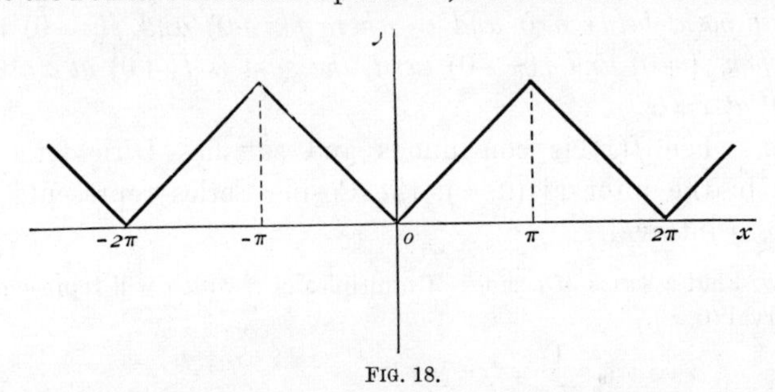

FIG. 18.

of x represents the ordinates of the lines shown in Fig. 18, the part from the interval $(-\pi, \pi)$ being repeated indefinitely in both directions.

The sum is continuous for all values of x.

Ex. 2. Find a series of cosines of multiples of x which will represent $f(x)$ in the interval $(0, \pi)$, where
$$\left.\begin{aligned} f(x)&=\tfrac{1}{4}\pi x, & 0&\leqq x\leqq \tfrac{1}{2}\pi,\\ f(x)&=\tfrac{1}{4}\pi(\pi-x), & \tfrac{1}{2}\pi&< x\leqq \pi.\end{aligned}\right\}$$

Here
$$a_0=\frac{1}{\pi}\int_0^{\frac{1}{2}\pi}\tfrac{1}{4}\pi x\,dx+\frac{1}{\pi}\int_{\frac{1}{2}\pi}^{\pi}\tfrac{1}{4}\pi(\pi-x)\,dx=\tfrac{1}{16}\pi^2,$$

and $$a_n = \frac{2}{\pi}\int_0^{\frac{1}{2}\pi} \tfrac{1}{4}\pi x \cos nx\, dx + \frac{2}{\pi}\int_{\frac{1}{2}\pi}^{\pi} \tfrac{1}{4}\pi(\pi - x)\cos nx\, dx$$
$$= \tfrac{1}{2}\int_0^{\frac{1}{2}\pi} x\cos nx\, dx + \tfrac{1}{2}\int_{\frac{1}{2}\pi}^{\pi}(\pi - x)\cos nx\, dx,$$

which gives $a_n = -\dfrac{1}{2n^2}[1 + \cos n\pi - 2\cos\tfrac{1}{2}n\pi] = \dfrac{2}{n^2}\cos\dfrac{n\pi}{2}\sin^2\dfrac{n\pi}{4}$.

Thus a_n vanishes when n is odd or a multiple of 4.

Also $$f(x) = \frac{\pi^2}{16} - 2\left[\frac{1}{2^2}\cos 2x + \frac{1}{6^2}\cos 6x + \dots\right], \quad 0 \leqq x \leqq \pi.$$

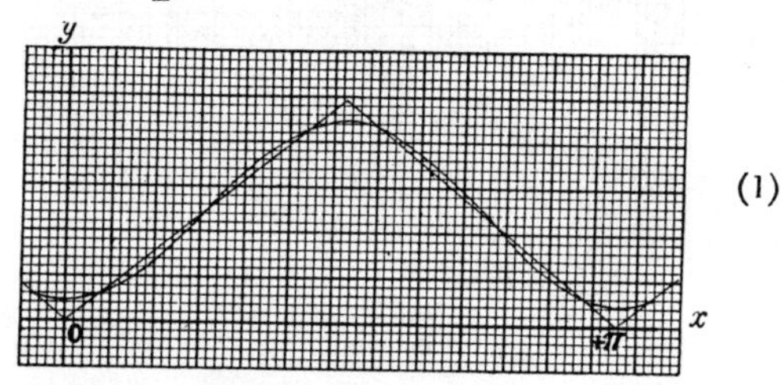

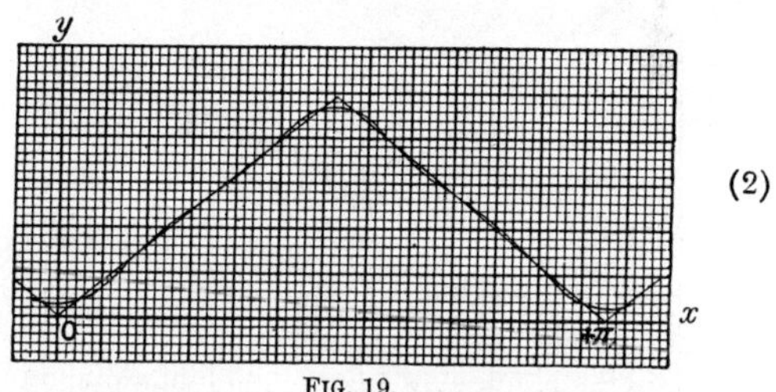

FIG. 19.

In Fig. 19, the lines
$$\left.\begin{aligned} y &= \tfrac{1}{4}\pi x, & 0 \leqq x \leqq \tfrac{1}{2}\pi, \\ y &= \tfrac{1}{4}\pi(\pi - x), & \tfrac{1}{2}\pi \leqq x \leqq \pi, \end{aligned}\right\}$$
and the approximation curves
$$\left.\begin{aligned} y &= \tfrac{1}{16}\pi^2 - \tfrac{1}{2}\cos 2x, \\ y &= \tfrac{1}{16}\pi^2 - \tfrac{1}{2}\cos 2x - \tfrac{1}{18}\cos 6x, \end{aligned}\right\} \quad 0 \leqq x \leqq \pi,$$
are drawn.

It will be noticed that the approximation curve, corresponding to the

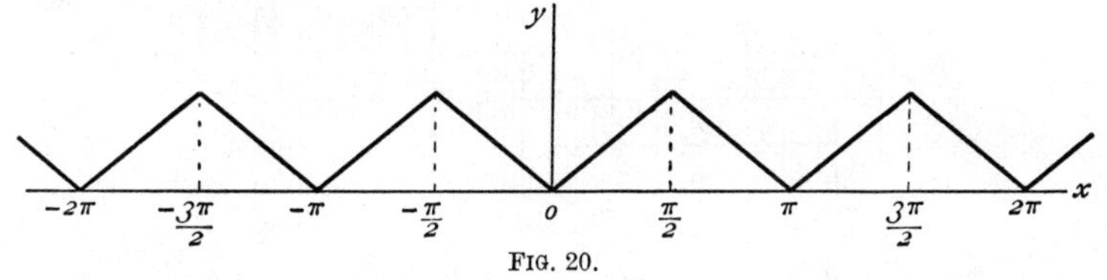

FIG. 20.

terms up to and including cos $6x$, approaches the given lines closely, except at the sharp corner, right through the interval $(0, \pi)$.

For unrestricted values of x the series represents the ordinates of the lines shown in Fig. 20, the part from $-\pi$ to π being repeated indefinitely in both directions.

The sum is continuous for all values of x.

Ex. 3. Find a series of cosines of multiples of x which will represent $f(x)$ in the interval $(0, \pi)$, where

$$\left.\begin{aligned} f(x)&=0, \quad 0 \leqq x < \tfrac{1}{2}\pi. \\ f(\tfrac{1}{2}\pi)&=\tfrac{1}{4}\pi, \\ f(x)&=\tfrac{1}{2}\pi, \quad \tfrac{1}{2}\pi < x \leqq \pi. \end{aligned}\right\}$$

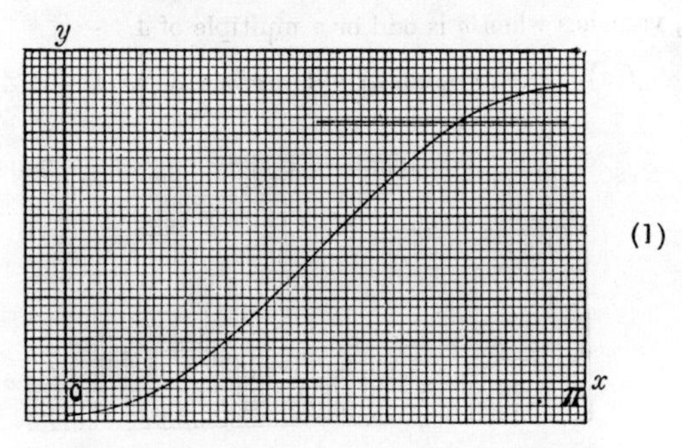

(1)

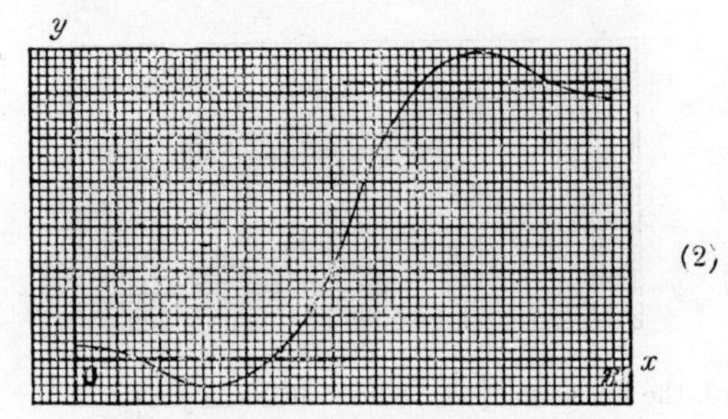

(2)

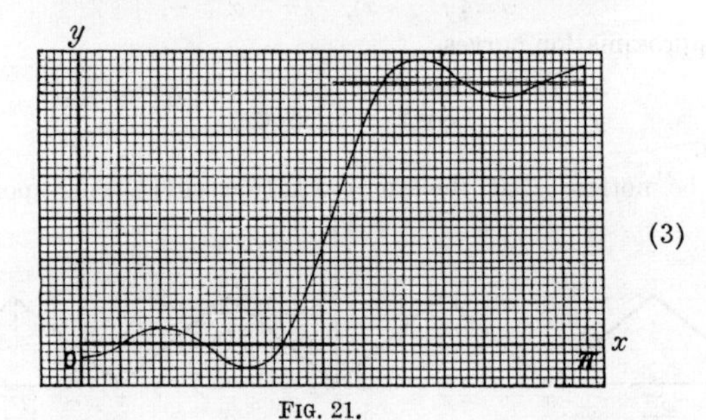

(3)

FIG. 21.

Here $\qquad a_0=\dfrac{1}{\pi}\displaystyle\int_0^\pi f(x)\,dx=\dfrac{1}{\pi}\int_{\frac{1}{2}\pi}^\pi \tfrac{1}{2}\pi\,dx=\tfrac{1}{4}\pi.$

Also $\qquad a_n=\dfrac{2}{\pi}\displaystyle\int_0^\pi f(x)\cos nx\,dx=\int_{\frac{1}{2}\pi}^\pi \cos nx\,dx=-\dfrac{1}{n}\sin\tfrac{1}{2}n\pi.$

Thus $\qquad f(x)=\tfrac{1}{4}\pi-[\cos x-\tfrac{1}{3}\cos 3x+\tfrac{1}{5}\cos 5x-\ldots], \quad 0 \leqq x \leqq \pi,$

since, when $x=\frac{1}{2}\pi$, the sum of the Fourier's Series is $\frac{1}{2}[f(\frac{1}{2}\pi+0)+f(\frac{1}{2}\pi-0)]$.

From the values at $x=0$ and $x=\pi$, we have the well-known result,

$$\frac{\pi}{4}=1-\tfrac{1}{3}+\tfrac{1}{5}-\dots.$$

In Fig. 21, the graph of the given function, and the approximation curves

$$\left.\begin{aligned} y&=\tfrac{1}{4}\pi-\cos x,\\ y&=\tfrac{1}{4}\pi-\cos x+\tfrac{1}{3}\cos 3x,\\ y&=\tfrac{1}{4}\pi-\cos x+\tfrac{1}{3}\cos 3x-\tfrac{1}{5}\cos 5x,\end{aligned}\right\}\quad 0\leqq x\leqq\pi,$$

are drawn.

The points $x=0$ and $x=\pi$ are points of continuity in the sum of the series: the point $x=\frac{1}{2}\pi$ is a point of discontinuity.

The behaviour of the approximation curves at a point of discontinuity, when n is large, will be treated fully in Chapter IX. It will be sufficient to say now that it is proved in § 117 that just before $x=\frac{1}{2}\pi$ the approximation curve for a large value of n will have a minimum at a depth nearly 0·14 below $y=0$: that it will then ascend at a steep gradient, passing near the point $(\frac{1}{2}\pi, \frac{1}{2}\pi)$, and rising to a maximum just after $x=\frac{1}{2}\pi$ at a height nearly 0·14 above $\frac{1}{2}\pi$.

Ex. 4. Find a series of cosines of multiples of x which will represent $f(x)$ in the interval $(0, \pi)$, where

$$\left.\begin{aligned} f(x)&=\tfrac{1}{3}\pi, & 0&\leqq x<\tfrac{1}{3}\pi,\\ f(x)&=0, & \tfrac{1}{3}\pi&<x<\tfrac{2}{3}\pi,\\ f(x)&=-\tfrac{1}{3}\pi, & \tfrac{2}{3}\pi&<x\leqq\pi.\end{aligned}\right\}$$

Also $\quad f(\frac{1}{3}\pi)=\frac{1}{6}\pi, \qquad f(\frac{2}{3}\pi)=-\frac{1}{6}\pi.$

Here
$$a_0=\tfrac{1}{3}\int_0^{\frac{1}{3}\pi}dx-\tfrac{1}{3}\int_{\frac{2}{3}\pi}^{\pi}dx=0,$$

and
$$\begin{aligned} a_n&=\tfrac{2}{3}\int_0^{\frac{1}{3}\pi}\cos nx\,dx-\tfrac{2}{3}\int_{\frac{2}{3}\pi}^{\pi}\cos nx\,dx\\ &=\frac{2}{3n}[\sin\tfrac{1}{3}n\pi+\sin\tfrac{2}{3}n\pi]\\ &=\frac{4}{3n}\sin\tfrac{1}{2}n\pi\cos\tfrac{1}{6}n\pi.\end{aligned}$$

Thus a_n vanishes when n is even or a multiple of 3.

And $f(x)=\dfrac{2\sqrt{3}}{3}[\cos x-\frac{1}{5}\cos 5x+\frac{1}{7}\cos 7x-\frac{1}{11}\cos 11x+\dots], \quad 0\leqq x\leqq\pi.$

The points $x=0$ and $x=\pi$ are points of continuity in the sum of the series. The points $x=\frac{1}{3}\pi$ and $x=\frac{2}{3}\pi$ are points of discontinuity.

Fig. 22 contains the graph of the given function, and the approximation curves

$$\left.\begin{aligned} y&=\frac{2\sqrt{3}}{3}\cos x,\\ y&=\frac{2\sqrt{3}}{3}[\cos x-\tfrac{1}{5}\cos 5x],\\ y&=\frac{2\sqrt{3}}{3}[\cos x-\tfrac{1}{5}\cos 5x+\tfrac{1}{7}\cos 7x],\\ y&=\frac{2\sqrt{3}}{3}[\cos x-\tfrac{1}{5}\cos 5x+\tfrac{1}{7}\cos 7x-\tfrac{1}{11}\cos 11x],\end{aligned}\right\}\quad 0\leqq x\leqq\pi.$$

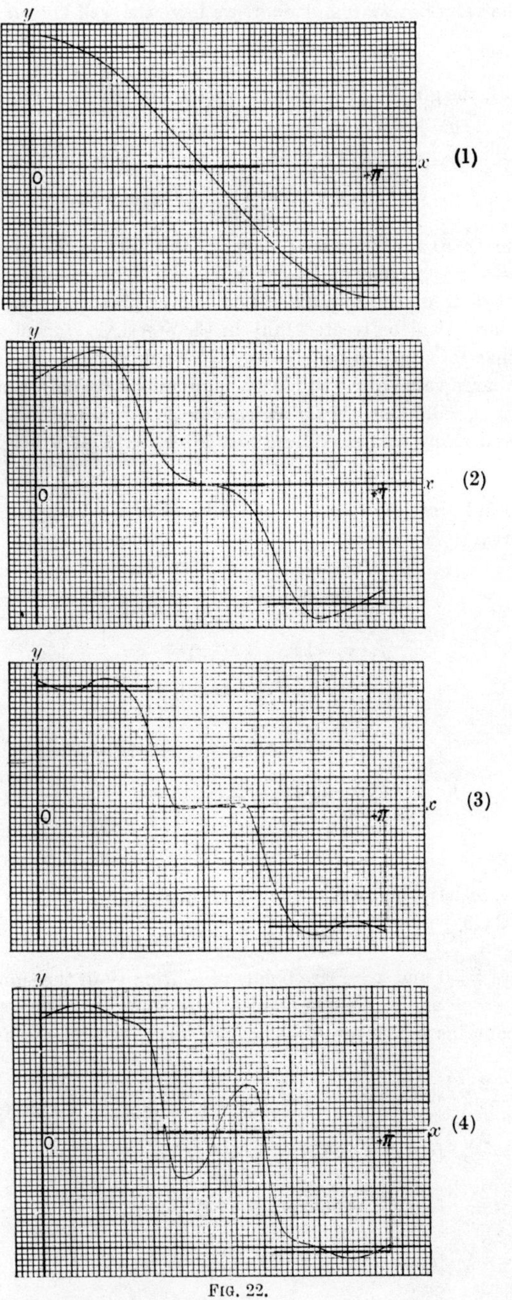

FIG. 22.

Ex. 5. Find a series of cosines of multiples of x which will represent $\log(2\sin\frac{1}{2}x)$* in the interval $(0, \pi)$.

Here
$$a_0=\frac{1}{\pi}\int_0^\pi \log(2\sin\tfrac{1}{2}x)\,dx$$
$$=\log 2+\frac{2}{\pi}\int_0^{\frac{1}{2}\pi}\log\sin x\,dx$$
$$=0, \text{ by Ex. 4, p. 131.}$$

And
$$a_n=\frac{2}{\pi}\int_0^\pi \cos nx \log(2\sin\tfrac{1}{2}x)\,dx$$
$$=\frac{4}{\pi}\int_0^{\frac{1}{2}\pi}\cos 2nx\log(2\sin x)\,dx$$
$$=-\frac{1}{n}, \text{ by Ex. 5, p. 131.}$$

Thus $\log(2\sin\frac{1}{2}x)=-[\cos x+\frac{1}{2}\cos 2x+\frac{1}{3}\cos 3x+\ldots]$, when $0<x\leqq\pi$.

It follows from this—or may be obtained independently—that

$$\log(2\cos\tfrac{1}{2}x)=[\cos x-\tfrac{1}{2}\cos 2x+\tfrac{1}{3}\cos 3x-\ldots], \text{ when } 0\leqq x<\pi.$$

These expansions have been obtained otherwise in § 70. 1. [See footnote on p. 159.]

97. The Sine Series. Again let $f(x)$ be given in the interval $(0, \pi)$, and satisfy Dirichlet's Conditions in that interval. Define $f(x)$ in $-\pi\leqq x<0$ by the equation $f(-x)=-f(x)$. The function thus defined for $(-\pi, \pi)$ satisfies Dirichlet's Conditions in this interval, and we can apply to it the results of § 95.

But it is clear that in this case

$$b_n=\frac{1}{\pi}\int_{-\pi}^{\pi} f(x')\sin nx'\,dx' \text{ leads to } b_n=\frac{2}{\pi}\int_0^\pi f(x')\sin nx'\,dx',$$

and that $a_n=0$ when $n\geqq 0$.

Thus the cosine terms disappear from this Fourier's Series.

Since all the terms of the series

$$b_1\sin x+b_2\sin 2x+\ldots$$

vanish when $x=0$ and $x=\pi$, the sum of the series is zero at these points.

It follows that, *when $f(x)$ is an arbitrary function satisfying Dirichlet's Conditions*† *in the interval* $(0, \pi)$, *the sum of the Sine Series,*

$$\frac{2}{\pi}\sum_1^\infty \sin nx\int_0^\pi f(x')\sin nx'\,dx',$$

is equal to
$$\tfrac{1}{2}[f(x+0)+f(x-0)]$$

---

*This function is infinite at $x=0$, but it satisfies Dirichlet's Conditions.

†See footnote, p. 230.

at every point between 0 *and* π *where* $f(x+0)$ *and* $f(x-0)$ *exist; and, when* $x=0$ *and* $x=\pi$, *the sum is zero.*

It will be noticed that, when $f(x)$ is continuous at the end-points $x=0$ and $x=\pi$, the Cosine Series gives the value of the function at these points. The Sine Series only gives the value of $f(x)$ at these points if $f(x)$ is zero there.

Ex. 1. Find a series of sines of multiples of x which will represent x in the interval $0<x<\pi$.

Here $$b_n=\frac{2}{\pi}\int_0^\pi x\sin nx\,dx=(-1)^{n-1}\frac{2}{n}.$$

Therefore $$x=2[\sin x-\tfrac{1}{2}\sin 2x+\tfrac{1}{3}\sin 3x-\dots],\quad 0\leqq x<\pi.$$

At $x=\pi$ the sum is discontinuous.

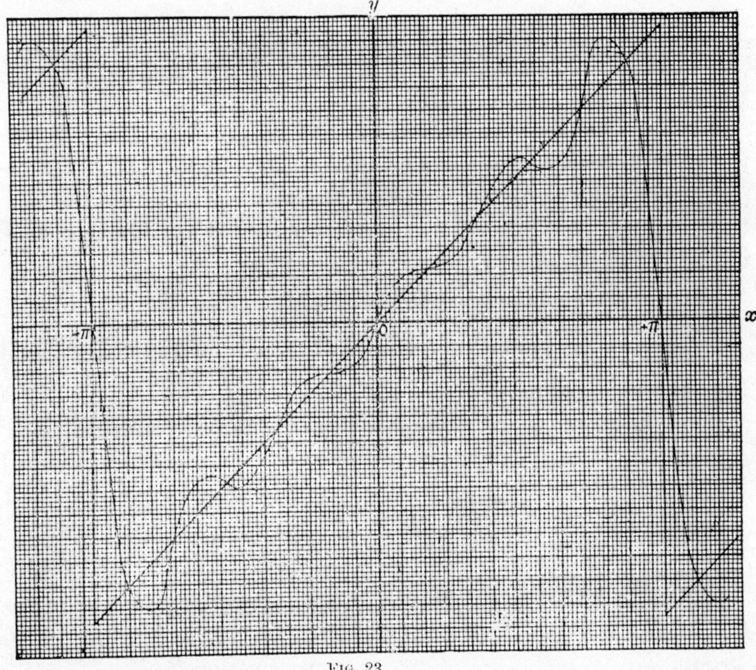

FIG. 23.

In Fig. 23, the line $$y=x.\quad -\pi\leqq x\leqq\pi,$$
and the approximation curve
$$y=2[\sin x-\tfrac{1}{2}\sin 2x+\tfrac{1}{3}\sin 3x-\tfrac{1}{4}\sin 4x+\tfrac{1}{5}\sin 5x],\quad -\pi\leqq x\leqq\pi,$$
are drawn.

The convergence of the series is so slow that this curve does not approach $y=x$ between $-\pi$ and π nearly as closely as the corresponding approximation curve in the cosine series approached $y=\pm x$. If n is taken large enough, the curve $y=s_n(x)$ will be a wavy curve oscillating about the line $y=x$ from $-\pi$ to $+\pi$, but we would be wrong if we were to say that it descends at a steep gradient from $x=-\pi$ to the end of $y=x$, and again descends from the other end of $y=x$ to $x=\pi$ at a steep gradient. As a matter of fact the summit of the first wave is some distance below $y=x$ at $x=-\pi$, and the summit of the last wave a corresponding distance above $y=x$ at $x=\pi$ when n is large.

To this question we return in Chapter IX.

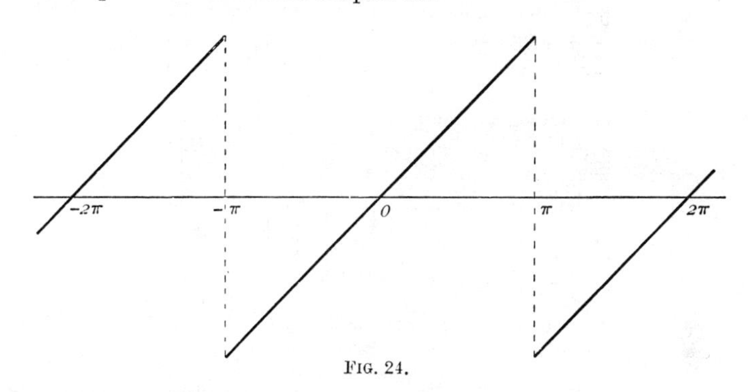

FIG. 24.

Since the Fourier's Series has a period 2π, this series for unrestricted values of x represents the ordinates of the lines shown in Fig. 24, the part from the open interval $(-\pi, \pi)$ being repeated indefinitely in both directions. The points $\pm\pi$, $\pm 3\pi$, ... are points of discontinuity. At these the sum is zero.

Ex. 2. Find a series of sines of multiples of x which will represent $f(x)$ in the interval $0 \leqq x \leqq \pi$, where

$$\left.\begin{aligned} f(x)&=\tfrac{1}{4}\pi x, & 0 &\leqq x \leqq \tfrac{1}{2}\pi, \\ f(x)&=\tfrac{1}{4}\pi(\pi-x), & \tfrac{1}{2}\pi &\leqq x \leqq \pi. \end{aligned}\right\}$$

Here $$b_n=\frac{2}{\pi}\int_0^{\frac{1}{2}\pi} \tfrac{1}{4}\pi x \sin nx\, dx+\frac{2}{\pi}\int_{\frac{1}{2}\pi}^{\pi} \tfrac{1}{4}\pi(\pi-x) \sin nx\, dx$$

$$=\tfrac{1}{2}\int_0^{\frac{1}{2}\pi} x \sin nx\, dx+\tfrac{1}{2}\int_{\frac{1}{2}\pi}^{\pi} (\pi-x) \sin nx\, dx,$$

which gives $$b_n=\frac{1}{n^2}\sin\frac{n\pi}{2}.$$

Thus $$f(x)=\sin x-\frac{1}{3^2}\sin 3x+\frac{1}{5^2}\sin 5x-\dots,\quad 0\leqq x\leqq\pi.$$

Fig. 25 contains the lines $$\left.\begin{aligned} y&=\tfrac{1}{4}\pi x, & 0 &\leqq x \leqq \tfrac{1}{2}\pi, \\ y&=\tfrac{1}{4}\pi(\pi-x), & \tfrac{1}{2}\pi &\leqq x \leqq \pi, \end{aligned}\right\}$$

and the approximation curves

$$\left.\begin{aligned} y&=\sin x, \\ y&=\sin x-\frac{1}{3^2}\sin 3x, \\ y&=\sin x-\frac{1}{3^2}\sin 3x+\frac{1}{5^2}\sin 5x, \end{aligned}\right\}\quad 0\leqq x\leqq\pi.$$

It will be noticed that the last of these curves approaches the given lines closely, except at the sharp corner, right through the interval.

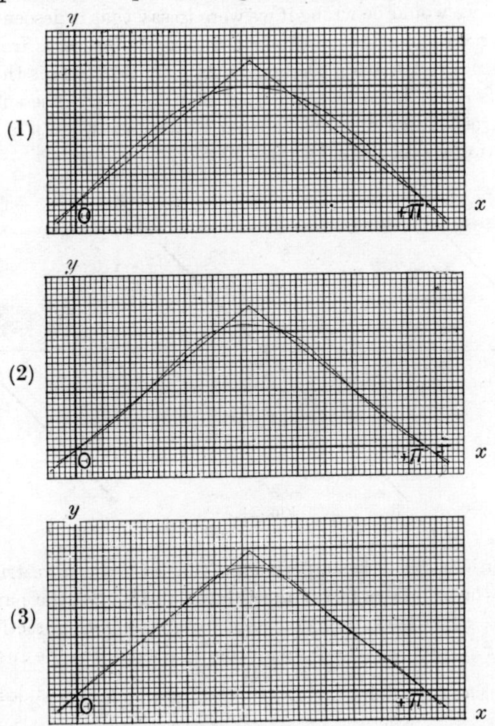

FIG. 25.

For unrestricted values of x the series represents the ordinates of the lines shown in Fig. 26, the part from $-\pi$ to $+\pi$ being repeated indefinitely in both directions.

The sum is continuous for all values of x.

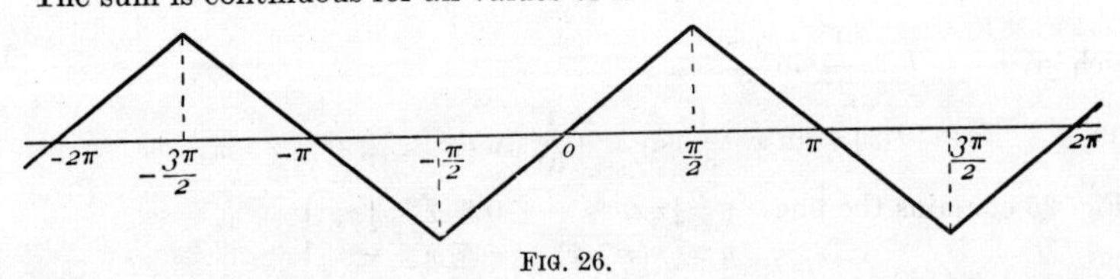

FIG. 26.

Ex. 3. Find a series of sines of multiples of x which will represent $f(x)$ in the interval $(0, \pi)$, where

$$\left.\begin{aligned} f(x)&=0, \quad 0 \leqq x < \tfrac{1}{2}\pi, \\ f(\tfrac{1}{2}\pi)&=\tfrac{1}{4}\pi, \\ f(x)&=\tfrac{1}{2}\pi, \quad \tfrac{1}{2}\pi < x < \pi, \\ f(\pi)&=0. \end{aligned}\right\}$$

Here

$$b_n = \int_{\frac{1}{2}\pi}^{\pi} \sin nx \, dx$$

$$= \frac{1}{n}\left(\cos \frac{n\pi}{2} - \cos n\pi\right)$$

$$= \frac{2}{n} \sin \frac{3n\pi}{4} \sin \frac{n\pi}{4}.$$

Therefore b_n vanishes when n is a multiple of 4.

And $f(x) = \sin x - \sin 2x + \frac{1}{3}\sin 3x + \frac{1}{5}\sin 5x - \frac{1}{3}\sin 6x + \dots, \quad 0 \leqq x \leqq \pi.$

Fig. 27 contains the graph of the given function, and the approximation curves

$$\left.\begin{aligned} y &= \sin x, \\ y &= \sin x - \sin 2x, \\ y &= \sin x - \sin 2x + \tfrac{1}{3}\sin 3x, \\ y &= \sin x - \sin 2x + \tfrac{1}{3}\sin 3x + \tfrac{1}{5}\sin 5x, \end{aligned}\right\} \quad 0 \leqq x \leqq \pi.$$

The points $x = \frac{1}{2}\pi$ and $x = \pi$ are points of discontinuity in the sum of the series. The behaviour of the approximation curves for large values of n at these points will be examined in Chapter IX.

Ex. 4. Find a series of sines of multiples of x which will represent $f(x)$ in the interval $(0, \pi)$, where

$$\left.\begin{aligned} &f(x) = \tfrac{1}{3}\pi, && 0 < x < \tfrac{1}{3}\pi, \\ &f(x) = 0, && \tfrac{1}{3}\pi < x < \tfrac{2}{3}\pi, \\ &f(x) = -\tfrac{1}{3}\pi, && \tfrac{2}{3}\pi < x < \pi. \end{aligned}\right.$$

Also $\quad f(0) = f(\pi) = 0;\ f(\frac{1}{3}\pi) = \frac{1}{6}\pi;\ f(\frac{2}{3}\pi) = -\frac{1}{6}\pi.$

Here

$$b_n = \frac{2}{3}\int_{\pi}^{\frac{1}{3}\pi} \sin nx \, dx - \frac{2}{3}\int_{\frac{2}{3}\pi}^{\pi} \sin nx \, dx$$

$$= \frac{2}{3n}\left[1 - \cos \tfrac{1}{3}n\pi - \cos \tfrac{2}{3}n\pi + \cos n\pi\right]$$

$$= \frac{8}{3n} \cos^2 \frac{n\pi}{2} \sin^2 \frac{n\pi}{6}.$$

Therefore a_n vanishes when n is odd or a multiple of 6.

And $\quad f(x) = \sin 2x + \frac{1}{2}\sin 4x + \frac{1}{4}\sin 8x + \frac{1}{5}\sin 10x + \dots, \quad 0 \leqq x \leqq \pi.$

The points $x = 0$, $x = \frac{1}{3}\pi$, $x = \frac{2}{3}\pi$ and $x = \pi$ are points of discontinuity in the sum of the series.

Fig. 28 contains the graph of the given function, and the approximation curves

$$\left.\begin{aligned} y &= \sin 2x, \\ y &= \sin 2x + \tfrac{1}{2}\sin 4x, \\ y &= \sin 2x + \tfrac{1}{2}\sin 4x + \tfrac{1}{4}\sin 8x, \\ y &= \sin 2x + \tfrac{1}{2}\sin 4x + \tfrac{1}{4}\sin 8x + \tfrac{1}{5}\sin 10x, \end{aligned}\right\} \quad 0 \leqq x \leqq \pi.$$

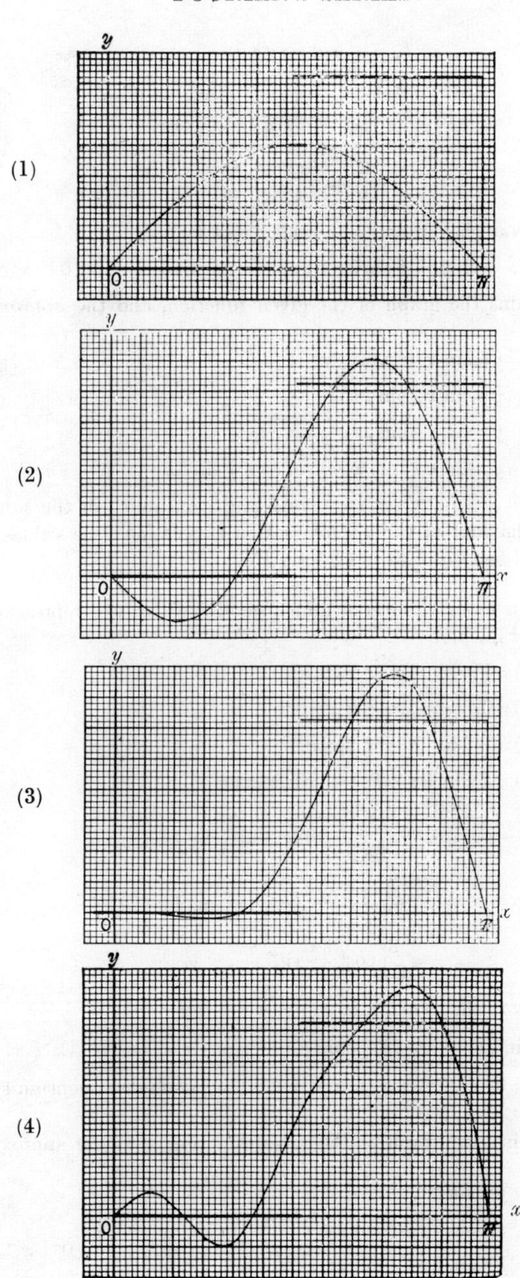

FIG. 27.

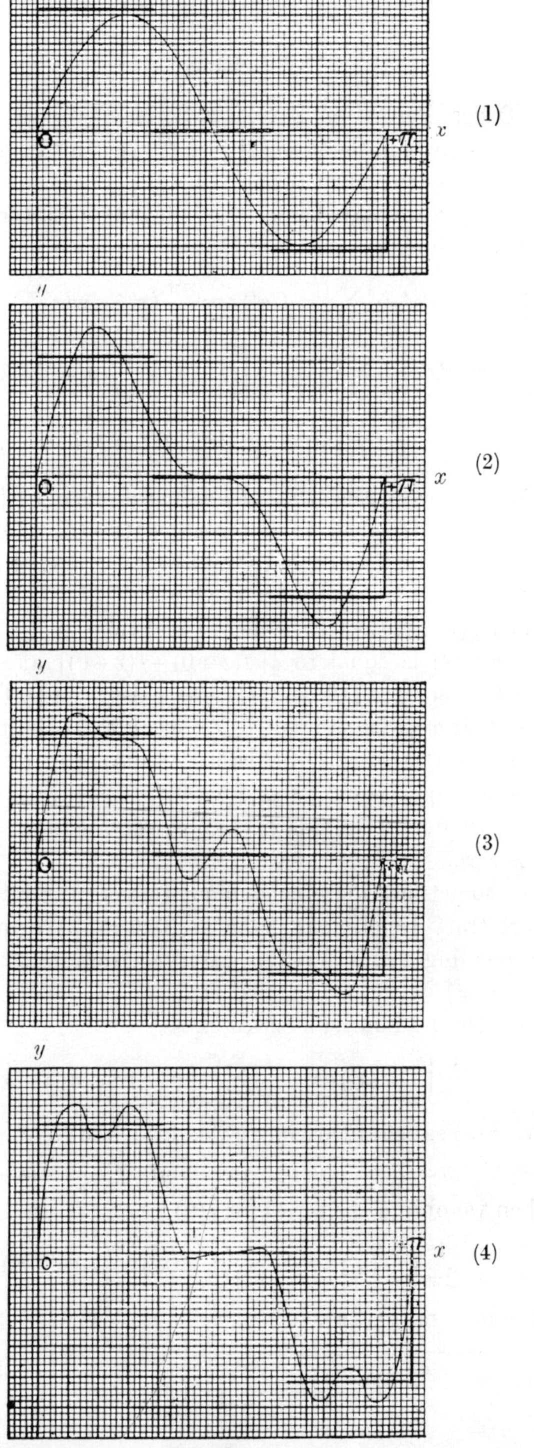

FIG. 28.

98. Other Forms of Fourier's Series.* When the arbitrary function is given in the interval $(-l, l)$, we can change this interval to $(-\pi, \pi)$ by the substitution $u = \pi x/l$.

In this way we may deduce the following expansions from those already obtained:

$$\frac{1}{2l}\int_{-l}^{l} f(x')\,dx' + \frac{1}{l}\sum_{1}^{\infty}\int_{-l}^{l} f(x')\cos\frac{n\pi}{l}(x'-x)\,dx', \quad -l \leqq x \leqq l, \ldots(1)$$

$$\frac{1}{l}\int_{0}^{l} f(x')\,dx' + \frac{2}{l}\sum_{1}^{\infty}\cos\frac{n\pi}{l}x\int_{0}^{l} f(x')\cos\frac{n\pi}{l}x'\,dx', \quad 0 \leqq x \leqq l, \ldots(2)$$

$$\frac{2}{l}\sum_{1}^{\infty}\sin\frac{n\pi}{l}x\int_{0}^{l} f(x')\sin\frac{n\pi}{l}x'\,dx', \quad 0 \leqq x \leqq l. \quad \ldots\ldots(3)$$

When $f(x)$ satisfies Dirichlet's Conditions in $(-l, l)$, the sum of the series (1) is equal to $\frac{1}{2}[f(x+0)+f(x-0)]$ at every point in $-l<x<l$ where $f(x+0)$ and $f(x-0)$ exist; and at $x = \pm l$ its sum is $\frac{1}{2}[f(-l+0)+f(l-0)]$, when the limits $f(-l+0)$ and $f(l-0)$ exist.

When $f(x)$ satisfies Dirichlet's Conditions in $(0, l)$, the sum of the series (2) is equal to $\frac{1}{2}[f(x+0)+f(x-0)]$ at every point in $0<x<l$ where $f(x+0)$ and $f(x-0)$ exist; and at $x=0$ its sum is $f(+0)$, at $x=l$ its sum is $f(l-0)$, when these limits exist.

When $f(x)$ satisfies Dirichlet's Conditions in $(0, l)$ the sum of the series (3) is equal to $\frac{1}{2}[f(x+0)+f(x-0)]$ at every point in $0<x<l$ where $f(x+0)$ and $f(x-0)$ exist; and at $x=0$ and $x=l$ its sum is zero.

It is sometimes more convenient to take the interval in which the arbitrary function is given as $(0, 2\pi)$. We may deduce the corresponding series for this interval from that already found for $(-\pi, \pi)$.

Consider the Fourier's Series

$$\frac{1}{2\pi}\int_{-\pi}^{\pi} F(x')\,dx' + \frac{1}{\pi}\sum_{1}^{\infty}\int_{-\pi}^{\pi} F(x')\cos n(x'-x)\,dx',$$

where $F(x)$ satisfies Dirichlet's Conditions in $(-\pi, \pi)$.

Let $\quad u = \pi + x, \quad u' = \pi + x' \quad$ and $\quad f(u) = F(u-\pi)$.

Then we obtain the series for $f(u)$,

$$\frac{1}{2\pi}\int_{0}^{2\pi} f(u')\,du' + \frac{1}{\pi}\sum_{1}^{\infty}\int_{0}^{2\pi} f(u')\cos n(u'-u)\,du',$$

for the interval $(0, 2\pi)$.

---

* See footnote, p. 230.

On changing u into x, we have the series for $f(x)$,

$$\frac{1}{2\pi}\int_0^{2\pi} f(x')\,dx' + \frac{1}{\pi}\sum_1^\infty \int_0^{2\pi} f(x')\cos n(x'-x)\,dx', \quad 0 \leqq x \leqq 2\pi. \quad \text{....(4)}$$

The sum of the series (4) is $\frac{1}{2}[f(x+0)+f(x-0)]$ at every point between 0 and 2π where $f(x+0)$ and $f(x-0)$ exist; and at $x=0$ and $x=2\pi$ its sum is

$$\tfrac{1}{2}[f(+0)+f(2\pi-0)],$$

when these limits exist.

In (4), it is assumed that $f(x)$ satisfies Dirichlet's Conditions in the interval $(0, 2\pi)$.

This reduces to a Cosine Series if $f(x)=f(2\pi-x)$ and to a Sine Series if $f(x)=-f(2\pi-x)$.

If the interval is (0, 1), we have instead of (4)

$$\int_0^1 f(x')\,dx' + 2\sum_1^\infty \int_0^1 f(x')\cos 2n\pi(x-x')\,dx', \quad 0 \leqq x \leqq 1, \quad \text{....(5)}$$

a series with period unity.

Again, it is sometimes convenient to take the interval in which the function is defined as (a, b). We can deduce the corresponding series for this interval from the result just obtained.

Taking the series

$$\frac{1}{2\pi}\int_0^{2\pi} F(x')\,dx' + \frac{1}{\pi}\sum_1^\infty \int_0^{2\pi} F(x')\cos n(x'-x)\,dx', \quad 0 \leqq x \leqq 2\pi,$$

we write $\quad u = a + \dfrac{(b-a)x}{2\pi} \quad$ and $\quad f(u) = F\left\{\dfrac{2\pi(u-a)}{b-a}\right\}.$

Then in the interval $a \leqq u \leqq b$ we have the series for $f(u)$,

$$\frac{1}{b-a}\int_a^b f(u')\,du' + \frac{2}{b-a}\sum_1^\infty \int_a^b f(u')\cos\frac{2n\pi}{b-a}(u'-u)\,du'.$$

On changing u into x, we obtain the series for $f(x)$ in $a \leqq x \leqq b$, namely,

$$\frac{1}{b-a}\int_a^b f(x')\,dx' + \frac{2}{b-a}\sum_1^\infty \int_a^b f(x')\cos\frac{2n\pi}{b-a}(x'-x)\,dx' \text{......(6)}$$

The sum of the series (6) is $\frac{1}{2}[f(x+0)+f(x-0)]$ at every point in $a<x<b$ where $f(x+0)$ and $f(x-0)$ exist; and at $x=a$ and $x=b$ its sum is $\frac{1}{2}[f(a+0)+f(b-0)]$, when these limits exist.

Of course $f(x)$ is again subject to Dirichlet's Conditions in the interval (a, b).

The corresponding Cosine Series and Sine Series are, respectively,

$$\frac{1}{b-a}\int_a^b f(x')dx' + \frac{2}{b-a}\sum_1^\infty \cos\frac{n\pi}{b-a}(x-a)\int_a^b f(x')\cos\frac{n\pi}{b-a}(x'-a)\,dx',$$
$$a \leqq x \leqq b, \quad \text{.........(7)}$$

and $$\frac{2}{b-a}\sum_1^\infty \sin\frac{n\pi}{b-a}(x-a)\int_a^b f(x')\sin\frac{n\pi}{b-a}(x'-a)\,dx',$$
$$a \leqq x \leqq b. \quad \text{.........(8)}$$

Ex. 1. Show that the series

$$\frac{4}{\pi}\left(\sin\frac{\pi x}{l} + \frac{1}{3}\sin\frac{3\pi x}{l} + \ldots\right)$$

is equal to 1 when $0 < x < l$.

Ex. 2. Show that the series

$$\frac{ca}{l} + \frac{2c}{\pi}\left(\sin\frac{\pi a}{l}\cos\frac{\pi x}{l} + \frac{1}{2}\sin\frac{2\pi a}{l}\cos\frac{2\pi x}{l} + \ldots\right)$$

is equal to c when $0 < x < a$ and to zero when $a < x < l$.

Ex. 3. Show that the series

$$\frac{v_1+v_2+v_3}{3} + \frac{1}{\pi}\sum_1^\infty \frac{1}{n}\sin\frac{n\pi}{3}\left\{2(v_3-v_1)\sin\frac{2n\pi}{3}\sin\frac{n\pi x}{l} + (2v_2-v_1-v_3)\cos\frac{n\pi x}{l}\right\}$$

is equal to $\quad v_1$ when $-l < x < -\frac{l}{3}$,

v_2 when $-\frac{l}{3} < x < \frac{l}{3}$,

v_3 when $\frac{l}{3} < x < l$.

Ex. 4. Show that the series

$$2\left[\sin x + \frac{\sin 2x}{2} + \frac{\sin 3x}{3} + \ldots\right]$$

represents $(\pi - x)$ in the interval $0 < x < 2\pi$.

99. Poisson's Discussion of Fourier's Series. As has been mentioned in the introduction, within a few years of Fourier's discovery of the possibility of representing an arbitrary function by what is now called its Fourier's Series, Poisson discussed the subject from a quite different standpoint.

He began with the equation

$$\frac{1-r^2}{1-2r\cos(x'-x)+r^2} = 1 + 2\sum_1^\infty r^n\cos n(x'-x),$$

where $|r|<1$, and he obtained, by integration,

$$\frac{1}{2\pi}\int_{-\pi}^{\pi}\frac{1-r^2}{1-2r\cos(x'-x)+r^2}f(x')\,dx'$$
$$=\frac{1}{2\pi}\int_{-\pi}^{\pi}f(x')dx'+\frac{1}{\pi}\sum_1^\infty r^n\int_{-\pi}^{\pi}f(x')\cos n(x'-x)\,dx'.$$

Poisson proceeded to show that, as $r\to 1$, the integral on the left-hand side of this equation has the limit $f(x)$, supposing $f(x)$ continuous at that point, and he argued that $f(x)$ must then be the sum of the series on the right-hand side when $r=1$. Apart from the incompleteness of his discussion of the questions connected with the limit of the integral as $r\to 1$, the conclusion he sought to draw is invalid until it is shown that the series does converge when $r=1$, and this, in fact, is the real difficulty. In accordance with Abel's Theorem on the Power Series (§ 72), if the series converges, when $r=1$, its sum is continuous up to and including $r=1$. In other words, if we write

$$F(r,x)=\frac{1}{2\pi}\int_{-\pi}^{\pi}f(x')\,dx'+\frac{1}{\pi}\sum_1^\infty r^n\int_{-\pi}^{\pi}f(x')\cos n(x'-x)\,dx',$$

we know that, if $F(1,x)$ converges, then

$$\lim_{r\to 1}F(r,x)=F(1,x).$$

But we have no right to assume, from the convergence of

$$\lim_{r\to 1}F(r,x),$$

that $F(1,x)$ does converge.

Poisson's method, however, has a definite value in the treatment of Fourier's Series, and we shall now give a presentation of it on the more exact lines which we have followed in the discussion of series and integrals in the previous pages of this book.

100. Poisson's Integral. The integral

$$\frac{1}{2\pi}\int_{-\pi}^{\pi}\frac{1-r^2}{1-2r\cos(x'-x)+r^2}f(x')dx', \quad |r|<1,$$

is called Poisson's Integral.

We shall assume that $f(x)$ is either bounded and integrable in the interval $(-\pi,\pi)$, or that the infinite integral $\int_{-\pi}^{\pi}f(x)dx$ is absolutely convergent.

Now we know that $\dfrac{1-r^2}{1-2r\cos\theta+r^2}=1+2\sum_1^\infty r^n\cos n\theta$

when $|r|<1$, and that this series is uniformly convergent for any interval of θ, when r has any given value between -1 and $+1$.

It follows that $$\frac{1}{2\pi}\int_{-\pi}^{\pi}\frac{1-r^2}{1-2r\cos(x'-x)+r^2}dx'=1,$$

and that

$$\frac{1}{2\pi}\int_{-\pi}^{\pi}\frac{1-r^2}{1-2r\cos(x'-x)+r^2}f(x')\,dx'$$

$$=\frac{1}{2\pi}\int_{-\pi}^{\pi}f(x')dx'+\frac{1}{\pi}\sum_{1}^{\infty}r^n\int_{-\pi}^{\pi}f(x')\cos n(x'-x)dx' \quad\text{.........(1)}$$

under the limitations above imposed upon $f(x)$. (Cf. § 70. 1, Cor. II, and § 74, I.)

Now let us choose a number x between $-\pi$ and π for which we wish the sum of this series, or, what is the same thing, the value of Poisson's Integral,

$$\frac{1}{2\pi}\int_{-\pi}^{\pi}\frac{1-r^2}{1-2r\cos(x'-x)+r^2}f(x')dx'.$$

Denote this sum, or the integral, by $F(r, x)$.

Let us assume that, for the value of x chosen,

$$\lim_{t\to 0}[f(x+t)+f(x-t)]$$

exists.

Also, let the function $\phi(x')$ be defined when $-\pi\leqq x'\leqq\pi$ by the equation

$$\phi(x')=f(x')-\tfrac{1}{2}\lim_{t\to 0}[f(x+t)+f(x-t)].$$

Then

$$F(r, x)-\tfrac{1}{2}\lim_{t\to 0}[f(x+t)+f(x-t)]$$

$$=\frac{1}{2\pi}\int_{-\pi}^{\pi}\frac{1-r^2}{1-2r\cos(x'-x)+r^2}\{f(x')-\tfrac{1}{2}\lim_{t\to 0}[f(x+t)+f(x-t)]\}\,dx'$$

$$=\frac{1}{2\pi}\int_{-\pi}^{\pi}\frac{1-r^2}{1-2r\cos(x'-x)+r^2}\phi(x')dx'. \quad\text{.................................(2)}$$

But we are given that $\lim_{t\to 0}[f(x+t)+f(x-t)]$ exists.

Let the arbitrary positive number ϵ be chosen, as small as we please. Then to $\tfrac{1}{2}\epsilon$ there will correspond a positive number η such that

$$|f(x+t)+f(x-t)-\lim_{t\to 0}[f(x+t)+f(x-t)]|<\tfrac{1}{2}\epsilon, \quad\text{............(3)}$$

when $0<t\leqq\eta$.

The number η fixed upon will be such that $(x-\eta, x+\eta)$ does not go beyond $(-\pi, \pi)$.

Then

$$\frac{1}{2\pi}\int_{x-\eta}^{x+\eta}\frac{1-r^2}{1-2r\cos(x'-x)+r^2}\phi(x')dx'$$

$$=\frac{1}{2\pi}\int_{0}^{\eta}\frac{1-r^2}{1-2r\cos t+r^2}[\phi(x+t)+\phi(x-t)]\,dt$$

$$=\frac{1}{2\pi}\int_{0}^{\eta}\frac{1-r^2}{1-2r\cos t+r^2}\{f(x+t)+f(x-t)-\lim_{t\to 0}[f(x+t)+f(x-t)]\}\,dt.$$

It follows that

$$\left|\frac{1}{2\pi}\int_{x-\eta}^{x+\eta}\frac{1-r^2}{1-2r\cos(x'-x)+r^2}\phi(x')dx'\right|<\frac{\epsilon}{4\pi}\int_0^{\eta}\frac{1-r^2}{1-2r\cos t+r^2}dt$$

$$<\frac{\epsilon}{4\pi}\int_{-\pi}^{\pi}\frac{1-r^2}{1-2r\cos t+r^2}dt$$

$$<\tfrac{1}{2}\epsilon. \qquad \text{.............................(4)}$$

Also, when $0<r<1$,

$$\left|\frac{1}{2\pi}\left(\int_{-\pi}^{x-\eta}+\int_{x+\eta}^{\pi}\right)\frac{1-r^2}{1-2r\cos(x'-x)+r^2}\phi(x')dx'\right|$$

$$<\frac{1}{2\pi}\frac{1-r^2}{1-2r\cos\eta+r^2}\int_{-\pi}^{\pi}|\phi(x')|dx'$$

$$<\frac{1-r^2}{1-2r\cos\eta+r^2}\left(\frac{1}{2\pi}\int_{-\pi}^{\pi}|f(x')|dx'+\tfrac{1}{2}\left|\lim_{t\to 0}[f(x+t)+f(x-t)]\right|\right)$$

$$<\frac{1-r^2}{1-2r\cos\eta+r^2}\times A, \text{ say.} \qquad \text{.............................(5)}$$

But
$$\frac{1-r^2}{1-2r\cos\eta+r^2}<\frac{2(1-r)}{(1-r)^2+4r\sin^2\dfrac{\eta}{2}}, \text{ if } 0<r<1,$$

$$<\frac{1-r}{2r\sin^2\dfrac{\eta}{2}}.$$

And
$$\frac{1-r}{2r\sin^2\dfrac{\eta}{2}}<\frac{\epsilon}{2A},$$

provided that
$$r>\frac{1}{1+\dfrac{\epsilon}{A}\sin^2\dfrac{\eta}{2}}.$$

It follows that

$$\left|\frac{1}{2\pi}\left(\int_{-\pi}^{x-\eta}+\int_{x+\eta}^{\pi}\right)\frac{1-r^2}{1-2r\cos(x'-x)+r^2}\phi(x')\,dx'\right|<\frac{\epsilon}{2},$$

if
$$1>r>\frac{1}{1+\dfrac{\epsilon}{A}\sin^2\dfrac{\eta}{2}}. \qquad \text{.............................(6)}$$

Combining (4) and (6), it will be seen that when any positive number ϵ has been chosen, as small as we please there is a positive number ρ such that

$$\left|F(r,x)-\tfrac{1}{2}\lim_{t\to 0}[f(x+t)+f(x-t)]\right|<\epsilon,$$

when $\rho\leqq r<1$, provided that for the value of x considered $\lim\limits_{t\to 0}[f(x+t)+f(x-t)]$ exists.

We have thus established the following theorem:

Let $f(x)$, given in the interval $(-\pi, \pi)$, be bounded and integrable, or have an absolutely convergent infinite integral, in this range. Then for any value of x in $-\pi<x<\pi$ for which

$$\lim_{t\to 0}[f(x+t)+f(x-t)]$$

exists, Poisson's Integral converges to that limit as $r\to 1$ from below.

In particular, at a point of ordinary discontinuity of $f(x)$, *Poisson's Integral converges to*

$$\tfrac{1}{2}[f(x+0)+f(x-0)],$$

and, at a point where $f(x)$ *is continuous, it converges to* $f(x)$.

It has already been pointed out that no conclusion can be drawn from this as to the convergence, or non-convergence, of the Fourier's Series at this point. But if we know that the Fourier's Series does converge, it follows from Abel's Theorem that it must converge to the limit to which Poisson's Integral converges as $r \to 1$.

We have thus the following theorem:

If $f(x)$ *is any function, given in* $(-\pi, \pi)$, *which is either bounded and integrable, or has an absolutely convergent infinite integral* $\int_{-\pi}^{\pi} f(x)\,dx$, *then, at any point* x *in* $-\pi < x < \pi$ *at which the Fourier's Series is convergent, its sum must be equal to*

$$\lim_{t\to 0}\,[f(x+t)+f(x-t)],$$

provided that this limit exists.

With certain obvious modifications these theorems can be made to apply to the points $-\pi$ and π as well as points between $-\pi$ and π.

It follows immediately from this theorem that:

If all the Fourier's Constants are zero for a function, continuous in the interval $(-\pi, \pi)$, *then the function vanishes identically.*

If the constants vanish but the function only satisfies the conditions ascribed to $f(x)$ in the earlier theorems of this section, we can only infer that the function must vanish at all points where it is continuous, and that at points where $\lim_{t\to 0}[f(x+t)+f(x-t)]$, exists, this limit must be zero.

Further, if (a, b) is an interval in which $f(x)$ is continuous, the same number ρ, corresponding to the arbitrary ϵ, may be chosen to serve all the values of x in the interval (a, b); for this is true, first of the number η in (3), then of A in (5), and thus finally of ρ.

It follows that *as* $r \to 1$ *Poisson's Integral converges uniformly to the value* $f(x)$ *in any interval* (a, b) *in which* $f(x)$ *is continuous.**

This last theorem has an important application in connection with the approximate representation of functions by finite trigonometrical series.†

101. Fejér's Theorem.‡

Let $f(x)$ *be given in the interval* $(-\pi, \pi)$. *If bounded, let it be*

---

* It is assumed in this that $f(a-0)=f(a)=f(a+0)$ and $f(b-0)=f(b)=f(b+0)$. Also $f(x)$ is subject to the conditions given at the beginning of this section. Cf. § 107.

† Cf. Picard, *Traité d'Analyse*, 1 (2ᵉ éd., 1905), 275; Bôcher, *Annals of Math.*, (2), 7 (1906), 102; Hobson, *Theory of Functions of a Real Variable*, 2 (2nd ed., 1926), 637.

‡Cf. *Math. Annalen*, 58 (1904), 51.

integrable; if unbounded, let the infinite integral $\int_{-\pi}^{\pi} f(x)dx$ *be absolutely convergent. Denote by* s_n *the sum of the* $(n+1)$ *terms*

$$\frac{1}{2\pi}\int_{-\pi}^{\pi} f(x')dx' + \frac{1}{\pi}\sum_{1}^{n}\int_{-\pi}^{\pi} f(x')\cos r(x'-x)\,dx'.$$

Also let

$$\sigma_n(x) = \frac{s_0 + s_1 + \dots + s_{n-1}}{n}.$$

Then at every point x *in the interval* $-\pi < x < \pi$ *at which* $f(x+0)$ *and* $f(x-0)$ *exist,*

$$\lim_{n\to\infty} \sigma_n(x) = \tfrac{1}{2}[f(x+0) + f(x-0)].$$

With the above notation,

$$s_n = \frac{1}{2\pi}\int_{-\pi}^{\pi} f(x')\frac{\cos n(x'-x) - \cos(n+1)(x'-x)}{1-\cos(x'-x)}dx'.$$

Therefore

$$\begin{aligned}\sigma_n(x) &= \frac{1}{2n\pi}\int_{-\pi}^{\pi} f(x')\frac{1-\cos n(x'-x)}{1-\cos(x'-x)}dx' \\ &= \frac{1}{2n\pi}\int_{-\pi}^{\pi} f(x')\frac{\sin^2\frac{1}{2}n(x'-x)}{\sin^2\frac{1}{2}(x'-x)}dx' \\ &= \frac{1}{2n\pi}\int_{-\pi+x}^{\pi+x} f(x')\frac{\sin^2\frac{1}{2}n(x'-x)}{\sin^2\frac{1}{2}(x'-x)}dx', \quad \text{......................(1)}\end{aligned}$$

if $f(x)$ is defined outside the interval $(-\pi, \pi)$ by the equation

$$f(x+2\pi) = f(x).$$

Dividing the range of integration into $(-\pi+x, x)$ and $(x, \pi+x)$, and substituting $x' = x - 2a$ in the first, and $x' = x + 2a$ in the second, we obtain

$$\sigma_n(x) = \frac{1}{n\pi}\int_0^{\frac{1}{2}\pi} f(x-2a)\frac{\sin^2 na}{\sin^2 a}da + \frac{1}{n\pi}\int_0^{\frac{1}{2}\pi} f(x+2a)\frac{\sin^2 na}{\sin^2 a}da. \quad \text{......(2)}$$

Now suppose that x is a point in $(-\pi, \pi)$ at which $f(x+0)$ and $f(x-0)$ exist.

Let ϵ be any positive number, chosen as small as we please.

Then to ϵ there corresponds a positive number η chosen less than $\frac{1}{2}\pi$ such that

$$|f(x+2a)| - f(x+0)| < \epsilon \text{ when } 0 < a \leqq \eta.$$

Also

$$\frac{1}{n\pi}\int_0^{\frac{1}{2}\pi} f(x+2a)\frac{\sin^2 na}{\sin^2 a}\,da = \frac{1}{n\pi}\int_0^{\eta}\{f(x+2a)-f(x+0)\}\frac{\sin^2 na}{\sin^2 a}\,da$$

$$+\frac{1}{n\pi}f(x+0)\int_0^{\frac{1}{2}\pi}\frac{\sin^2 na}{\sin^2 a}\,da$$

$$+\frac{1}{n\pi}\int_{\eta}^{\frac{1}{2}\pi} f(x+2a)\frac{\sin^2 na}{\sin^2 a}\,da$$

$$-\frac{1}{n\pi}f(x+0)\int_{\eta}^{\frac{1}{2}\pi}\frac{\sin^2 na}{\sin^2 a}\,da$$

$$= I_1+I_2+I_3+I_4, \text{ say.*} \qquad \ldots\ldots\ldots\ldots(3)$$

Putting $C_{n-1}=\frac{1}{2}+\cos 2a+\cos 4a+\ldots+\cos 2(n-1)a$,

we have $$C_{n-1}=\frac{1}{2}\frac{(\cos 2(n-1)a-\cos 2na)}{4(1-\cos 2a)},$$

and $$C_0+C_1+\ldots+C_{n-1}=\frac{1}{2}\frac{(1-\cos 2na)}{(1-\cos 2a)}=\frac{1}{2}\frac{\sin^2 na\dagger}{\sin^2 a}.$$

Thus $$\int_0^{\frac{1}{2}\pi}\frac{\sin^2 na}{\sin^2 a}\,da = 2\int_0^{\frac{1}{2}\pi}(C_0+C_1+\ldots+C_{n-1})\,da$$

$$=\tfrac{1}{2}n\pi,$$

since all the terms on the right-hand side disappear on integration except the first in each of the C's.

It follows that $$I_2=\tfrac{1}{2}f(x+0).$$

Also $$|I_1| \leqq \frac{1}{n\pi}\int_0^{\eta}|f(x+2a)-f(x+0)|\frac{\sin^2 na}{\sin^2 a}\,da$$

$$<\frac{\epsilon}{n\pi}\int_0^{\eta}\frac{\sin^2 na}{\sin^2 a}\,da$$

$$<\frac{\epsilon}{n\pi}\int_0^{\frac{1}{2}\pi}\frac{\sin^2 na}{\sin^2 a}\,da$$

$$<\tfrac{1}{2}\epsilon. \qquad \ldots\ldots\ldots\ldots(4)$$

Further, $$|I_3| \leqq \frac{1}{n\pi}\int_{\eta}^{\frac{1}{2}\pi}|f(x+2a)|\frac{\sin^2 na}{\sin^2 a}\,da$$

$$<\frac{1}{n\pi\sin^2\eta}\int_{\eta}^{\frac{1}{2}\pi}|f(x+2a)|\,da$$

$$<\frac{1}{2n\pi\sin^2\eta}\int_{x+2\eta}^{x+\pi}|f(x')|\,dx'. \qquad \ldots\ldots\ldots\ldots(5)$$

---

*This discussion also applies when the upper limit of the integral on the left is any positive number less than $\frac{1}{2}\pi$.

†This integral can be obtained at once from (1) by putting $f(x)=1$ in $(-\pi, \pi)$.

But we are given that $\int_{-\pi}^{\pi} |f(x')|dx'$ converges, and we have defined $f(x)$ outside the interval $(-\pi, \pi)$ by the equation $f(x+2\pi)=f(x)$.

Let $$\int_{-\pi}^{\pi} |f(x')|dx' = \pi J, \text{ say.}$$

Then we have $$|I_3| < \frac{J}{2n \sin^2 \eta}.$$

Also $$|I_4| < \frac{1}{2n \sin^2 \eta} |f(x+0)|. \qquad \text{.........................(6)}$$

Combining these results, it follows from (3) that

$$\left| \frac{1}{n\pi} \int_0^{\frac{1}{2}\pi} f(x+2\alpha) \frac{\sin^2 n\alpha}{\sin^2 \alpha} d\alpha - \tfrac{1}{2} f(x+0) \right|$$
$$< \tfrac{1}{2}\epsilon + \frac{1}{2n \sin^2 \eta} \{J + |f(x+0)|\}.$$

Now let ν be a positive integer such that

$$\frac{1}{\nu \sin^2 \eta} \{J + |f(x+0)|\} < \epsilon. \qquad \text{.......................(7)}$$

Then

$$\left| \frac{1}{n\pi} \int_0^{\frac{1}{2}\pi} f(x+2\alpha) \frac{\sin^2 n\alpha}{\sin^2 \alpha} d\alpha - \tfrac{1}{2} f(x+0) \right| < \tfrac{1}{2}\epsilon + \tfrac{1}{2}\epsilon$$
$$< \epsilon, \text{ when } n \geqq \nu.$$

In other words,

$$\lim_{n\to\infty} \frac{1}{n\pi} \int_0^{\frac{1}{2}\pi} f(x+2\alpha) \frac{\sin^2 n\alpha}{\sin^2 \alpha} d\alpha = \tfrac{1}{2} f(x+0),$$

when $f(x+0)$ exists.

In precisely the same way we find that

$$\lim_{n\to\infty} \frac{1}{n\pi} \int_0^{\frac{1}{2}\pi} f(x-2\alpha) \frac{\sin^2 n\alpha}{\sin^2 \alpha} d\alpha = \tfrac{1}{2} f(x-0),$$

when $f(x-0)$ exists.

Then, returning to (2), we have

$$\lim_{n\to\infty} \sigma_n(x) = \tfrac{1}{2}[f(x+0) + f(x-0)],$$

when $f(x \pm 0)$ exist.

This proof applies also to the points $x = \pm\pi$, when $f(\pi - 0)$ and $f(-\pi+0)$ exist. Since we have defined $f(x)$ outside the interval $(-\pi, \pi)$ by the equation $f(x+2\pi)=f(x)$, it is clear that

$$f(-\pi+0) = f(\pi+0) \quad \text{and} \quad f(-\pi-0) = f(\pi-0).$$

In this way we obtain

$$\lim_{n\to\infty} \sigma_n(\pm\pi) = \tfrac{1}{2}[f(-\pi+0) + f(\pi-0)],$$

when $f(-\pi+0)$ and $f(\pi-0)$ exist.

Corollary. If $f(x)$ is continuous in $a \leqq x \leqq b$, including the end-points, when the arbitrary positive number ϵ is chosen, the same η will do for all values of x from a to b, including the end-points. Then, from (7), it follows that *the sequence of arithmetic means*

$$\sigma_1, \quad \sigma_2, \quad \sigma_3, \ldots$$

converges uniformly to the sum $f(x)$ in the interval (a, b).

It is assumed in this statement that $f(x)$ is continuous at $x=a$ and $x=b$ as well as in the interval (a, b); *i.e.* $f(a-0)=f(a)=f(a+0)$, and

$$f(b-0)=f(b)=f(b+0).$$

102. Two Theorems on the Arithmetic Means. Before applying this very important theorem to the discussion of Fourier's Series, we shall prove two theorems regarding the sequence of arithmetic means for any series

$$u_1+u_2+u_3+\ldots.$$

In this connection we adopt the notation

$$s_n=u_1+u_2+\ldots+u_n,$$

$$\sigma_n=\frac{s_1+s_2+\ldots+s_n}{n}.$$

Theorem I. *If the series $u_1+u_2+u_3+\ldots$ converges and its sum is s, then the sequence of arithmetic means σ_n also converges to s.**

(i) First, we assume that $\lim_{n\to\infty} s_n=0$, and we prove that $\lim_{n\to\infty}\sigma_n=0$.

Take the arbitrary positive number ϵ.

Then there is a positive integer N, such that

$$|s_n|<\tfrac{1}{2}\epsilon, \text{ when } n \geqq N.$$

Also $$|\sigma_n| \leqq \frac{|s_1+s_2+\ldots+s_N|}{n}+\frac{|s_{N+1}|+|s_{N+2}|+\ldots+|s_n|}{n}.$$

But we can choose $\nu>N$, so that

$$\frac{|s_1+s_2+\ldots+s_N|}{n}<\tfrac{1}{2}\epsilon, \text{ when } n \geqq \nu.$$

Therefore $$|\sigma_n|<\tfrac{1}{2}\epsilon+\tfrac{1}{2}\left(1-\frac{N}{n}\right)\epsilon, \text{ when } n \geqq \nu$$

$$<\epsilon, \text{ when } n \geqq \nu.$$

Thus $$\lim_{n\to\infty}\sigma_n=0.$$

(ii) Let $$\lim_{n\to\infty} s_n=s\neq 0.$$

---

* *If the series $\sum_1^\infty u_r$ diverges to $+\infty$ (or to $-\infty$), then* $\lim_{n\to\infty}\sigma_n=+\infty$ (or $-\infty$).

For, in the case of divergence to $+\infty$, however large K may be, we know that there is a positive integer N such that $s_n>K$, when $n \geqq N$.

But $$\sigma_n=\frac{s_1+s_2+\ldots+s_N}{n}+\frac{s_{N+1}+s_{N+2}+\ldots+s_n}{n}.$$

The first part tends to zero when $n\to\infty$, and the second part is greater than $\left(1-\frac{N}{n}\right)K$, which tends to K ,when $n\to\infty$.

Then $$0=(u_1-s)+u_2+u_3+\dots .$$

The arithmetic mean for n terms of this series is equal to $\frac{1}{n}\left(\sum_1^n s_n - ns\right)$, where $s_n=\sum_1^n u_r$; and thus it is equal to $(\sigma_n - s)$.

But by (i) the limit of this sequence of arithmetic means is zero.

Therefore $\lim_{n\to\infty} \sigma_n = s$.

THEOREM II. *Let the sequence of arithmetic means σ_n of the series*

$$u_1+u_2+u_3+\dots$$

converge to σ: and **either** $n(s_n - s_{n+1})<K$ **or** $n(s_{n+1}-s_n)<K$, *where K is some positive constant.*

Then $$\lim_{n\to\infty} s_n = \sigma.\text{*}$$

We may, as above, without loss of generality, take

$$\sigma=0,\ K=1, \quad \text{and} \quad n(s_n-s_{n+1})<1.\dagger$$

Suppose we are given $\quad \lim_{n\to\infty}\sigma_n=0 \quad$ and $\quad n(s_n-s_{n+1})<1.$

It is clear that $\lim_{n\to\infty} s_n$ is not equal to $+\infty$ (or $-\infty$), because if this limit were $+\infty$ (or $-\infty$) we would have $\lim_{n\to\infty}\sigma_n=+\infty$ (or $-\infty$). [Cf. footnote on p. 258].

(i) If possible, let $\overline{\lim_{n\to\infty}}\, s_n=\Lambda$, where Λ is $+\infty$, or a finite positive number.

Then, if A is any positive number $<\Lambda$,

$s_n>A$, for an infinite number of values of n, say $M_1, M_2, M_3, \dots$.

But to the arbitrary positive number ϵ, there corresponds a positive integer μ, such that

$$|\sigma_n|<\epsilon, \text{ when } n \geqq \mu.$$

Let M be the first of the sequence $M_1, M_2, M_3, \dots$ which is greater than μ, and such that $MA \geqq$ an even positive integer.

Let $2p$ be the largest even positive integer not greater than MA.

Then $$2p \leqq MA < 4p.$$

---

*This proof of the Hardy-Landau Theorem is due to Professor A. E. Jolliffe. This theorem, in a less general form, was given by Hardy in *Proc. London Math. Soc.* (2), 8 (1910), 302. In the earlier edition of this book Littlewood's proof, given in Whittaker and Watson's *Modern Analysis* is followed. Cf. also de la Vallée Poussin, *loc. cit.* 2 (4e éd., 1922), § 93; and Bromwich, *loc. cit.* (2nd ed., 1926), § 151, where further references will be found.

†For if we put $U_1=\frac{u_1-\sigma}{K}$, $U_2=\frac{u_2}{K}$, etc., and $S_n=\sum_1^n U_r$,

we have $$S_n=\frac{s_n-\sigma}{K}, \quad \text{and} \quad \frac{1}{n}\sum_1^n S_r=\frac{\sigma_n-\sigma}{K}.$$

Thus $$n(S_n-S_{n+1})<1, \quad \text{if} \quad n(s_n-s_{n+1})<K.$$

In the other case, we put $U_1=\frac{\sigma-u_1}{K}$, $U_2=-\frac{u_2}{K}$, etc.

Also
$$s_{M+1} > s_M - \frac{1}{M},$$
$$s_{M+2} > s_{M+1} - \frac{1}{M+1} > s_M - \left(\frac{1}{M} + \frac{1}{M+1}\right) > s_M - \frac{2}{M},$$
$$s_{M+3} > \quad \dots\dots\dots\dots \quad > s_M - \frac{3}{M},$$
$$\dots\dots\dots\dots\dots\dots$$
$$s_{M+p} > \quad \dots\dots\dots\dots \quad > s_M - \frac{p}{M},$$

and each of these $> A - \dfrac{p}{M}$.

But
$$\sigma_{M+p} = \frac{M\sigma_M + (s_{M+1} + s_{M+2} + \dots + s_{M+p})}{M+p}$$
$$> \frac{M}{M+p}\sigma_M + \frac{p}{M+p}\left(A - \frac{p}{M}\right)$$
$$> \frac{a}{a+1}(A-a) - \epsilon, \quad \text{where } a = \frac{p}{M} \leqq \tfrac{1}{2}A.$$

Now $A - a \geqq \frac{1}{2}A$ and $\dfrac{a}{a+1} > \dfrac{\frac{A}{4}}{\frac{A}{4}+1}$, since $\dfrac{p}{M} = a > \frac{1}{4}A$.

Therefore
$$\frac{a}{a+1}(A-a) > \frac{A^2}{2(A+4)},$$

and
$$\sigma_{M+p} > \frac{A^2}{2(A+4)} - \epsilon.$$

If we take $\epsilon = \dfrac{A^2}{4(A+4)}$, we have $\sigma_{M+p} > \epsilon$, which is impossible.

Thus
$$\overline{\lim}\, s_n < 0.$$

(ii) We shall now prove that $\underline{\lim}\, s_n \geqq 0$.

For, if possible, let $\underline{\lim}\, s_n = \lambda < 0$.

Take B any positive number $< -\lambda$.

Then $s_n < -B$, for an infinite number of values of n, say $N_1, N_2, N_3, \dots$.

And
$$|\sigma_n| < \epsilon, \text{ when } n \geqq \nu.$$

Let N be the first of the sequence $N_1, N_2, N_3, \dots$ such that $\dfrac{N}{1+\frac{1}{2}B} > \nu$ and such that between $\dfrac{N}{1+\frac{1}{2}B}$ and $\dfrac{N}{1+\frac{1}{4}B}$ there shall be at least one positive integer.

Let the integer next above $\dfrac{N}{1+\frac{1}{2}B}$ be q, and write $\dfrac{N}{q} = 1 + b$.

Then since
$$\frac{N}{1+\frac{1}{2}B} < q < \frac{N}{1+\frac{1}{4}B},$$

we have
$$\tfrac{1}{4}B < b < \tfrac{1}{2}B.$$

Also as before
$$\sigma_N = \frac{q\sigma_q + s_{q+1} + \dots + s_N}{N}.$$

But
$$s_{N-1} < s_N + \frac{1}{N-1},$$
$$s_{N-2} < s_{N-1} + \frac{1}{N-2} < s_N + \left(\frac{1}{N-1} + \frac{1}{N-2}\right).$$
$$\cdots\cdots\cdots\cdots\cdots\cdots\cdots\cdots\cdots\cdots\cdots\cdots$$
$$s_{q+1} < \ldots < s_N + \left(\frac{1}{N-1} + \frac{1}{N-2} + \ldots + \frac{1}{q+1} + \frac{1}{q}\right).$$

Therefore each of these $< - B + \dfrac{N-q}{q} < - B + b.$

Also
$$\sigma_N < \frac{q}{N}\sigma_q + \frac{N-q}{N}(-B+b)$$
$$< -\frac{b}{1+b}(B-b) + \epsilon.$$

But
$$B - b > \tfrac{1}{2}B \quad \text{and} \quad \frac{b}{1+b} > \frac{B}{(B+4)}.$$

Therefore
$$\frac{b}{1+b}(B-b) > \frac{B^2}{2(B+4)},$$

and
$$\sigma_N < -\frac{B^2}{2(B+4)} + \epsilon.$$

If we take $\epsilon = \dfrac{B^2}{4(B+4)}$, we have $\sigma_N < -\epsilon$, which is impossible.

Thus
$$\varliminf_{n\to\infty} s_n \geqq 0.$$

But we have seen that
$$\varlimsup_{n\to\infty} s_n \leqq 0.$$

It follows that
$$\lim s_n = 0.$$

COROLLARY I. *Let the sequence of arithmetic means σ_n for the series*
$$u_1 + u_2 + u_3 + \ldots$$
converge to σ.

If a positive integer n_0 exists such that $|u_n| < \dfrac{K}{n}$, when $n \geqq n_0$, where K is a positive number independent of n,

then the series $\sum_1^\infty u_n$ converges and its sum is σ.

This is a special case of Theorem II.

COROLLARY II. *Let $u_1(x) + u_2(x) + \ldots$ be a series whose terms are functions of x, and let the sequence of arithmetic means $\sigma_n(x)$ converge uniformly to $\sigma(x)$ in an interval (a, b).*

Then, if **either** $n[s_n(x) - s_{n+1}(x)] < K$ **or** $n[s_{n+1}(x) - s_n(x)] > K$, *when $n \geqq n_0$, where K is independent of x and n, and the same n_0 serves for all values of x in (a, b),*
$$\lim_{n\to\infty} s_n(x) = \sigma(x)$$
uniformly.

As before we may, without loss of generality, put
$$K = 1,\ \sigma(x) = 0, \quad \text{and} \quad n[s_n(x) - s_{n+1}(x)] < 1.$$

Then we have
$$|\sigma_n| < \epsilon, \text{ when } n \geqq \nu,$$
the same ν serving for all values of x in (a, b).

If $s_n(x)$ does not converge uniformly to zero, there must be a positive number ϵ_0 such that an infinite number of the set

$$|s_1(x)|, \quad |s_2(x)|, \quad |s_3(x)|, \ldots .$$

are greater than or equal to ϵ_0, each for some value of x in (a, b); that is, there must be an infinite set of positive integers $N_1, N_2, N_3, \ldots$ such that

$$|s_{N_1}(x_{N_1})|, \quad |s_{N_2}(x_{N_2})|, \ldots$$

are each greater than or equal to ϵ_0, $x_{N_1}, x_{N_2}, \ldots$ being points of (a, b), corresponding to $N_1, N_2, \ldots$.

Let N be the first of this set $N_1, N_2, \ldots$ which is greater than ν_0 and n_0, and such that $N\epsilon_0 \geqq$ an even positive integer, say $2p$.

Then

$$s_{N+1}(x_N) > s_N(x_N) - \frac{1}{N},$$

$$s_{N+2}(x_N) > s_N(x_N) - \frac{2}{N},$$

$$\cdots\cdots\cdots\cdots\cdots\cdots$$

$$\cdots\cdots\cdots\cdots\cdots\cdots$$

$$s_{N+p}(x_N) > s_N(x_N) - \frac{p}{N}.$$

And the argument proceeds as before.

103. Fejér's Theorem and Fourier's Series.* We shall now use Fejér's Theorem (§ 101) to establish the convergence of Fourier's Series under the limitations imposed in our previous discussion; that is we shall show that: *When $f(x)$ satisfies Dirichlet's Conditions in the interval $(-\pi, \pi)$, and*

$$a_0 = \frac{1}{2\pi}\int_{-\pi}^{\pi} f(x')\,dx', \quad a_n = \frac{1}{\pi}\int_{-\pi}^{\pi} f(x') \cos nx'\,dx',$$

$$b_n = \frac{1}{\pi}\int_{-\pi}^{\pi} f(x') \sin nx'\,dx' \quad (n \geqq 1),$$

the sum of the series

$$a_0 + (a_1 \cos x + b_1 \sin x) + (a_2 \cos 2x + b_2 \sin 2x) + \ldots$$

is $\frac{1}{2}[f(x+0) + f(x-0)]$ at every point in $-\pi < x < \pi$ where $f(x+0)$ and $f(x-0)$ exist; and at $x = \pm\pi$ the sum is $\frac{1}{2}[f(-\pi+0) + f(\pi-0)]$, when these limits exist.

I. First, let $f(x)$ be bounded in $(-\pi, \pi)$ and otherwise satisfy Dirichlet's Conditions in this interval.†

If the interval $(-\pi, \pi)$ can be broken up into a finite number (say p) of open partial intervals in which $f(x)$ is monotonic, it follows at once from the Second Theorem of Mean Value that each of these intervals contributes to $|a_n|$ or $|b_n|$ a part less than $4M/n\pi$, where $|f(x)| < M$ in $(-\pi, \pi)$.

Thus we have $\qquad |a_n \cos nx + b_n \sin nx| < 8pM/n\pi,$

where M is independent of n.

It follows from Fejér's Theorem, combined with Theorem II, Cor. I of § 102, that the Fourier's Series

$$a_0 + (a_1 \cos x + b_1 \sin x) + (a_2 \cos 2x + b_2 \sin 2x) + \ldots$$

converges, and its sum is $\frac{1}{2}[f(x+0) + f(x-0)]$ at every point in $-\pi < x < \pi$

---

* Cf. Whittaker and Watson, *loc. cit.* (2nd ed., 1915), 167.

† If the more general condition that the function is of bounded variation in $(-\pi, \pi)$ is taken, then $f(x)$ is the difference of two positive, bounded and monotonic functions, and a similar argument applies.

at which $f(x \pm 0)$ exist, and at $x = \pm\pi$ its sum is $\frac{1}{2}[f(-\pi+0)+f(\pi-0)]$, provided that $f(-\pi+0)$ and $f(\pi-0)$ exist.

II. Next, let there be a finite number of points of infinite discontinuity in $(-\pi, \pi)$, but, when arbitrarily small neighbourhoods of these points are excluded, let $f(x)$ be bounded in the remainder of the interval, which can be broken up into a finite number of open partial intervals in each of which $f(x)$ is monotonic. In addition, let the infinite integral $\int_{-\pi}^{\pi} f(x')dx'$ be absolutely convergent.

In this case, let x be a point between $-\pi$ and π at which $f(x+0)$ and $f(x-0)$ exist. Then we may suppose it an internal point of an interval (a, b), where $b - a < \pi$, and $f(x)$ is bounded in (a, b) and otherwise satisfies Dirichlet's Conditions therein.

Let
$$\left.\begin{aligned} \pi a_n' &= \int_a^b f(x') \cos nx'\, dx' \\ \pi b_n' &= \int_a^b f(x') \sin nx'\, dx' \end{aligned}\right\} n \geqq 1,$$
and

while
$$2\pi a_0' = \int_a^b f(x')dx'.$$

Then, forming the arithmetic means for the series
$$a_0' + (a_1' \cos x + b_1' \sin x) + (a_2' \cos 2x + b_2' \sin 2x) + \dots,$$
we have, with the notation of § 101,
$$\begin{aligned} \sigma_n(x) &= \frac{s_0 + s_1 + \dots + s_{n-1}}{n} \\ &= \frac{1}{2n\pi}\int_a^b f(x') \frac{\sin^2 \frac{1}{2}n(x'-x)}{\sin^2 \frac{1}{2}(x'-x)} dx' \\ &= \frac{1}{n\pi}\left\{\int_0^{\frac{1}{2}(x-a)} f(x-2\alpha) \frac{\sin^2 n\alpha}{\sin^2\alpha} d\alpha + \int_a^{\frac{1}{2}(b-x)} f(x+2\alpha) \frac{\sin^2 n\alpha}{\sin^2\alpha} d\alpha\right\}, \end{aligned}$$
where $\frac{1}{2}(x-a)$ and $\frac{1}{2}(b-x)$ are each positive and less than $\frac{1}{2}\pi$.

But it will be seen that the argument used in Fejér's Theorem with regard to the integrals
$$\frac{1}{n\pi}\int_0^{\frac{1}{2}\pi} f(x \pm 2\alpha) \frac{\sin^2 n\alpha}{\sin^2\alpha} d\alpha$$
applies equally well when the upper limits of the integrals are positive and less than $\frac{1}{2}\pi$.*

Therefore, in this case.
$$\lim_{n\to\infty} \sigma_n(x) = \tfrac{1}{2}[f(x+0)+f(x-0)].$$

And, as the terms $(a_n' \cos nx + b_n' \sin nx)$
satisfy the condition of Theorem II of § 102, it follows that the series
$$a_0' + (a_1' \cos x + b_1' \sin x) + (a_2' \cos 2x + b_2' \sin 2x) + \dots$$
converges and that its sum is $\lim_{n\to\infty} S_n(x)$.

---

*Cf. footnote, p. 256.

But $(a_0 - a_0') + \sum_1^n \{(a_n - a_n')\cos nx + (b_n - b_n')\sin nx\}$

$$= \frac{1}{\pi}\left\{\int_{-\pi}^{a} + \int_{b}^{\pi}\right\} f(x')\{\tfrac{1}{2} + \sum_1^n \cos n(x'-x)\}\, dx'$$

$$= \frac{1}{2\pi}\left\{\int_{-\pi+x}^{a} + \int_{b}^{\pi+x}\right\} f(x') \frac{\sin \frac{1}{2}(2n+1)(x'-x)}{\sin \frac{1}{2}(x'-x)}\, dx' \, *$$

$$= \frac{1}{\pi}\int_{\frac{1}{2}(x-a)}^{\frac{1}{2}\pi} f(x-2\alpha)\frac{\sin(2n+1)\alpha}{\sin\alpha}\, d\alpha + \frac{1}{\pi}\int_{\frac{1}{2}(b-x)}^{\frac{1}{2}\pi} f(x+2\alpha)\frac{\sin(2n+1)\alpha}{\sin\alpha}\, d\alpha.$$

By § 94 both of these integrals vanish in the limit as $n \to \infty$.

It follows that the series

$$(a_0 - a_0') + \sum_1^\infty \{(a_n - a_n')\cos nx + (b_n - b_n')\sin nx\}$$

converges, and that its sum is zero.

But we have already shown that

$$a_0' + \sum_1^n (a_n' \cos nx + b_n' \sin nx)$$

converges, and that its sum is

$$\tfrac{1}{2}[f(x+0) + f(x-0)].$$

It follows, by adding the two series, that

$$a_0 + \sum_1^\infty (a_n \cos nx + b_n \sin nx)$$

converges, and that its sum is

$$\tfrac{1}{2}[f(x+0) + f(x-0)]$$

at any point between $-\pi$ and π at which these limits exist.

When the limits $f(-\pi+0)$ and $f(\pi-0)$ exist, we can reduce the discussion of the sum of the series for $x = \pm\pi$ to the above argument, using the equation $f(x+2\pi) = f(x)$.

We can then treat $x = \pm\pi$ as inside an interval (a, b), as above.

REFERENCES.

BÔCHER, "Introduction to the Theory of Fourier's Series," *Annals of Math.*, (2), 7 (1906).

DE LA VALLÉE POUSSIN, *loc. cit.*, 2 (4e éd., 1922), Ch. IV.

DINI, *Serie di Fourier e altre rappresentazioni analitiche delle funzioni di una variabile reale* (Pisa, 1880), Cap. IV.

FOURIER, *Théorie analytique de la chaleur*, Ch. III.

GOURSAT, *loc. cit.*, T. I (3e éd.), Ch. IX.

HOBSON, *Theory of Functions of a Real Variable*, 2 (2nd ed., 1926), Ch. VIII.

JORDAN, *Cours d'Analyse*, 2 (3e éd., Paris, 1913), Ch. V.

LEBESGUE, *Leçons sur les séries trigonométriques* (Paris, 1906), Ch. II.

NEUMANN, *Über d. nach Kreis-, Kugel- u. Cylinder-Functionen fortschreitenden Entwickelungen* (Leipzig, 1881), Kap. II.

* $f(x)$ is defined outside the interval $(-\pi, \pi)$ by the equation $f(x+2\pi) = f(x)$.

RIEMANN, *loc. cit.*

WEBER-RIEMANN, *Die partiellen Differential-gleichungen der mathematischen Physik*, 1 (2 Aufl., Braunschweig, 1910).

WHITTAKER AND WATSON, *loc. cit.* (5th ed., 1928), Ch. IX.

EXAMPLES ON CHAPTER VII.

1. In the interval $\qquad 0<x<\frac{l}{2},\ f(x)=\frac{1}{4}l-x,$

and in the interval $\qquad \frac{l}{2}<x<l,\ f(x)=x-\frac{3}{4}l.$

Prove that $\qquad f(x)=\frac{2l}{\pi^2}\left(\cos\frac{2\pi x}{l}+\frac{1}{9}\cos\frac{6\pi x}{l}+\frac{1}{25}\cos\frac{10\pi x}{l}+\dots\right).$

2. The function $f(x)$ is defined as follows for the interval $(0, \pi)$;

$$f(x)=\tfrac{3}{2}x, \quad \text{when} \quad 0\leqq x\leqq \tfrac{1}{3}\pi,$$
$$f(x)=\tfrac{1}{2}\pi, \quad \text{when} \quad \tfrac{1}{3}\pi<x<\tfrac{2}{3}\pi,$$
$$f(x)=\tfrac{3}{2}(\pi-x), \quad \text{when} \quad \tfrac{2}{3}\pi\leqq x\leqq \pi.$$

Show that

$$f(x)=\frac{6}{\pi}\sum_1^\infty \frac{\sin\frac{1}{3}(2n-1)\pi\sin(2n-1)x}{(2n-1)^2}, \text{ when } 0\leqq x\leqq \pi.$$

3. Expand $f(x)$ in a series of sines of multiples of $\pi x/a$, given that

$$f(x)=mx, \qquad \text{when} \quad 0\leqq x\leqq \tfrac{1}{2}a,$$
$$f(x)=m(a-x), \quad \text{when} \quad \tfrac{1}{2}a\leqq x\leqq a.$$

4. Prove that

$$\tfrac{1}{2}l-x=\frac{l}{\pi}\sum_1^\infty \frac{\sin\dfrac{2n\pi x}{l}}{n}, \text{ when } 0<x<l,$$

and

$$(\tfrac{1}{2}l-x)^2=\frac{l^2}{12}+\frac{l^2}{\pi^2}\sum_1^\infty \frac{\cos\dfrac{2n\pi x}{l}}{n^2}, \text{ when } 0\leqq x\leqq l.$$

5. Obtain an expansion in a mixed series of sines and cosines of multiples of x which is zero between $-\pi$ and 0, and is equal to e^x between 0 and π, and give its values at the three limits.

6. Show that between the values $-\pi$ and $+\pi$ of x the following expansions hold:

$$\sin mx=\frac{2}{\pi}\sin m\pi\left(\frac{\sin x}{1^2-m^2}-\frac{2\sin 2x}{2^2-m^2}+\frac{3\sin 3x}{3^2-m^2}+\dots\right),$$

$$\cos mx=\frac{2}{\pi}\sin m\pi\left(\frac{1}{2m}+\frac{m\cos x}{1^2-m^2}-\frac{m\cos 2x}{2^2+m^2}+\frac{m\cos 3x}{3^2+m^2}-\dots\right),$$

$$\frac{\cosh mx}{\sinh m\pi}=\frac{2}{\pi}\left(\frac{1}{2m}-\frac{m\cos x}{1^2+m^2}+\frac{m\cos 2x}{2^2+m^2}-\frac{m\cos 3x}{3^2+m^2}+\dots\right).$$

7. Express x^2 for values of x between $-\pi$ and π as the sum of a constant and a series of cosines of multiples of x.

Prove that the locus represented by

$$\sum_1^\infty \frac{(-1)^{n-1}}{n^2}\sin nx\sin ny=0$$

is two systems of lines at right angles dividing the plane of x, y into squares of area π^2.

8. Prove that

$$y^2 = \frac{2}{3}\frac{c^3}{d} + \sum_1^\infty \frac{4d}{n^3\pi^3}\left(d \sin\frac{n\pi c}{d} - n\pi c \cos\frac{n\pi c}{d}\right)\cos\frac{n\pi}{d}x$$

represents a series of circles of radius c with their centres on the axis of x at distances $2d$ apart, and also the portions of the axis exterior to the circles, one circle having its centre at the origin.

9. A polygon is inscribed in a circle of radius a, and is such that the alternate sides beginning at $\theta=0$ subtend angles α and β at the centre of the circle. Prove that the first, third, ... pairs of sides of the polygon may be represented, except at angular points, by the polar equation

$$r = a_1 \sin\frac{\pi\theta}{\alpha+\beta} + a_2 \sin\frac{2\pi\theta}{\alpha+\beta} + \dots,$$

where
$$\frac{a_n(\alpha+\beta)}{4a} = \sin\frac{n\pi\alpha}{2(\alpha+\beta)}\cos\frac{\alpha}{2}\int_0^{\frac{1}{2}\alpha}\cos\frac{n\pi\phi}{\alpha+\beta}\sec\phi\, d\phi$$
$$+\sin\frac{n\pi(2\alpha+\beta)}{2(\alpha+\beta)}\cos\frac{\beta}{2}\int_0^{\frac{1}{2}\beta}\cos\frac{n\pi\phi}{\alpha+\beta}\sec\phi\, d\phi.$$

Find a similar equation to represent the other sides.

10. A regular hexagon has a diagonal lying along the axis of x. Investigate a trigonometrical series which shall represent the value of the ordinate of any point of the perimeter lying above the axis of x.

11. If $0<x<2\pi$, prove that

$$\frac{\pi}{2}\frac{\sinh a(\pi-x)}{\sinh a\pi} = \frac{\sin x}{a^2+1^2} + \frac{2\sin 2x}{a^2+2^2} + \frac{3\sin 3x}{a^2+3^2} + \dots.$$

12. Prove that the equation in rectangular coordinates

$$y = \frac{2}{3}h + \frac{4h}{\pi^2}\left(\cos\frac{\pi x}{\kappa} + \frac{1}{2^2}\cos\frac{2\pi x}{\kappa} + \frac{1}{3^2}\cos\frac{3\pi x}{\kappa} - \dots\right)$$

represents a series of equal and similar parabolic arcs of height h and span 2κ standing in contact along the axis of x.

13. The arcs of equal parabolas cut off by the latera recta of length $4a$ are arranged alternately on opposite sides of a straight line formed by placing the latera recta end to end, so as to make an undulating curve. Prove that the equation of the curve can be written in the form

$$\frac{\pi^3 y}{64a} = \sin\frac{\pi x}{4a} + \frac{1}{3^3}\sin\frac{3\pi x}{4a} + \frac{1}{5^3}\sin\frac{5\pi x}{4a} + \dots.$$

14. If circles be drawn on the sides of a square as diameters, prove that the polar equation of the quatrefoil formed by the external semicircles, referred to the centre as origin, is

$$\frac{\pi r}{4a\sqrt{2}} = \tfrac{1}{2} + \tfrac{1}{15}\cos 4\theta - \tfrac{1}{63}\cos 8\theta + \tfrac{1}{143}\cos 12\theta + \dots,$$

where a is the side of the square.

15. On the sides of a regular pentagon remote from the centre are described segments of circles which contain angles equal to that of the pentagon; prove that the equation to the cinquefoil thus obtained is

$$\pi r = 5a \tan\frac{\pi}{5}\left[1 - 2\sum_1^\infty (-1)^n \frac{\cos 5n\theta}{25n^2-1}\right],$$

a being the radius of the circle circumscribing the pentagon.

16. In the interval $\quad 0 < x < \frac{l}{2}, \quad f(x) = x^2;$

and in the interval $\quad \frac{l}{2} < x < l, \quad f(x) = 0.$

Express the function by means of a series of sines and also by means of a series of cosines of multiples of $\frac{\pi x}{l}$. Draw figures showing the functions represented by the two series respectively for all values of x not restricted to lie between 0 and l. What are the sums of the series for the value $x = \frac{l}{2}$?

17. A point moves in a straight line with a velocity which is initially u, and which receives constant increments each equal to u at equal intervals τ. Prove that the velocity at any time t after the beginning of the motion is

$$v = \frac{1}{2}u + \frac{ut}{\tau} + \frac{u}{\pi}\sum_{n=1}^\infty \frac{1}{n}\sin\frac{2n\pi}{\tau}t,$$

and that the distance traversed is

$$\frac{ut}{2\tau}(t+\tau) + \frac{u\tau}{12} - \frac{u\tau}{2\pi^2}\sum_1^\infty \frac{1}{n^2}\cos\frac{2n\pi}{\tau}t. \quad \text{[See Ex. 4 above.]}$$

18. A curve is formed by the positive halves of the circles

$$(x-(4n+1)a)^2 + y^2 = a^2$$

and the negative halves of the circles

$$(x-(4n-1)a)^2 + y^2 = a^2,$$

n being an integer. Prove that the equation for the complete curve obtained by Fourier's method is

$$y = \frac{2}{a}\sum_{\kappa=1}^{\kappa=\infty}(-1)^{\kappa-1}\sin\left(\kappa-\frac{1}{2}\right)\frac{\pi x}{a}\int_0^a \sin\left(\kappa-\frac{1}{2}\right)\frac{\pi x'}{a}\sqrt{a^2-x'^2}\,dx'.$$

19. Having given the form of the curve $y = f(x)$, trace the curves

$$y = \frac{2}{\pi}\sum_1^\infty \sin rx\int_0^{\frac{\pi}{2}} f(t)\sin rt\,dt,$$

$$y = \frac{4}{\pi}\sum_1^\infty \sin(2r-1)x\int_0^{\frac{\pi}{2}} f(t)\sin(2r-1)t\,dt,$$

and show what these become when the upper limit is $\frac{\pi}{4}$ instead of $\frac{\pi}{2}$.

20. Prove that for all values of t between 0 and $\frac{b}{a}$ the sum of the series

$$\sum_1^\infty \frac{1}{r}\sin\frac{r\pi b}{l}\sin\frac{r\pi x}{l}\sin\frac{r\pi at}{l}$$

is zero for all values of x between 0 and $b - at$ and between $at + b$ and l, and is $\frac{\pi}{4}$ for all values of x between $b - at$ and $at + b$, when $b < \frac{l}{2}$.

21. Find the sum of the series

$$u=\frac{1}{\pi}\sum_1^\infty \frac{\sin 2n\pi x}{n},$$

$$v=\frac{1}{\pi}\sum_1^\infty \frac{\sin (2n-1)\pi x}{2n-1},$$

and hence prove that the greatest integer in the positive number x is represented by $x+u-8v^2$. [See Ex. 1, p. 250, and Ex. 4 above.]

22. If x, y, z are the rectangular coordinates of a point which moves so that from $y=0$ to $y=x$ the value of z is $\kappa(a^2-x^2)$, and from $y=x$ to $y=a$ the value of z is $\kappa(a^2-y^2)$, show that for all values of x and y between 0 and a, z may be expressed by a series in the form

$$\Sigma A_{p,q} \sin\frac{p\pi}{2}\left(\frac{x-a}{a}\right)\sin\frac{q\pi}{2}\left(\frac{y+a}{a}\right),$$

and find the values of $A_{p,q}$ for the different types of terms.

23. If $f(x)=\frac{1}{2}\pi \sin x$, when $0 \leqq x \leqq \frac{1}{2}\pi$,

and $f(x)=\frac{1}{2}\pi$, when $\frac{1}{2}\pi < x \leqq \pi$,

prove that, when $0 \leqq x < \pi$,

$$f(x)=\tfrac{1}{4}\pi \sin x + S_1 + S_2 - S_3,$$

where

$$S_1=\frac{2}{1\,.\,3}\sin 2x-\frac{4}{3\,.\,5}\sin 4x+\frac{6}{5\,.\,7}\sin 6x-\dots,$$

$$S_2=\sin x+\tfrac{1}{3}\sin 3x+\tfrac{1}{5}\sin 5x+\dots,$$

$$S_3=\sin 2x+\tfrac{1}{3}\sin 6x+\tfrac{1}{5}\sin 10x+\dots,$$

and find the values of S_1, S_2, and S_3 separately for values of x lying within the assigned interval. [Cf. Ex. 1, p. 250.]

24. If

$$f(x)=\frac{4}{\pi^2}\left(\sin x-\frac{\sin 3x}{3^2}+\frac{\sin 5x}{5^2}-\dots\right)$$

$$+\frac{2}{\pi}\left(\sin x-\frac{\sin 2x}{2}+\frac{\sin 3x}{3}-\dots\right),$$

show that $f(x)$ is continuous between 0 and π, and that $f(\pi-0)=1$. Also show that $f'(x)$ has a sudden change of value $\frac{2}{\pi}$ at the point $\frac{\pi}{2}$. [See Ex. 1, 2, pp. 242-3.]

25. Let

$$f(x)=\sum_1^\infty\frac{\sin 3(2n-1)x}{2n-1}-2\sum_1^\infty\frac{\sin (2n-1)x}{2n-1}+\frac{6}{\pi}\sum_1^\infty\frac{\sin \frac{1}{3}(2n-1)\pi \sin (2n-1)x}{(2n-1)^2},$$

when $0 \leqq x \leqq \pi$.

Show that

$$f(+0)=f(\pi-0)=-\tfrac{1}{4}\pi,$$

$$f(\tfrac{1}{3}\pi+0)-f(\tfrac{1}{3}\pi-0)=-\tfrac{1}{2}\pi,$$

$$f(\tfrac{2}{3}\pi+0)-f(\tfrac{2}{3}\pi-0)=\tfrac{1}{2}\pi;$$

also that

$$f(0)=f(\tfrac{1}{3}\pi)=f(\tfrac{2}{3}\pi)=f(\pi)=0.$$

Draw the graph of $f(x)$ in the interval $(0, \pi)$. [See Ex. 1, p. 250, and Ex. 2 above.]

CHAPTER VIII

THE NATURE OF THE CONVERGENCE OF FOURIER'S SERIES AND SOME PROPERTIES OF FOURIER'S CONSTANTS

104. The Order of the Terms. Before entering upon the discussion of the nature of the convergence of the Fourier's Series for a function satisfying Dirichlet's Conditions, we shall show that in certain cases the order of the terms may be determined easily.

I. *If $f(x)$ is bounded and otherwise satisfies Dirichlet's Conditions in the interval $(-\pi, \pi)$, the coefficients in the Fourier's Series for $f(x)$ are less in absolute value than K/n, where K is some positive number independent of n.*

If the interval $(-\pi, \pi)$ can be broken up into a finite number of open partial intervals (c_r, c_{r+1}) in which $f(x)$ is monotonic, it follows, from the Second Theorem of Mean Value, that

$$\pi a_n = \Sigma \int_{c_r}^{c_{r+1}} f(x) \cos nx\, dx$$

$$= \Sigma \left\{ f(c_r + 0) \int_{c_r}^{\xi_r} \cos nx\, dx + f(c_{r+1} - 0) \int_{\xi_r}^{c_{r+1}} \cos nx\, dx \right\},$$

where ξ_r is some definite number in (c_r, c_{r+1}).

Thus
$$\pi |a_n| < \frac{2}{n} \Sigma \{ |f(c_r + 0)| + |f(c_{r+1} - 0)| \}$$

$$< \frac{4pM}{n},$$

where p is the number of partial intervals and M is the upper bound of $|f(x)|$ in the interval $(-\pi, \pi)$.

Therefore
$$|a_n| < K/n,$$

where K is some positive number independent of n.

And similarly we obtain

$$|b_n| < K/n.$$

We may speak of the terms of this series as of the order $1/n$. When the terms are of the order $1/n$, the series will, in general, be only conditionally convergent, the convergence being due to the presence of both positive and negative terms.

II. If we are given that $f(x)$ is of bounded variation in $(-\pi, \pi)$· the same result follows at once, since $f(x) = F(x) - G(x)$, where $F(x)$ and $G(x)$ are bounded and monotonic functions.

III. *If $f(x)$ is bounded and continuous, and otherwise satisfies Dirichlet's Conditions in $-\pi < x < \pi$, while $f(\pi - 0) = f(-\pi + 0)$, and if $f'(x)$ is bounded and otherwise satisfies Dirichlet's Conditions in the same interval, the coefficients in the Fourier's Series for $f(x)$ are less in absolute value than K/n^2, where K is some positive number independent of n.*

In this case we can make $f(x)$ continuous in the closed interval $(-\pi, \pi)$ by giving to it the values $f(-\pi + 0)$ and $f(\pi - 0)$ at $x = -\pi$ and π respectively.

Then $$\pi a_n = \int_{-\pi}^{\pi} f(x) \cos nx \, dx$$
$$= \frac{1}{n}\Big[f(x) \sin nx \Big]_{-\pi}^{\pi} - \frac{1}{n}\int_{-\pi}^{\pi} f'(x) \sin nx \, dx$$
$$= -\frac{1}{n}\int_{-\pi}^{\pi} f'(x) \sin nx \, dx.$$

But we have just seen that with the given conditions

$$\int_{-\pi}^{\pi} f'(x) \sin nx \, dx$$

is of the order $1/n$.

It follows that $$|a_n| < K/n^2,$$

where K is some positive number independent of n.

A similar argument, in which it will be seen that the condition $f(\pi - 0) = f(-\pi + 0)$ is used, shows that

$$|b_n| < K/n^2.$$

Since the terms of this Fourier's Series are of the order $1/n^2$, it follows that it is absolutely convergent, and also uniformly convergent in any interval.

The above result can be generalised as follows: *If the function $f(x)$ and its differential coefficients, up to the $(p-1)^{th}$, are bounded,* *continuous and otherwise satisfy Dirichlet's Conditions in the interval* $-\pi<x<\pi$, *and*

$$f^{(r)}(-\pi+0)=f^{(r)}(\pi-0), \quad [r=0, 1, \dots (p-1)],$$

and if the pth *differential coefficient is bounded and otherwise satisfies Dirichlet's Conditions in the same interval, the coefficients in the Fourier's Series for* $f(x)$ *will be less in absolute value than* K/n^{p+1}, *where* K *is some positive number independent of* n.

105. The Riemann-Lebesgue Theorem,* and its Consequences. *Let* $f(x)$ *be bounded and integrable in* (a, b), *or, if* $f(x)$ *is unbounded, let* $\int_a^b f(x)\,dx$ *be absolutely convergent.*

Then
$$\lim_{n\to\infty}\int_a^b f(x) \begin{matrix}\sin\\ \cos\end{matrix}\, nx\,dx=0.$$

(α) Let $|f(x)|$ be less than A in (a, b), and ϵ the usual arbitrary positive number.

There is a mode of division of (a, b), say $a=x_0, x_1, x_2, \dots x_{m-1}, x_m=b$, such that $S-s$ for it is less than $\frac{1}{2}\epsilon$ (§ 42).

Thus
$$\int_a^b f(x)\sin nx\,dx=\sum_1^m\int_{x_{r-1}}^{x_r}[f(x_r)+(f(x)-f(x_r))]\sin nx\,dx$$

$$\leqq \sum_1^m\left\{|f(x_r)|\left|\int_{x_{r-1}}^{x_r}\sin nx\,dx\right|+\int_{x_{r-1}}^{x_r}|f(x)-f(x_r)||\sin nx|\,dx\right\}$$

$$<\frac{2mA}{n}+\sum_1^m(M_r-m_r)(x_r-x_{r-1}), \text{ with the notation of § 39}$$

$$<\frac{2mA}{n}+(S-s)$$

$$<\frac{2mA}{n}+\tfrac{1}{2}\epsilon$$

$$<\epsilon, \text{ when } n\geqq\frac{4mA}{\epsilon}.$$

Hence
$$\lim_{n\to\infty}\int_a^b f(x)\sin nx\,dx=0.$$

And in the same way,
$$\lim_{n\to\infty}\int_a^b f(x)\cos nx\,dx=0.$$

(β) If $\int_a^b f(x)dx$ is an absolutely convergent infinite integral, according to the definition of § 51, we have only a finite number of points of infinite discontinuity in (a, b).

As we can treat these separately, it is clear that we need only discuss the case when a or b is a point of infinite discontinuity. We take the latter alternative.

* This theorem was proved by Riemann [*Math. Werke*, 1 (2 Aufl., 1892) 254] for the functions stated in the text, and extended by Lebesgue to functions with a Lebesgue Integral [*Ann. sc. de l'École normale* (3), 20 (1903), 471].

In this case, there is a point β, between a and b, such that

$$\int_\beta^b |f(x)|dx < \tfrac{1}{2}\epsilon.$$

Also
$$\left|\int_a^b f(x)\sin nx\,dx\right| = \left|\int_a^\beta f(x)\sin nx\,dx + \int_\beta^b f(x)\sin nx\,dx\right|$$
$$\leqq \left|\int_a^\beta f(x)\sin nx\,dx\right| + \int_\beta^b |f(x)|\,dx.$$

But $f(x)$ is bounded in (a, β): and thus, by the above, we know that

$$\left|\int_a^\beta f(x)\sin nx\,dx\right| < \tfrac{1}{2}\epsilon \text{ when } n \geqq \nu.$$

Therefore
$$\left|\int_a^b f(x)\sin nx\,dx\right| < \tfrac{1}{2}\epsilon + \tfrac{1}{2}\epsilon, \text{ when } n \geqq \nu,$$

and the theorem is proved.

The following results can be deduced almost immediately from the Riemann-Lebesgue Theorem. In all of them x_0 is a point of the interval $(-\pi, \pi)$; and $f(x)$ is subject to the conditions named in that theorem: it is bounded and integrable in $(-\pi, \pi)$, or, if unbounded, the integral $\int_{-\pi}^{\pi} f(x)dx$ is absolutely convergent.

(i) *The Fourier's Constants a_n and b_n of $f(x)$ tend to zero when $n \to \infty$.*

(ii) *The behaviour of the Fourier's Series corresponding to $f(x)$, as to convergence, divergence, or oscillation at a point x_0, depends only on the values of $f(x)$ in the neighbourhood of x_0.*

Here, with the notation of § 95, we have

$$s_n(x_0) = \frac{1}{2\pi}\left(\int_{-\pi}^{x_0-2\eta} + \int_{x_0-2\eta}^{x_0+2\eta} + \int_{x_0+2\eta}^{\pi}\right) f(x')\,\frac{\sin\frac{1}{2}(2n+1)(x'-x_0)}{\sin\frac{1}{2}(x'-x_0)}\,dx'$$

where $(x_0 - 2\eta, x_0 + 2\eta)$ is a neighbourhood of x_0.

Thus

$$s_n(x_0) = \frac{1}{\pi}\int_\eta^{\frac{1}{2}(\pi+x_0)} f(x_0 - 2\alpha)\,\frac{\sin(2n+1)\alpha}{\sin\alpha}\,d\alpha$$
$$+ \frac{1}{\pi}\int_\eta^{\frac{1}{2}(\pi-x_0)} f(x_0+2\alpha)\frac{\sin(2n+1)\alpha}{\sin\alpha}\,d\alpha + \frac{1}{\pi}\int_{-\eta}^{\eta} f(x_0+2\alpha)\,\frac{\sin(2n+1)\alpha}{\sin\alpha}\,d\alpha.$$

By the Riemann-Lebesgue Theorem, the first and second integrals vanish in the limit when $n \to \infty$: and the result follows.

(iii) *The behaviour of the Fourier's Series corresponding to $f(x)$ in an interval (a, b), where $-\pi < a < b < \pi$, as to convergence, divergence, or oscillation, depends only on the values of $f(x)$ in $(a - \delta, b + \delta)$, where δ is an arbitrarily small positive number.*

This is proved as in (ii).

(iv) *If $f(x)$ is of bounded variation in a neighbourhood of x_0, the series converges at x_0 to $\frac{1}{2}[f(x_0+0) + f(x_0-0)]$.*

As above,
$$\lim_{n\to\infty} s_n(x_0) = \frac{1}{\pi}\lim_{n\to\infty}\int_{-\eta}^{\eta} f(x_0+2\alpha)\,\frac{\sin(2n+1)\alpha}{\sin\alpha}\,d\alpha,$$

and the result follows from Dirichlet's Integral (§ 94).

If $\lim_{h \to 0} [f(x_0+h)+f(x_0-h)]$ exists, and $2f(x_0)$ is taken equal to this limit, two important sufficient conditions for the convergence of the Fourier's Series at x_0 to the value $f(x_0)$ are given in (v) and (vi). They are usually called Dini's Condition and Lipschitz's Condition. In both

$$\phi(a)=f(x_0+2a)+f(x_0-2a)-2f(x_0).$$

(v) **Dini's Condition.*** *The Fourier's Series corresponding to $f(x)$ has $f(x_0)$ for its sum when $x=x_0$, if there is a positive number η such that $\int_0^\eta \frac{|\phi(a)|}{a}\,da$ is a convergent integral.*

For, we see from (ii), that $\lim_{n\to\infty} s_n(x_0)=f(x_0)$, provided that

$$\lim_{n\to\infty}\int_0^\eta \left\{f(x_0+2a)+f(x_0-2a)-2f(x_0)\right\}\frac{\sin(2n+1)a}{\sin a}\,da=0;$$

i.e. if

$$\lim_{n\to\infty}\int_0^\eta \phi(a)\,\frac{\sin(2n+1)a}{\sin a}\,da=0.$$

Also

$$\left|\int_0^\eta \phi(a)\,\frac{\sin(2n+1)a}{\sin a}\,da\right| \leqq \tfrac{1}{2}\pi \int_0^\eta \frac{|\phi(a)|}{a}\,da, \text{ if } 0<\eta<\tfrac{1}{2}\pi.$$

(vi) **Lipschitz's Condition.†** *The Fourier's Series corresponding to $f(x)$ has $f(x_0)$ for its sum when $x=x_0$, if positive numbers C and k exist such that, $|f(x_0+t)-f(x_0)|<C|t|^k$, when $|t| \leqq$ some fixed positive number.*

In this case, there is a value of η for which

$$|f(x_0+2a)+f(x_0-2a)-2f(x_0)|<2^{k+1}Ca^k, \text{ when } 0 \leqq a \leqq \eta.$$

And

$$\left|\int_0^\eta \phi(a)\,\frac{\sin(2n+1)a}{\sin a}\,da\right| < \pi 2^k C\int_0^\eta a^{k-1}\,da,$$

which can be made as small as we please by taking η small enough.

This condition is a special case of the preceding.

106. Discussion of a case in which f(x) satisfies Dirichlet's Conditions and has an infinity in $(-\pi, \pi)$.‡ We have seen that Dirichlet's Conditions include the possibility of $f(x)$ having a certain number of points of infinite discontinuity in the interval, subject to the condition that the infinite integral $\int_{-\pi}^{\pi} f(x)dx$ is absolutely convergent.

Let us suppose that near the point x_0, where $-\pi<x_0<\pi$, the function $f(x)$ is such that

$$f(x)=\frac{\phi(x)}{(x-x_0)^\nu},$$

where $0<\nu<1$ and $\phi(x)$ is monotonic to the right and left of x_0, while $\phi(x_0\pm 0)$ do not both vanish.

In this case the condition for absolute convergence is satisfied.

---

*Dini, *Serie di Fourier e altre rappresentazioni analitiche delle funzioni di una variabile reale* (Pisa, 1880), p. 102.

†Lipschitz, *Journal für Math.*, 63 (1864), 296.

‡See also Ex. 5, p. 241.

Then, in determining a_n and b_n, where

$$\pi a_n = \int_{-\pi}^{\pi} f(x) \cos nx \, dx \quad \text{and} \quad \pi b_n = \int_{-\pi}^{\pi} f(x) \sin nx \, dx,$$

we break up the interval into

$$(-\pi, a), \quad (a, x_0), \quad (x_0, \beta), \quad \text{and} \quad (\beta, \pi),$$

where (a, β) is the interval in which $f(x)$ has the given form. In $(-\pi, a)$ and (β, π) it is supposed that $f(x)$ is bounded and otherwise satisfies Dirichlet's Conditions, and we know from § 104 that these partial intervals give to a_n and b_n contributions of the order $1/n$.

The remainder of the integral, *e.g.* in a_n, is given by the sum of

$$\lim_{\delta \to 0} \int_a^{x_0 - \delta} \frac{\phi(x)}{(x - x_0)^\nu} \cos nx \, dx \quad \text{and} \quad \lim_{\delta \to 0} \int_{x_0 + \delta}^{\beta} \frac{\phi(x)}{(x - x_0)^\nu} \cos nx \, dx,$$

these limits being known to exist.

We take the second of these integrals, and apply to it the Second Theorem of Mean Value.

Thus we have

$$\int_{x_0+\delta}^{\beta} \frac{\phi(x)}{(x - x_0)^\nu} \cos nx \, dx = \phi(x_0 + \delta) \int_{x_0+\delta}^{\xi} \frac{\cos nx}{(x - x_0)^\nu} dx + \phi(\beta) \int_{\xi}^{\beta} \frac{\cos nx}{(x - x_0)^\nu} dx,$$

where $x_0 + \delta \leqq \xi \leqq \beta$.

Putting $n(x - x_0) = y$, we obtain

$$\int_{x_0+\delta}^{\beta} \frac{\phi(x)}{(x - x_0)^\nu} \cos nx \, dx = \frac{\phi(x_0 + \delta)}{n^{1-\nu}} \int_{n\delta}^{n(\xi - x_0)} \frac{\cos(y + nx_0)}{y^\nu} dy$$
$$+ \frac{\phi(\beta)}{n^{1-\nu}} \int_{n(\xi - x_0)}^{n\beta} \frac{\cos(y + nx_0)}{y^\nu} dy$$

But
$$\int_a^b \frac{\cos(y + nx_0)}{y^\nu} dy = \cos nx_0 \int_a^b \frac{\cos y}{y^\nu} dy - \sin nx_0 \int_a^b \frac{\sin y}{y^\nu} dy.$$

Also when a, b are positive.

$$\left| \int_a^b \frac{\cos y}{y^\nu} dy \right|, \quad \left| \int_a^b \frac{\sin y}{y^\nu} dy \right|$$

are both less than definite numbers independent of a and b, when $0 < \nu < 1$.

Thus, whatever positive integer n may be, and whatever value δ may have, subject to $0 < \delta < \beta - x_0$,

$$\left| \int_{x_0+\delta}^{\beta} \frac{\phi(x)}{(x - x_0)^\nu} \cos nx \, dx \right| < K'/n^{1-\nu},$$

where K' is some positive number independent of n and ξ.

It follows that

$$\left| \lim_{\delta \to 0} \int_{x_0+\delta}^{\beta} \frac{\phi(x)}{(x - x_0)^\nu} \cos nx \, dx \right| < K'/n^{1-\nu}.$$

A similar argument applies to the integral

$$\int_a^{x_0 - \delta} \frac{\phi(x)}{(x - x_0)^\nu} \cos nx \, dx.$$

It thus appears that the coefficient of $\cos nx$ in the Fourier's Series for the given function $f(x)$ is less in absolute value than $K/n^{1-\nu}$, where K is some positive number independent of n; and a corresponding result holds with regard to the coefficient of $\sin nx$.

It is easy to modify the above argument so that it will apply to the case when the infinity occurs at $\pm\pi$.

107. The Uniform Convergence of Fourier's Series.* We shall deal, first of all, with the case of the Fourier's Series for $f(x)$, when $f(x)$ is bounded in $(-\pi, \pi)$ and otherwise satisfies Dirichlet's Conditions. Later we shall discuss the case where a finite number of points of infinite discontinuity are admitted.

It is clear that the Fourier's Series for $f(x)$ cannot be uniformly convergent in any interval which contains a point of discontinuity; since uniform convergence, in the case of series whose terms are continuous, involves continuity in the sum.

Let $f(x)$ be bounded in the interval $(-\pi, \pi)$, and otherwise satisfy Dirichlet's Conditions in that interval. Then the Fourier's Series for $f(x)$ converges uniformly to $f(x)$ in any interval which contains neither in its interior nor at an end any point of discontinuity of the function.†

As before the bounded function $f(x)$, satisfying Dirichlet's Conditions in $(-\pi, \pi)$, is defined outside that interval by the equation

$$f(x+2\pi)=f(x).$$

Then we can express $f(x)$ in any interval—*e.g.* $(-2\pi, 2\pi)$—as the difference of two functions, which we shall denote by $F(x)$ and $G(x)$, where $F(x)$ and $G(x)$ are bounded, positive and monotonic increasing. They are also continuous at all points where $f(x)$ is continuous [§ 36.1 or § 36.2].

Let $f(x)$ be continuous at a and b‡ and at all points in $a<x<b$, where, to begin with, we shall assume $-\pi<a$ and $b<\pi$.

Also let x be any point in (a, b).

Then with the notation of § 95,

$$s_n(x)=\frac{1}{2\pi}\int_{-\pi}^{\pi} f(x')\frac{\sin\frac{1}{2}(2n+1)(x'-x)}{\sin\frac{1}{2}(x'-x)}\,dx'$$

$$=\frac{1}{\pi}\int_{-\frac{1}{2}\pi}^{\frac{1}{2}\pi} f(x+2\alpha)\frac{\sin m\alpha}{\sin\alpha}\,d\alpha, \text{ where } m=2n+1.\S$$

---

*See footnote, p. 230.

†It will be seen from § 105, (ii), that if (a, b) be any interval contained in $(-\pi, \pi)$ such that $f(x)$ is continuous in (a, b), including the end-points, the answer to the question whether the Fourier's Series converges uniformly in (a, b), or not, depends only upon the nature of $f(x)$ in an interval (a', b'), which includes (a, b) in its interior and exceeds it in length by an arbitrarily small amount.

‡Thus $f(a+0)=f(a)=f(a-0)$ and $f(b+0)=f(b)=f(b-0)$.

§We have replaced the limits $-\pi$, π by $-\pi+x$, $\pi+x$ in the integral before changing the variable from x' to α by the substitution $x'=x+2\alpha$. Cf. § 101.

Thus

$$s_n(x) = \frac{1}{\pi}\int_{-\frac{1}{2}\pi}^{\frac{1}{2}\pi} F(x+2\alpha)\frac{\sin m\alpha}{\sin\alpha}\,d\alpha$$

$$-\frac{1}{\pi}\int_{-\frac{1}{2}\pi}^{\frac{1}{2}\pi} G(x+2\alpha)\frac{\sin m\alpha}{\sin\alpha}\,d\alpha. \quad \text{......................(1)}$$

We shall now discuss the first integral in (1),

$$\int_{-\frac{1}{2}\pi}^{\frac{1}{2}\pi} F(x+2\alpha)\frac{\sin m\alpha}{\sin\alpha}\,d\alpha,$$

i.e.
$$\int_0^{\frac{1}{2}\pi} F(x+2\alpha)\frac{\sin m\alpha}{\sin\alpha}\,d\alpha + \int_0^{\frac{1}{2}\pi} F(x-2\alpha)\frac{\sin m\alpha}{\sin\alpha}\,d\alpha.$$

Let μ be any number such that $0<\mu<\frac{1}{2}\pi$.

Then

$$\int_0^{\frac{1}{2}\pi} F(x+2\alpha)\frac{\sin m\alpha}{\sin\alpha}\,d\alpha = F(x+0)\int_0^{\frac{1}{2}\pi}\frac{\sin m\alpha}{\sin\alpha}\,d\alpha$$

$$+\int_0^{\mu}\{F(x+2\alpha)-F(x+0)\}\frac{\sin m\alpha}{\sin\alpha}\,d\alpha$$

$$+\int_{\mu}^{\frac{1}{2}\pi}\{F(x+2\alpha)-F(x+0)\}\frac{\sin m\alpha}{\sin\alpha}\,d\alpha$$

$$= I_1 + I_2 + I_3, \text{ say.} \quad \text{......................(2)}$$

We can replace $F(x+0)$ by $F(x)$, since $F(x)$ is continuous at x.

But
$$\int_0^{\frac{1}{2}\pi}\frac{\sin(2n+1)\alpha}{\sin\alpha}\,d\alpha = \int_0^{\frac{1}{2}\pi}\left(1+2\sum_1^n \cos 2r\alpha\right)d\alpha = \tfrac{1}{2}\pi.$$

Thus
$$I_1 = \tfrac{1}{2}\pi F(x). \quad \text{............................(3)}$$

Also $\{F(x+2\alpha)-F(x)\}$ is bounded, positive and monotonic increasing in any interval; and $\dfrac{\alpha}{\sin\alpha}$ is also bounded, positive and monotonic increasing in $0<\alpha\leqq\frac{1}{2}\pi$.

Therefore we can apply the Second Theorem of Mean Value to the integral

$$\int_0^{\mu}\phi(\alpha)\frac{\sin m\alpha}{\alpha}\,d\alpha,$$

where
$$\phi(\alpha) = \{F(x+2\alpha)-F(x)\}\frac{\alpha}{\sin\alpha}.$$

It follows that

$$I_2 = \{F(x+2\mu)-F(x)\}\frac{\mu}{\sin\mu}\int_{\xi}^{\mu}\frac{\sin m\alpha}{\alpha}\,d\alpha,$$

where $0\leqq\xi\leqq\mu$.

But we know that

$$\left|\int_{\xi}^{\mu} \frac{\sin m\alpha}{\alpha}\,d\alpha\right| = \left|\int_{m\xi}^{m\mu} \frac{\sin \alpha}{\alpha}\,d\alpha\right| < \pi. \quad (\S\ 91.)$$

Therefore $$|I_2| < \{F(x+2\mu) - F(x)\}\,\frac{\mu\pi}{\sin\mu}. \quad \text{..................(4)}$$

Finally $$I_3 = \{F(x+2\mu) - F(x)\}\int_{\mu}^{\xi'} \frac{\sin m\alpha}{\sin\alpha}\,d\alpha$$
$$+\{F(x+\pi) - F(x)\}\int_{\xi'}^{\frac{1}{2}\pi} \frac{\sin m\alpha}{\sin\alpha}\,d\alpha,$$

where $\mu \leqq \xi' \leqq \frac{1}{2}\pi$.

But, if $0 < \theta < \phi \leqq \frac{1}{2}\pi$,

$$\int_{\theta}^{\phi} \frac{\sin m\alpha}{\sin\alpha}\,d\alpha = \frac{1}{\sin\theta}\int_{\theta}^{\chi} \sin m\alpha\,d\alpha + \frac{1}{\sin\phi}\int_{\chi}^{\phi} \sin m\alpha\,d\alpha,$$

where $\theta \leqq \chi \leqq \phi$.

Therefore $$\left|\int_{\theta}^{\phi} \frac{\sin m\alpha\,d\alpha}{\sin\alpha}\right| < \frac{2}{m}\{\operatorname{cosec}\theta + \operatorname{cosec}\phi\}$$
$$< \frac{4}{m}\operatorname{cosec}\theta.$$

It follows that

$$|I_3| < \frac{4}{m\sin\mu}\left[\{F(x+2\mu) - F(x)\} + \{F(x+\pi) - F(x)\}\right]$$
$$< \frac{4K}{m\sin\mu}, \quad \text{..(5)}$$

where K is some positive number, independent of m, and depending on the upper bound of $|f(x)|$ in $(-\pi, \pi)$.

Combining (3), (4) and (5), we see from (2) that

$$\left|\frac{1}{\pi}\int_0^{\frac{1}{2}\pi} F(x+2\alpha)\,\frac{\sin m\alpha}{\sin\alpha}\,d\alpha - \tfrac{1}{2}F(x)\right|$$
$$< \{F(x+2\mu) - F(x)\}\frac{\mu}{\sin\mu} + \frac{4K}{m\pi\sin\mu}. \quad \text{......(6)}$$

A similar argument applies to the integral

$$\int_0^{\frac{1}{2}\pi} F(x-2\alpha)\,\frac{\sin m\alpha}{\sin\alpha}\,d\alpha,$$

but in this case it has to be remembered that $F(x-2\alpha)$ is monotonic decreasing as α increases from 0 to $\frac{1}{2}\pi$.

The corresponding result for this integral is that

$$\left|\frac{1}{\pi}\int_0^{\frac{1}{2}\pi} F(x-2\alpha)\,\frac{\sin m\alpha}{\sin\alpha}\,d\alpha - \tfrac{1}{2}F(x)\right|$$
$$< |F(x-2\mu) - F(x)|\,\frac{\mu}{\sin\mu} + \frac{4K}{m\pi\sin\mu}, \quad \text{......(7)}$$

K as before being some positive number independent of m, and depending on the upper bound of $|f(x)|$ in $(-\pi, \pi)$.

Without loss of generality we can take K the same in (6) and (7).

From (6) and (7) we obtain at once

$$\left|\frac{1}{\pi}\int_{-\frac{1}{2}\pi}^{\frac{1}{2}\pi} F(x+2a)\frac{\sin ma}{\sin a}\,da - F(x)\right|$$
$$< \frac{\mu}{\sin\mu}\{|F(x+2\mu)-F(x)| + |F(x-2\mu)-F(x)|\} + \frac{8K}{m\pi\sin\mu}. \quad (8)$$

Similarly we find that

$$\left|\frac{1}{\pi}\int_{-\frac{1}{2}\pi}^{\frac{1}{2}\pi} G(x+2a)\frac{\sin ma}{\sin a}\,da - G(x)\right|$$
$$< \frac{\mu}{\sin\mu}\{|G(x+2\mu)-G(x)| + |G(x-2\mu)-G(x)|\} + \frac{8K}{m\pi\sin\mu}. \quad (9)$$

Thus, from (1),

$$\left|\frac{1}{\pi}\int_{-\frac{1}{2}\pi}^{\frac{1}{2}\pi} f(x+2a)\frac{\sin ma}{\sin a}\,da - f(x)\right|$$
$$< \frac{\mu}{\sin\mu}\{|F(x+2\mu)-F(x)| + |F(x-2\mu)-F(x)|$$
$$+ |G(x+2\mu)-G(x)| + |G(x-2\mu)-G(x)|\}$$
$$+ \frac{16K}{m\pi\sin\mu}. \quad \text{......(10)}$$

Now $F(x)$ and $G(x)$ are continuous in $a<x<b$, and also when $x=a$ and $x=b$.

Thus, to the arbitrary positive number ϵ, there will correspond a positive number μ_0 (which can be taken less than $\frac{1}{2}\pi$) such that

$$|F(x+2\mu)-F(x)|<\epsilon, \quad |G(x+2\mu)-G(x)|<\epsilon,$$

when $|\mu| \leqq \mu_0$, the same μ_0 serving for all values of x in $a \leqq x \leqq b$.

Also we know that $\mu \operatorname{cosec} \mu$ increases continuously from unity to $\frac{1}{2}\pi$ as μ passes from 0 to $\frac{1}{2}\pi$.

Choose μ_0, as above, less than $\frac{1}{2}\pi$, and put $\mu=\mu_0$ in the argument of (1) to (10). This is allowable,. as the only restriction upon μ was that it must lie between 0 and $\frac{1}{2}\pi$.

Then the terms on the right-hand side of (10), not including $16K/m\pi \sin \mu_0$, are together less than 8ϵ for all values of x in (a, b).

So far nothing has been said about the number m, except that it is an odd positive integer $(2n+1)$.

Let n_0 be the smallest positive integer which satisfies the inequality

$$\frac{16K}{(2n_0+1)\pi \sin \mu_0} < \epsilon.$$

As K, μ_0 and ϵ are independent of x, so also is n_0. We now choose m (*i.e.* $2n+1$) so that $n \geqq n_0$.

Then it follows from (10) that

$$|s_n(x) - f(x)| < 9\epsilon, \text{ when } n \geqq n_0,$$

the same n_0 serving for every x in $a \leqq x \leqq b$.

In other words, we have shown that the Fourier's Series converges uniformly to $f(x)$, under the given conditions, in the interval (a, b).*

If $f(-\pi+0) = f(\pi-0)$, we can regard the points $\pm\pi$, $\pm 2\pi$, etc., as points at which $f(x)$, extended beyond $(-\pi, \pi)$ by the equation $f(x+2\pi) = f(x)$, is continuous, for we can give to $f(\pm\pi)$ the common value of $f(-\pi+0)$ and $f(\pi-0)$.

108. The Uniform Convergence of Fourier's Series † (*continued*). By argument similar to that employed in the preceding section, it can be proved that when $f(x)$, bounded or not, satisfies Dirichlet's Conditions in the interval $(-\pi, \pi)$, and $f(x)$ is bounded in the interval (a', b') contained within $(-\pi, \pi)$, then the Fourier's Series for $f(x)$ converges uniformly in any interval (a, b) in the interior of (a', b'), provided $f(x)$ is continuous in (a, b), including its end-points.

But, instead of developing the discussion on these lines, we shall now show

* It may help the reader to follow the argument of this section if we take a special case:

$$E.g.\quad \left.\begin{array}{ll} f(x)=0, & -\pi \leqq x \leqq 0, \\ f(x)=1, & 0 \leqq x \leqq \pi. \end{array}\right\}$$

Then we have:

Interval.	$f(x)$	$F(x)$	$G(x)$
$-2\pi < x < -\pi$	1	1	0
$-\pi \leqq x \leqq 0$	0	1	1
$0 < x \leqq \pi$	1	2	1
$\pi < x \leqq 2\pi$	0	2	2

If $0 < a \leqq x \leqq b < \pi$, the interval (a, b) is an interval in which $f(x)$ is continuous.

The argument of the preceding section will then apply to the case in which $-\pi$ or π is an end-point of the interval (a, b) inside and at the ends of which $f(x)$ is continuous.

† See footnote, p. 230.

how the question can be treated by Fejér's Arithmetic Means (cf. § 101), and we shall prove the following theorem:

Let $f(x)$, bounded or not, satisfy Dirichlet's Conditions in the interval $(-\pi, \pi)$, and let it be continuous at a and b and in (a, b), where $-\pi \leqq a$ and $b \leqq \pi$. Then the Fourier's Series for $f(x)$ converges uniformly to $f(x)$ in any interval $(a+\delta, b-\delta)$ contained within (a, b).

Without loss of generality we may assume $b-a \leqq \pi$, for a greater interval could be treated as the sum of two such intervals.

Let
$$2\pi a_0' = \int_a^b f(x')dx',$$

and
$$\left.\begin{aligned} \pi a_n' &= \int_a^b f(x') \cos nx'\, dx' \\ \pi b_n' &= \int_a^b f(x') \sin nx'\, dx', \end{aligned}\right\} n \geqq 1.$$

Since $f(x)$ is continuous at a and b and in (a, b), it is also bounded in (a, b), and we can use the Corollary to Fejér's Theorem (§ 101) and assert that the sequence of Arithmetic Means for the series

$$a_0' + \sum_1^\infty (a_n' \cos nx + b_n' \sin nx)$$

converges *uniformly* to $f(x)$ in (a, b).

Also
$$|a_n' \cos nx + b_n' \sin nx| \leqq (a_n'^2 + b_n'^2)^{\frac{1}{2}}.$$

But $f(x)$ is bounded in (a, b) and satisfies Dirichlet's Conditions therein.

Thus we can write
$$f(x) = F(x) - G(x),$$

where $F(x)$ and $G(x)$ are bounded, positive and monotonic increasing functions in (a, b). It follows that we can apply the Second Theorem of Mean Value to the integrals

$$\int_a^b F(x') \frac{\cos}{\sin} nx'\, dx', \quad \int_a^b G(x') \frac{\cos}{\sin} nx'\, dx',$$

and we deduce at once that $(a_n'^2 + b_n'^2)^{\frac{1}{2}} < K/n$,

where K is some positive number depending on the upper bound of $|f(x)|$ in (a, b).

Then we know, from Theorem II, Cor. II of § 102, that the series

$$a_0' + \sum_1^\infty (a_n' \cos nx + b_n' \sin nx)$$

converges uniformly to $f(x)$ in (a, b).

Let us now suppose x to be any point in the interval $(a+\delta, b-\delta)$ lying within the interval (a, b).

With the usual notation

$$2\pi a_0 = \int_{-\pi}^{\pi} f(x')dx',$$

$$\left.\begin{aligned} \pi a_n &= \int_{-\pi}^{\pi} f(x') \cos nx'\, dx', \\ \pi b_n &= \int_{-\pi}^{\pi} f(x') \sin nx'\, dx'. \end{aligned}\right\} n \geqq 1$$

It follows, as in § 103, that

$$(a_0 - a_0') + \sum_1^\infty \{(a_n - a_n') \cos nx + (b_n - b_n') \sin nx\}$$

is equal to

$$\frac{1}{\pi}\int_{\frac{1}{2}(x-a)}^{\frac{1}{2}\pi} f(x-2\alpha) \frac{\sin(2n+1)\alpha}{\sin \alpha} d\alpha + \frac{1}{\pi}\int_{\frac{1}{2}(b-x)}^{\frac{1}{2}\pi} f(x+2\alpha) \frac{\sin(2n+1)\alpha}{\sin \alpha} d\alpha,$$

$f(x)$ being defined outside the interval $(-\pi, \pi)$ by the equation

$$f(x+2\pi) = f(x).$$

Now $f(x)$ is supposed to have not more than a finite number of points (say m) of infinite discontinuity in $(-\pi, \pi)$, and $\int_{-\pi}^{\pi} |f(x')| \, dx'$ converges.

We can therefore take intervals $2\gamma_1, 2\gamma_2, \ldots 2\gamma_m$ enclosing these points, the intervals being so small that

$$\int_{2\gamma_r} |f(x')| \, dx' < 2\epsilon \sin \tfrac{1}{2}\delta, \quad [r = 1, 2, \ldots m]$$

ϵ being any given positive number.

Consider the integral

$$\int_{\frac{1}{2}(x-a)}^{\frac{1}{2}\pi} f(x-2\alpha) \frac{\sin(2n+1)\alpha}{\sin \alpha} d\alpha,$$

x, as already stated, being a point in $(a+\delta, b-\delta)$.

As α passes from $\frac{1}{2}(x-a)$ to $\frac{1}{2}\pi$, we may meet some or all of the m points of discontinuity of the given function in $f(x-2\alpha)$. Let these be taken as the centres of the corresponding intervals $\gamma_1, \gamma_2, \ldots \gamma_m$.

Also the smallest value of $(x-a)$ is δ.

Thus $$\left|\int_{\gamma_r} f(x-2\alpha) \frac{\sin(2n+1)\alpha}{\sin \alpha} d\alpha\right| < \frac{1}{\sin \frac{1}{2}\delta} \int_{\gamma_r} |f(x-2\alpha)| \, d\alpha$$

$$< \frac{1}{2 \sin \frac{1}{2}\delta} \int_{2\gamma_r} |f(x')| \, dx'$$

$$< \epsilon.$$

When these intervals, such of them as occur, have been cut out, the integral

$$\int_{\frac{1}{2}(x-a)}^{\frac{1}{2}\pi} f(x-2\alpha) \frac{\sin(2n+1)\alpha}{\sin \alpha} d\alpha$$

will at most consist of $(m+1)$ separate integrals (I_r). $[r = 1, 2, \ldots m+1.]$

In each of these integrals (I_r) we can take $f(x-2\alpha)$ as the difference of two bounded, positive and monotonic increasing functions

$$F(x-2\alpha) \quad \text{and} \quad G(x-2\alpha).$$

Then, confining our attention to (I_r), we see that

$$\left|\int (F-G) \frac{\sin(2n+1)\alpha}{\sin \alpha} d\alpha\right|$$

$$= \left|\int (F-G)\{\operatorname{cosec} \tfrac{1}{2}\delta - (\operatorname{cosec} \tfrac{1}{2}\delta - \operatorname{cosec} \alpha)\} \sin(2n+1)\alpha \, d\alpha\right|$$

$$\leqq |J_1| + |J_2| + |J_3| + |J_4|,$$

where $$J_1 = \operatorname{cosec} \tfrac{1}{2}\delta \int F \sin(2n+1)\alpha \, d\alpha,$$

$$J_2 = \operatorname{cosec} \tfrac{1}{2}\delta \int G \sin(2n+1)\alpha \, d\alpha,$$

$$J_3 = \int F \{\operatorname{cosec} \tfrac{1}{2}\delta - \operatorname{cosec} \alpha\} \sin(2n+1)\alpha \, d\alpha,$$

$$J_4 = \int G \{\operatorname{cosec} \tfrac{1}{2}\delta - \operatorname{cosec} \alpha\} \sin(2n+1)\alpha \, d\alpha.$$

But we can apply the Second Theorem of Mean Value to each of these integrals, since the factor in each integrand which multiplies $\sin(2n+1)\alpha$ is monotonic.

It follows that $$|I_r| < \frac{K}{2n+1} \operatorname{cosec} \tfrac{1}{2}\delta,$$
where K is some positive number independent of n and x, and depending only on the values of $f(x)$ in $(-\pi, \pi)$, when the intervals $2\gamma_1, 2\gamma_2, \ldots$ have been removed from that interval.

Thus
$$\left| \int_{\frac{1}{2}(x-a)}^{\frac{1}{2}\pi} f(x-2\alpha) \frac{\sin(2n+1)\alpha}{\sin \alpha} d\alpha \right| < \frac{(m+1)K}{(2n+1)\sin\frac{1}{2}\delta} + m\epsilon$$
$$< (2m+1)\epsilon,$$
when $(2n+1)\epsilon > K \operatorname{cosec} \frac{1}{2}\delta$.

Since this choice of n is independent of x, the integral converges uniformly to zero as $n \to \infty$, when x lies in the interval $(a+\delta, b-\delta)$.

Similarly we find that
$$\int_{\frac{1}{2}(b-x)}^{\frac{1}{2}\pi} f(x+2\alpha) \frac{\sin(2n+1)\alpha}{\sin \alpha} d\alpha$$
converges uniformly to zero when x lies in this interval.

Thus the series
$$(a_0 - a_0') + \sum_1^\infty \{(a_n - a_n') \cos nx + (b_n - b_n') \sin nx\}$$
converges uniformly to zero in $(a+\delta, b-\delta)$.

But we have shown that the series
$$a_0' + \sum_1^\infty (a_n' \cos nx + b_n' \sin nx)$$
converges uniformly to $f(x)$ in (a, b).

Since the sum of two uniformly convergent series converges uniformly, we see that
$$a_0 + \sum_1^\infty (a_n \cos nx + b_n \sin nx)$$
converges uniformly to $f(x)$ in $(a+\delta, b-\delta)$.

109. Differentiation and Integration of Fourier's Series.

Differentiation. From the worked out examples in §§ 95-97 it is clear that term by term differentiation of Fourier's Series is not in general permissible, as the terms do not tend to zero sufficiently quickly.

This difficulty does not arise in the application of Fourier's Series to the solution of the Equation of Conduction of Heat—often written as $\dfrac{\partial v}{\partial t} = \kappa \dfrac{\partial^2 v}{\partial x^2}$.

Here * the terms in the appropriate series are multiplied by a factor (*e.g.* $e^{-\kappa n^2\pi^2 t/a^2}$), which may be called a convergency factor, as it increases the rapidity of the convergence of the series, and allows term by term differentiation both with respect to x and t.

Integration. Again we have seen that under certain conditions the Fourier's Series for $f(x)$ converges uniformly to $f(x)$ in the interior of any interval in which the function is continuous. In this case we know that we may integrate the series term by term within such an interval, and equate the result to the integral of $f(x)$ between the same limits. Also this operation can be repeated any number of times (§ 70. 1).

But such a simple series as the Sine Series for unity in $0 < x < \pi$, namely

$$1 = \frac{4}{\pi}\left(\sin x + \frac{1}{3}\sin 3x + \frac{1}{5}\sin 5x + \ldots\right), \quad 0 < x < \pi$$

is not uniformly convergent in the interval $0 \leqq x \leqq \pi$, as its sum is discontinuous at $x = 0$ and $x = \pi$.

However it can be integrated term by term between the limits 0 and x, where $x \leqq \pi$ (cf. § 70. 2). This can be verified at once by comparing the Cosine Series for x in $(0, \pi)$, namely

$$x = \frac{\pi}{2} - \frac{4}{\pi}\left(\cos x + \frac{1}{3^2}\cos 3x + \frac{1}{5^2}\cos 5x + \ldots\right), \quad 0 \leqq x \leqq \pi$$

with the series obtained from the above by integration.

In the days when Fourier's Series were first used, term by term integration was employed without any hesitation, both in the case of the Fourier's Series for $f(x)$, and when the series considered was that obtained by multiplying the Fourier's Series term by term by another function. Later it was seen that such a step required justification. Hence the importance attached to the question of the uniform convergence of Fourier's Series in certain cases. But the theorem that follows shows that the presence or absence of uniformity of convergence has little or no bearing on the subject of the integration of the Fourier's Series: and that, even the convergence of the series, is of secondary interest.

Let $f(x)$ be bounded and integrable in $(-\pi, \pi)$, or, if unbounded, let $\int_{-\pi}^{\pi} f(x)dx$ be absolutely convergent. Then, whether the Fourier's Series corresponding to $f(x)$, namely

$$a_0 + (a_1 \cos x + b_1 \sin x) + (a_2 \cos 2x + b_2 \sin 2x) + \ldots,$$

converges or not,

$$\int_{-\pi}^{x} f(x)dx = a_0(x+\pi) + \sum_{1}^{\infty} \frac{1}{n}(a_n \sin nx + b_n(\cos n\pi - \cos nx)),$$

when $$-\pi \leqq x \leqq \pi.$$

Let $y = F(x) = \int_{-\pi}^{x} f(x)dx - a_0 x.$

Then $F(\pi) = \int_{-\pi}^{\pi} f(x)dx - a_0\pi = a_0\pi,$ since $2\pi a_0 = \int_{-\pi}^{\pi} f(x)dx.$

and $F(-\pi) = F(\pi).$

---

*Carslaw, *Conduction of Heat* (2nd ed., 1921), Ch. IV. § 30.

Also $F(x)$ is continuous in $(-\pi, \pi)$ (§ 49).

And it is of bounded variation in this interval; for, with the notation of § 36. 2,

$$\sum_0^{n-1} \left| y_{r+1} - y_r \right| = \sum_0^{n-1} \left| \int_{x_r}^{x_{r+1}} f(x)dx \right| \leqq \sum_0^{n-1} \int_{x_r}^{x_{r+1}} |f(x)|dx \leqq \int_{-\pi}^{\pi} |f(x)|dx.$$

Hence, by § 95, the Fourier's Series for $F(x)$, which we write as

$$A_0 + (A_1 \cos x + B_1 \sin x) + (A_2 \cos 2x + B_2 \sin 2x) + \ldots$$

has $F(x)$ for its sum at every point of $-\pi \leqq x \leqq \pi$.

Also, when $n \geqq 1$,
$$A_n = \frac{1}{\pi}\int_{-\pi}^{\pi} F(x) \cos nx\, dx$$
$$= -\frac{1}{n\pi}\int_{-\pi}^{\pi} F'(x) \sin nx\, dx$$
$$= -\frac{1}{n\pi}\int_{-\pi}^{\pi} (f(x) - a_0) \sin nx\, dx^*$$
$$= -\frac{b_n}{n}.$$

Similarly
$$B_n = \frac{1}{\pi}\int_{-\pi}^{\pi} F(x) \sin nx\, dx = \frac{a_n}{n}.$$

Therefore
$$F(x) = A_0 + \sum_1^{\infty} \frac{1}{n}(a_n \sin nx - b_n \cos nx).$$

But since $F(\pi) = F(-\pi) = a_0\pi$, we have

$$a_0\pi = A_0 + \sum_1^{\infty} \frac{1}{n}(a_n \sin n\pi - b_n \cos n\pi).$$

Hence, on subtraction,

$$\int_{-\pi}^{x} f(x)dx = a_0(x+\pi) + \sum_1^{\infty} \frac{1}{n}(a_n \sin nx + b_n(\cos n\pi - \cos nx)), \quad -\pi \leqq x \leqq \pi.$$

It also follows that, when $-\pi < x_1 < x_2 < \pi$,

$$\int_{x_1}^{x_2} f(x)dx = a_0(x_2 - x_1) + \sum_1^{\infty} \frac{1}{n}(a_n(\sin nx_2 - \sin nx_1) + b_n(\cos nx_1 - \cos nx_2)).$$

Again since $F(x)$ is continuous in $-\pi \leqq x \leqq \pi$ and $F(\pi) = F(-\pi)$, we know that the Fourier's Series for $F(x)$ converges uniformly to $F(x)$ in this interval.

Hence term by term integration of $f(x)$ can be repeated any number of times.

110. Parseval's Theorem on Fourier's Constants.

In this section, as usual,

$$s_n = a_0 + \sum_1^n (a_r \cos rx + b_r \sin rx),$$

$$\sigma_n = \frac{1}{n}(s_0 + s_1 + \ldots + s_{n-1}),$$

and $a_0, a_1, \ldots b_1, \ldots$ are Fourier's Constants for the function $f(x)$.

---

*It is assumed that the rule for integration by parts can be applied, and that $F'(x) = f(x) - a_0$. This makes the condition attached to $f(x)$ less general than in the statement of the theorem. For another proof, see §**110**, IV.

We shall prove that under certain conditions

$$\frac{1}{\pi}\int_{-\pi}^{\pi}[f(x)]^2\,dx = 2a_0^2 + \sum_1^\infty (a_n^2 + b_n^2),$$

a result usually called Parseval's Theorem.*

I. *Let $f(x)$ be bounded and integrable in $(-\pi, \pi)$, or, if unbounded, let*

$$\int_{-\pi}^{\pi}[f(x)]^2\,dx$$

be convergent.

Then $$2a_0^2 + \sum_1^\infty (a_n^2 + b_n^2) \leqq \frac{1}{\pi}\int_{-\pi}^{\pi}[f(x)]^2\,dx.$$

$$\int_{-\pi}^{\pi}[f(x) - s_n]^2\,dx = \int_{-\pi}^{\pi}[f(x)]^2\,dx - 2\int_{-\pi}^{\pi} f(x)s_n\,dx + \int_{-\pi}^{\pi} s_n^2\,dx.$$

Substitute for s_n on the right-hand side, and we have at once

$$\int_{-\pi}^{\pi}[f(x) - s_n]^2\,dx = \int_{-\pi}^{\pi}[f(x)]^2\,dx - 2\pi[2a_0^2 + \sum_1^n (a_r^2 + b_r^2)] + \pi[2a_0^2 + \sum_1^n (a_r^2 + b_r^2)].$$

It follows that

$$\frac{1}{\pi}\int_{-\pi}^{\pi}[f(x) - s_n]^2\,dx = \frac{1}{\pi}\int_{-\pi}^{\pi}[f(x)]^2\,dx - [2a_0^2 + \sum_1^n (a_r^2 + b_r^2)].$$

Thus $$2a_0^2 + \sum_1^n (a_r^2 + b_r^2) \leqq \frac{1}{\pi}\int_{-\pi}^{\pi}[f(x)]^2\,dx. \quad \text{.....................(1)}$$

And $$2a_0^2 + \sum_1^\infty (a_n^2 + b_n^2) \leqq \frac{1}{\pi}\int_{-\pi}^{\pi}[f(x)]^2\,dx. \quad \text{.....................(2)}$$

But if this Fourier's Series had converged uniformly to the continuous function $f(x)$, we could have multiplied both sides of the equation

$$f(x) = a_0 + (a_1 \cos x + b_1 \sin x) + \dots$$

by $f(x)$, and, integrating term by term, we would have obtained

$$\frac{1}{\pi}\int_{-\pi}^{\pi}[f(x)]^2\,dx = 2a_0^2 + \sum_1^\infty (a_n^2 + b_n^2),$$

in this case.

We proceed to prove that this equality holds when the only condition attached to $f(x)$ is that it is bounded and integrable in $(-\pi, \pi)$.

II. *Let $f(x)$ be bounded and integrable in $(-\pi, \pi)$.*

Then $$\frac{1}{\pi}\int_{-\pi}^{\pi}[f(x)]^2\,dx = 2a_0^2 + \sum_1^\infty (a_n^2 + b_n^2).$$

Since $$s_n = \frac{1}{2\pi}\int_{-\pi}^{\pi} f(x')(1 + 2\sum_1^n \cos r(x' - x))\,dx',$$

---

*If the Lebesgue Integral is used, the following theorem also holds :

Any trigonometrical series

$$a_0 + (a_1 \cos x + b_1 \sin x) + (a_2 \cos 2x + b_2 \sin 2x) + \dots$$

for which $\sum_1^\infty (a_n^2 + b_n^2)$ converges is the Fourier's Series of a function whose square is integrable (L) in $(-\pi, \pi)$.

This is known as the Riesz-Fischer Theorem.

we have, as in § 101,

$$\sigma_n = \frac{1}{2n\pi}\int_{-\pi}^{\pi} f(x')\frac{\sin^2 \frac{1}{2}n(x'-x)}{\sin^2 \frac{1}{2}(x'-x)}\,dx'.$$

Also we know that

$$\int_{-\pi}^{\pi} \frac{\sin^2 \frac{1}{2}n(x'-x)}{\sin^2 \frac{1}{2}(x'-x)}\,dx' = 2n\pi \text{ (cf. p. 256).}$$

Therefore
$$\sigma_n - f(x) = \frac{1}{2n\pi}\int_{-\pi}^{\pi} [f(x')-f(x)]\frac{\sin^2 \frac{1}{2}n(x'-x)}{\sin^2 \frac{1}{2}(x'-x)}\,dx', \quad \text{............(3)}$$

and
$$|\sigma_n - f(x)| \leqq M - m, \quad \text{..............................(4)}$$

where M, m are the upper and lower bounds of $f(x)$ in $(-\pi, \pi)$ and x is any point in this interval.

Now let ϵ and κ be any arbitrary positive numbers.

We know from § 42, III, that there is a mode of division of $(-\pi, \pi)$ into a finite number of partial intervals such that the sum of these intervals for which the oscillation of $f(x)$ is greater than or equal to κ is less than ϵ.

Let Δ denote the intervals in this mode of division for which the oscillation is greater than or equal to κ, and δ the other intervals.

Cut off from the intervals of δ at each of their ends a part, so that the sum of these segments cut off is less than ϵ.

Let δ'' denote the segment cut off, and δ' the inner parts of δ which remain.

Then
$$\int_{-\pi}^{\pi} [\sigma_n - f(x)]^2\,dx = \left\{\sum_{\delta'}\int_{\delta'} + \sum_{\delta''}\int_{\delta''} + \sum_{\Delta}\int_{\Delta}\right\}[\sigma_n - f(x)]^2\,dx, \quad \text{.........(5)}$$

where, by this notation, we mean that the integrals are taken respectively over the intervals of δ', δ'' and Δ.

Now let (a, b) be one of the intervals of δ, and (a', b') the corresponding interval of δ'.

Also let x be a point of (a', b').

Then from (3),

$$\sigma_n - f(x) = \frac{1}{2n\pi}\left(\int_{-\pi}^{a} + \int_{a}^{b} + \int_{b}^{\pi}\right)[f(x')-f(x)]\frac{\sin^2 \frac{1}{2}n(x'-x)}{\sin^2 \frac{1}{2}(x'-x)}\,dx'. \quad \text{......(6)}$$

Taking these three integrals separately, we have

$$\left|\frac{1}{2n\pi}\int_{-\pi}^{a} [f(x')-f(x)]\frac{\sin^2 \frac{1}{2}n(x'-x)}{\sin^2 \frac{1}{2}(x'-x)}\,dx'\right| < \frac{M-m}{2n\pi}\int_{-\pi}^{a} \operatorname{cosec}^2 \tfrac{1}{2}(x'-x)\,dx'$$
$$< \frac{K}{n},$$

where K is a positive constant depending upon the position of (a, b) and (a', b').

Similarly we find that

$$\left|\frac{1}{2n\pi}\int_{b}^{\pi} [f(x')-f(x)]\frac{\sin^2 \frac{1}{2}n(x'-x)}{\sin^2 \frac{1}{2}(x'-x)}\,dx'\right| < \frac{M-m}{2n\pi}\int_{b}^{\pi} \operatorname{cosec}^2 \tfrac{1}{2}(x'-x)\,dx'$$
$$< \frac{K}{n},$$

where it is clear that we may take the same value for K in both, and replace it later by any larger value we please.

Finally

$$\left|\frac{1}{2n\pi}\int_a^b [f(x')-f(x)]\frac{\sin^2 \frac{1}{2}n(x'-x)}{\sin^2 \frac{1}{2}(x'-x)}dx'\right| < \frac{\kappa}{2n\pi}\int_a^b \frac{\sin^2 \frac{1}{2}n(x'-x)}{\sin^2 \frac{1}{2}(x'-x)}dx'$$

$$< \frac{\kappa}{2n\pi}\int_{-\pi}^{\pi} \frac{\sin^2 \frac{1}{2}n(x'-x)}{\sin^2 \frac{1}{2}(x'-x)}dx'$$

$$< \kappa.$$

Thus, from (6),

$$|\sigma_n - f(x)| \leqq \left(\kappa + 2\frac{K}{n}\right) \quad \text{.................................(7)}$$

But the sum of the intervals of δ' does not exceed 2π.

Therefore in (5), we have

$$\sum_{\delta'}\int_{\delta'}[\sigma_n - f(x)]^2\,dx \leqq 2\pi\left(\kappa + 2\frac{K}{n}\right)^2 \quad \text{..........................(8)}$$

Also $\sum_{\delta''}\int_{\delta''}[\sigma_n - f(x)]^2\,dx \leqq \epsilon(M-m)^2,$ ....(9)

by (4), since the sum of the intervals δ'' is less than ϵ.

And $\sum_{\Delta}\int_{\Delta}[\sigma_n - f(x)]^2\,dx \leqq \epsilon(M-m)^2,$ ....(10)

for a similar reason.

It follows from (5) and (8), (9), (10), that

$$\lim_{n\to\infty}\int_{-\pi}^{\pi}[\sigma_n - f(x)]^2\,dx = 0, \quad \text{..........................(11)}$$

since κ and ϵ are arbitrary positive numbers, which can be chosen as small as we please.

But $\sigma_n = a_0 + \sum_1^{n-1}\left(\frac{n-r}{n}\right)(a_r \cos rx + b_r \sin rx).$

Therefore, as in (I), we have

$$\frac{1}{\pi}\int_{-\pi}^{\pi}[\sigma_n - f(x)]^2\,dx = \frac{1}{\pi}\int_{-\pi}^{\pi}[f(x)]^2\,dx - \left[2a_0^2 + \sum_1^{n-1}\left(\frac{n^2-r^2}{n^2}\right)(a_r^2+b_r^2)\right]$$

$$= \left\{\frac{1}{\pi}\int_{-\pi}^{\pi}[f(x)]^2\,dx - [2a_0^2 + \sum_1^{n-1}(a_r^2+b_r^2)]\right\}$$

$$+\frac{1}{n^2}\sum_1^{n-1} r^2(a_r^2+b_r^2). \quad \text{..................................(12)}$$

But we know from (1) that

$$\frac{1}{\pi}\int_{-\pi}^{\pi}[f(x)]^2\,dx - [2a_0^2 + \sum_1^{n-1}(a_r^2+b_r^2)] \geqq 0.$$

Therefore from (11) and (12), we see that, under the conditions stated in the theorem,

$$\frac{1}{\pi}\int_{-\pi}^{\pi}[f(x)]^2\,dx = 2a_0^2 + \sum_1^{\infty}(a_r^2+b_r^2).$$

Also $$\lim_{n\to\infty}\left(\frac{1}{n^2}\sum_1^{n-1} r^2(a_r^2+b_r^2)\right) = 0.$$

III. *Let $f(x)$ and $g(x)$ be bounded and integrable in $(-\pi, \pi)$ and a_n, b_n the Fourier's Constants for $f(x)$, α_n, β_n the Fourier's Constants for $g(x)$.*

Then
$$\frac{1}{\pi}\int_{-\pi}^{\pi} f(x)g(x)dx = 2a_0\alpha_0 + \sum_1^{\infty}(a_n\alpha_n + b_n\beta_n).$$

We have from (II),
$$\frac{1}{\pi}\int_{-\pi}^{\pi}[f(x)+g(x)]^2dx = 2(a_0+\alpha_0)^2 + \sum_1^{\infty}[(a_n+\alpha_n)^2+(b_n+\beta_n)^2],$$
and
$$\frac{1}{\pi}\int_{-\pi}^{\pi}[f(x)-g(x)]^2dx = 2(a_0-\alpha_0)^2 + \sum_1^{\infty}[(a_n-\alpha_n)^2+(b_n-\beta_n)^2].$$

On subtraction, the result follows.

IV. We may put $g(x)=0$ in $(-\pi, x_1)$ and (x_2, π), where $-\pi \leqq x_1 < x_2 \leqq \pi$. Thus we have the following theorem:

Let $f(x)$ be bounded and integrable in $(-\pi, \pi)$ and $g(x)$ be bounded and integrable in (x_1, x_2) where $-\pi \leqq x_1 < x_2 \leqq \pi$.

Then from (III),
$$\int_{x_1}^{x_2} f(x)g(x)dx = a_0\int_{x_1}^{x_2} g(x)dx + \sum_1^{\infty}\left(a_n\int_{x_1}^{x_2} g(x)\cos nx\,dx + b_n\int_{x_1}^{x_2} g(x)\sin nx\,dx\right).$$

It will be seen that (IV) establishes the possibility of term by term integration of the Fourier's Series for the bounded and integrable function $f(x)$, and also shows that this can be done when $f(x)$ is multiplied by another function of the same class.

The argument in (II) is taken from Hurwitz' proof of Parseval's Theorem.*

REFERENCES.

BÔCHER, *loc. cit.*

DE LA VALLÉE POUSSIN, *loc. cit.* (2e éd., 1912), Ch. IV.

HOBSON, *loc. cit.*, 2 (2nd ed., 1926), Ch. VIII.

LEBESGUE, *Leçons sur les séries trigonométriques* (Paris, 1906), Ch. III.

WHITTAKER AND WATSON, *loc. cit.* (5th ed., 1928), Ch. IX.

HILB U. M. RIESZ, "Neuere Untersuchungen über trigonometrische Reihen," *Enc. d. math. Wiss.*, Bd. II, Tl. III, 2, p. 1189 *et seq.* (Leipzig, 1923).

---

* *Math. Annalen*, **57** (1903), 425. Another proof is given by de la Vallée Poussin in his *Cours d'Analyse* 2 (2e éd., 1912), § 158. In his proof $\int_{-\pi}^{\pi} f(x)dx$ is to be absolutely convergent: but the integration under the sign towards the beginning of his demonstration is not discussed. The question of the removal of the restrictions upon the functions involved is now of less interest, as these and other theorems concerning Fourier's Constants are discussed with Lebesgue's Integral instead of Riemann's.

CHAPTER IX

THE APPROXIMATION CURVES AND THE GIBBS PHENOMENON IN FOURIER'S SERIES

111. We have seen in § 104 that, when $f(x)$ is bounded and continuous, and otherwise satisfies Dirichlet's Conditions in $-\pi<x<\pi$, $f(\pi-0)$ being equal to $f(-\pi+0)$, and $f'(x)$ is bounded and otherwise satisfies Dirichlet's Conditions in the same interval, the coefficients in the Fourier's Series for $f(x)$ are of the order $1/n^2$, and the series is uniformly convergent in any interval.

In this case the approximation curves

$$y=s_n(x)$$

in the interval $-\pi<x<\pi$ will nearly coincide with

$$y=f(x),$$

when n is taken large enough.

As an example, let $f(x)$ be the odd function defined in $(-\pi, \pi)$ as follows:

$$\begin{aligned} f(x)&=-\tfrac{1}{4}\pi(\pi+x), & -\pi \leqq x \leqq -\tfrac{1}{2}\pi, \\ f(x)&=\tfrac{1}{4}\pi x, & -\tfrac{1}{2}\pi \leqq x \leqq \tfrac{1}{2}\pi, \\ f(x)&=\tfrac{1}{4}\pi(\pi-x), & \tfrac{1}{2}\pi \leqq x \leqq \pi. \end{aligned}$$

The Fourier's Series for $f(x)$ in this case is the Sine Series

$$\sin x-\frac{1}{3^2}\sin 3x+\frac{1}{5^2}\sin 5x-\ldots,$$

which is uniformly convergent in any interval.

The approximation curves

$$\begin{aligned} y&=\sin x, \\ y&=\sin x-\frac{1}{3^2}\sin 3x, \\ y&=\sin x-\frac{1}{3^2}\sin 3x+\frac{1}{5^2}\sin 5x, \end{aligned}$$

are given in Fig. 29, along with $y=f(x)$, for the interval $(0, \pi)$, and it will be seen how closely the last of these three approaches the sum of the series right through the interval.

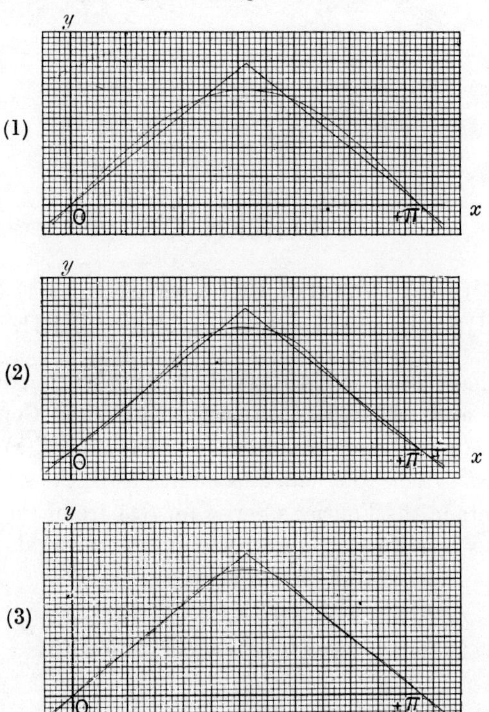

FIG. 29.

Again, let $f(x)$ be the corresponding even function :

$$f(x)=\tfrac{1}{4}\pi(\pi+x), \qquad -\pi \leqq x \leqq -\tfrac{1}{2}\pi,$$
$$f(x)=-\tfrac{1}{4}\pi x, \qquad \tfrac{1}{2}\pi \leqq x \leqq 0,$$
$$f(x)=\tfrac{1}{4}\pi x, \qquad 0 \leqq x \leqq \tfrac{1}{2}\pi,$$
$$f(x)=\tfrac{1}{4}\pi(\pi-x), \qquad \tfrac{1}{2}\pi \leqq x \leqq \pi.$$

The Fourier's Series for $f(x)$ in this case is the Cosine Series

$$\frac{1}{16}\pi^2-2\left\{\frac{1}{2^2}\cos 2x+\frac{1}{6^2}\cos 6x+\frac{1}{10^2}\cos 10x+\dots\right\},$$

which is again uniformly convergent in any interval.

The approximation curves

$$y=\tfrac{1}{16}\pi^2-\tfrac{1}{2}\cos 2x,$$
$$y=\tfrac{1}{16}\pi^2-\tfrac{1}{2}\cos 2x-\tfrac{1}{18}\cos 6x,$$

are given in Fig. 30, along with $y=f(x)$, for the interval $(0, \pi)$, and again it will be seen how closely the second of these curves approaches $y=f(x)$ right through the interval.

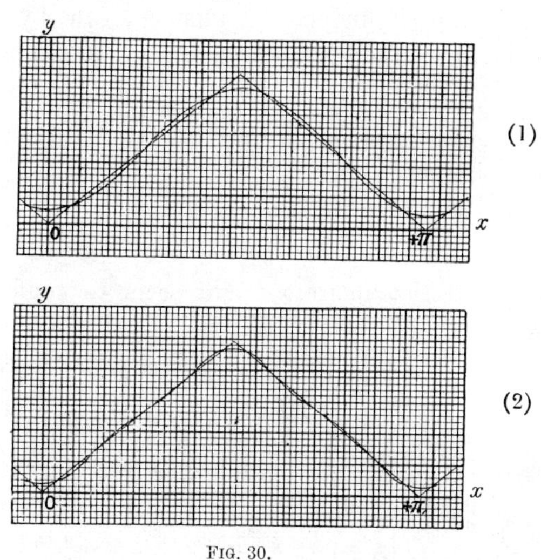

FIG. 30.

It will be noticed that in both these examples for large values of n the slope of $y=s_n(x)$ nearly agrees with that of $y=f(x)$, except at the corners, corresponding to $x=\pm\frac{1}{2}\pi$, where $f'(x)$ is discontinuous. This would lead us to expect that these series may be differentiated term by term, as in fact is the case.

112. When the function $f(x)$ is given by the equation

$$f(x)=x, \quad -\pi<x<\pi,$$

the corresponding Fourier's Series is the Sine Series

$$2\{\sin x-\tfrac{1}{2}\sin 2x+\tfrac{1}{3}\sin 3x-\ldots\}.$$

The sum of this series is x for all values in the open interval $-\pi<x<\pi$, and it is zero when $x=\pm\pi$.

This series converges uniformly in any interval $(-\pi+\delta, \pi-\delta)$ contained within $(-\pi, \pi)$ (cf. § 107), and in such an interval, by taking n large enough, we can make the approximation curves oscillate about $y=x$ as closely as we please.

Until recent years it was wrongly believed that, for large values of n, each approximation curve passed at a steep gradient from the point $(-\pi, 0)$ to a point near $(-\pi, -\pi)$, and then oscillated about $y=x$ till x approached the value π, when the curve passed at a steep gradient from a point near (π, π) to $(\pi, 0)$. And to those who did not properly understand what is meant by the sum of an infinite series, the difference between the approximation curves

$$y=s_n(x),$$

for large values of n, and the curve $y=\lim_{n\to\infty} s_n(x)$ offered considerable difficulty.

In Fig. 31, the line $y=x$ and the curve

$$y=2(\sin x-\tfrac{1}{2}\sin 2x+\tfrac{1}{3}\sin 3x-\tfrac{1}{4}\sin 4x+\tfrac{1}{5}\sin 5x)$$

are drawn, and the diagram might seem to confirm the above

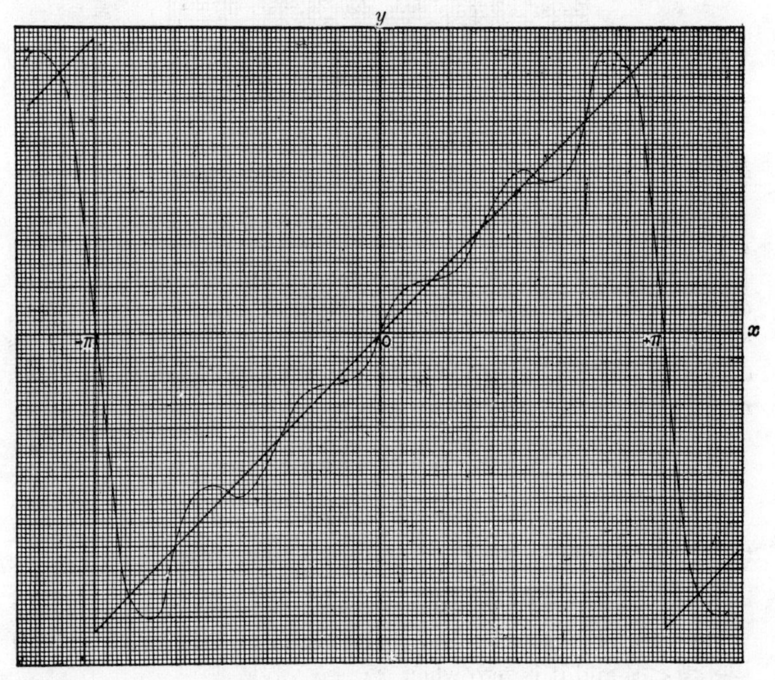

FIG. 31.

view of the matter—namely, that there will be a steep descent near one end of the line, from the point $(-\pi, 0)$ to near the point $(-\pi, -\pi)$, and a corresponding steep descent near the other end of the line. But it must be remembered that the convergence of this series is slow, and that $n=5$ would not count as a large value of n.

113. In 1899 Gibbs, in a letter to *Nature*,* pointed out that the approximation curves for this series do, in fact, behave in quite a different way at the points of discontinuity $\pm\pi$ in the sum. He stated, in effect, that the curve $y=s_n(x)$, for large values of n, falls from the point $(-\pi, 0)$ at a steep gradient to a point very nearly at a depth $2\int_0^\pi \frac{\sin x}{x} dx$ below the axis of x, then oscillates above and below $y=x$ close to this line until x approaches π, when it falls from a point very nearly at a height $2\int_0^\pi \frac{\sin x}{x} dx$ above the axis of x at a steep gradient to $(\pi, 0)$.

The approximation curves, for large values of n, would thus in $(-\pi, \pi)$ approach closely to the line $y=x$ of Fig. 32 with the lines parallel to the axis of y as drawn in that figure.

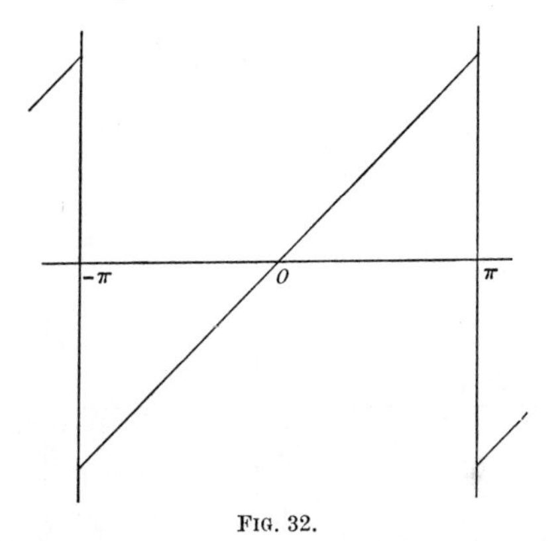

FIG. 32.

His statement was not accompanied by any proof. Though the remainder of the correspondence, of which his letter formed a part, attracted considerable attention, this remarkable observation passed practically unnoticed for several years. In 1906 Bôcher returned to the subject in a memoir † on Fourier's Series, and greatly extended Gibbs's result. He showed, among other things, that the phenomenon which Gibbs had observed in the case of this particular Fourier's Series holds in general at ordinary points of discontinuity. To quote his own words:‡

***Nature*, **59** (1899), 606.

†*Annals of Math.*, (2), **7** (1906), 81.

‡*loc. cit.*, p. 131.

If $f(x)$ has the period 2π and in any finite interval has no discontinuities other than a finite number of finite jumps, and if it has a derivative which in any finite interval has no discontinuities other than a finite number of finite discontinuities, then as n becomes infinite the approximation curve $y = s_n(x)$ approaches uniformly the continuous curve made up of

(a) *the discontinuous curve $y = f(x)$,*

(b) *an infinite number of straight lines of finite lengths parallel to the axis of y and passing through the points $a_1, a_2, \ldots$ on the axis of x where the discontinuities of $f(x)$ occur. If a is any one of these points, the line in question extends between the two points whose ordinates are*

$$f(a-0) + \frac{DP_1}{\pi}, \quad f(a+0) - \frac{DP_1}{\pi},$$

where D is the magnitude of the jump in $f(x)$ at a, and*

$$P_1 = \int_\pi^\infty \frac{\sin x}{x}\, dx = -0{\cdot}2811.$$

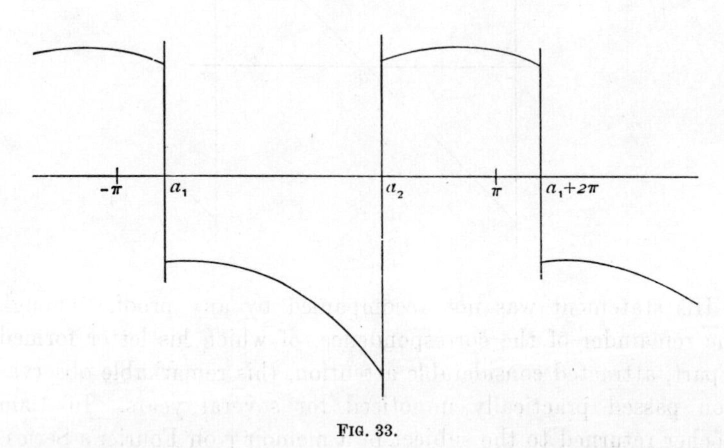

FIG. 33.

This theorem is illustrated in Fig. 33, where the amounts of the jumps at a_1, a_2 are respectively negative and positive. Both Gibbs and Bôcher thought they were describing properties previously unknown. However, in 1848, Wilbraham † had noticed its

* *i.e.* $D = f(a+0) - f(a-0)$.

† *Camb. and Dublin Math. Journal*, 3 (1848), 198.

occurrence in the approximation curves for the series

$$\sum_1^\infty (-1)^{r-1}\frac{\sin rx}{r} \quad \text{and} \quad \sum_1^\infty (-1)^{r-1}\frac{\cos(2r-1)x}{2r-1}.$$

And in 1874 du Bois Reymond * would have reached the same conclusion, both for Fourier's Series and Integrals, had he not made a curious slip in dealing with the integral $\int_0^{nx}\frac{\sin t}{t}\,dt$, when $n\to\infty$ and $x\to 0$ simultaneously. In recent years a number of other writers have dealt with the matter, and the property in Fourier's and other series is now well known as the Gibbs Phenomenon.†

114. 1. The series on which Bôcher founded his demonstration of this and other extensions of Gibbs's theorem is

$$\sin x+\tfrac{1}{2}\sin 2x+\tfrac{1}{3}\sin 3x+\dots,$$

which, in the interval $(0, 2\pi)$ represents the function $f(x)$ defined as follows:

$$f(0)=f(2\pi)=0,$$
$$f(x)=\tfrac{1}{2}(\pi-x), \quad 0<x<2\pi.$$

In this case

$$\begin{aligned} s_n(x)&=\sin x+\frac{1}{2}\sin 2x+\dots+\frac{1}{n}\sin nx \\ &=\int_0^x(\cos\alpha+\cos 2\alpha+\dots+\cos n\alpha)\,d\alpha \\ &=\tfrac{1}{2}\int_0^x\frac{\sin(n+\frac{1}{2})\alpha}{\sin\frac{1}{2}\alpha}\,d\alpha-\tfrac{1}{2}x. \end{aligned}$$

The properties of the maxima and minima of $s_n(x)$ are not so easy to obtain,‡ nor are they so useful in the argument, as those of

$$\begin{aligned} R_n(x)&=\tfrac{1}{2}(\pi-x)-s_n(x) \\ &=\tfrac{1}{2}\pi-\tfrac{1}{2}\int_0^x\frac{\sin(n+\frac{1}{2})\alpha}{\sin\frac{1}{2}\alpha}\,d\alpha. \end{aligned}$$

In his memoir Bôcher dealt with the maxima and minima of

---

* *Math. Annalen*, 7 (1874), 241.

†Cf. *Enc. d. math. Wiss.*, Bd. II, Tl. III, 2, p. 1203. In addition to the papers referred to on pp. 1203-4, the following may be named:

Weyl, *Rend. Circ. Mat. Palermo*, 29 (1910) and 30 (1910).

Cooke, *Proc. London Math. Soc.* (2) 22 (1928), 171.

Wilton, *Journal für Math.*, 159 (1928), 144.

And a historical note by Carslaw, *Bull. Amer. Math. Soc.*, 31 (1925), 420.

‡Gronwall discussed the properties of $s_n(x)$ for this series, and deduced the Gibbs Phenomenon for the first wave, and some other results. Cf. *Math. Annalen*, 72 (1912). Also Jackson, *Rend. Circ. Mat. Palermo*, 32 (1911).

$R_n(x)$, and he called attention more than once to the fact that the height of a wave from the curve $y=f(x)$ was measured parallel to the axis of y. This point has been lost sight of in some expositions of his work.

114. 2. However the series

$$2(\sin x+\tfrac{1}{3}\sin 3x+\tfrac{1}{5}\sin 5x+\ldots)$$

has certain advantages as an approach to the Gibbs Phenomenon.

This is the Fourier's Series for $f(x)$, when

$$\left.\begin{aligned} f(x)&=-\tfrac{1}{2}\pi \quad \text{for} \quad -\pi<x<0\\ &=\tfrac{1}{2}\pi \quad \text{for} \quad 0<x<\pi\\ \text{and} \qquad f(-\pi)&=f(\pi)=f(0)=0. \end{aligned}\right\}$$

Let $$s_n(x)=2\sum_1^n \frac{\sin(2r-1)x}{2r-1}.$$

Then $$\frac{d}{dx}s_n(x)=2\sum_1^n\cos(2r-1)x=\frac{\sin 2nx}{\sin x}$$

and $$s_n(x)=\int_0^x \frac{\sin 2n\alpha}{\sin\alpha}\,d\alpha. \qquad \text{.......(1)}$$

Also $$s_n(x)-\int_0^x\frac{\sin 2n\alpha}{\alpha}\,d\alpha=\int_0^x \sin 2n\alpha\left(\frac{1}{\sin\alpha}-\frac{1}{\alpha}\right)d\alpha.$$

Thus $$s_n(x)-\int_0^{2nx}\frac{\sin\alpha}{\alpha}\,d\alpha=\int_0^x\sin 2n\alpha\,\frac{\alpha}{\sin\alpha}\left(\frac{\alpha}{3!}-\frac{\alpha^3}{5!}+\ldots\right)d\alpha.$$

But $x\operatorname{cosec} x$ continually increases from 1 to $\frac{1}{2}\pi$ as x passes from 0 to $\frac{1}{2}\pi$ and $0<\frac{x}{3!}-\frac{x^3}{5!}+\ldots<\frac{x}{3!}$ when $0<x\leqq\frac{1}{2}\pi$.

Thus $$\left|s_n(x)-\int_0^{2nx}\frac{\sin\alpha}{\alpha}\,d\alpha\right|<\frac{\pi}{12}\int_0^x x\,dx$$

$$<\frac{\pi}{24}x^2 \text{ when } 0<x\leqq\tfrac{1}{2}\pi.$$

If we take the arbitrary positive number ϵ, there is a corresponding positive number η, such that

$$\left|s_n(x)-\int_0^{2nx}\frac{\sin\alpha}{\alpha}\,d\alpha\right|<\epsilon \text{ when } 0\leqq x\leqq\eta, \qquad \text{...........(2)}$$

and this holds for all values of n.

Thus, if we choose n so large that $\frac{\pi}{2n}<\eta$, we have

$$\left|s_n\left(\frac{\pi}{2n}\right)-\int_0^\pi\frac{\sin\alpha}{\alpha}\,d\alpha\right|<\epsilon;$$

that is $$\left| s_n\left(\frac{\pi}{2n}\right) - \left(\frac{\pi}{2} + \left| \int_\pi^\infty \frac{\sin a}{a}\, da \right| \right) \right| < \epsilon.$$

This shows that *the approximation curves* $y = s_n(x)$ *for this series rise above the sum in the right-hand neighbourhood of the origin by nearly* $\left| \int_\pi^\infty \frac{\sin a}{a}\, da \right|$, *when* n *is sufficiently large.*

The integral $\int_\pi^\infty \frac{\sin x}{x}\, dx$ is known to be approximately $-0{\cdot}2811$.*

Or, we may put the matter as follows:

Let $$y = \int_0^x \frac{\sin a}{a}\, da = \phi(x).$$

The turning points of this curve occur when $x = r\pi$; the odd multiples of π give maximum ordinates; the even multiples give minimum ordinates.

The maxima continually decrease from $\int_0^\pi \frac{\sin a}{a}\, da$ towards $\frac{1}{2}\pi$ when $x \to \infty$.

The minima continually increase from $\int_0^{2\pi} \frac{\sin a}{a}\, da$ towards $\frac{1}{2}\pi$ when $x \to \infty$.

If the abscissae of this curve are reduced in the ratio $1 : 2n$, we obtain the curve
$$y = \int_0^{2nx} \frac{\sin a}{a}\, da,$$
and, when n increases, the turning points of the curve come closer and closer to $x = 0$, the first (and largest) ordinate being always at $x = \frac{\pi}{2n}$, and of height $\int_0^\pi \frac{\sin a}{a}\, da = \frac{\pi}{2} + \left| \int_\pi^\infty \frac{\sin a}{a}\, da \right|$.

By (2), the curve $y = s_n(x)$ in the neighbourhood of the origin differs by the arbitrarily small ϵ from the curve $y = \int_0^{2nx} \frac{\sin a}{a}\, da.$

THE TRIGONOMETRICAL SUM $2\sum_1^n \frac{\sin(2r-1)x}{2r-1}$

115. The discussion in § 114. 2 establishes the existence of the Gibbs Phenomenon in the Fourier's Series $2\sum_1^\infty \frac{\sin(2n-1)x}{2n-1}$, and, as we shall see in § 117, it can at once be deduced that the phenomenon will appear at $x = a$ in the approximation curves for the

*Cf. Bôcher, *loc. cit.*, p. 124.

Fourier's Series for any function $f(x)$ with an ordinary discontinuity at $x=a$, if $f(x)$ satisfies Dirichlet's Conditions in $(-\pi, \pi)$.

But the approximation curves

$$y=2\left(\sin x+\frac{1}{3}\sin 3x+\dots+\frac{1}{2n-1}\sin(2n-1)x\right)=s_n(x)$$

are worth a more detailed examination, for it will be found that from the properties of their turning points and their graphs, the Gibbs Phenomenon is exhibited in the clearest possible manner.*

In this section we proceed to obtain the properties of the maxima and minima of these approximation curves $y=2\sum_1^n\frac{\sin(2r-1)x}{2r-1}$, when $0<x<\pi$.

I. Since, for any integer m,

$$\sin(2m-1)(\tfrac{1}{2}\pi+x')=\sin(2m-1)(\tfrac{1}{2}\pi-x'),$$

it follows from the series that $s_n(x)$ *is symmetrical about* $x=\frac{1}{2}\pi$, *and when* $x=0$ *and* $x=\pi$ *it is zero.*

II. *When* $0<x<\pi$, $s_n(x)$ *is positive.*

From (I) we need only consider $0<x\leqq\frac{1}{2}\pi$. We have, by § 114. 2 (1)

$$s_n(x)=\int_0^x\frac{\sin 2na}{\sin a}\,da=\frac{1}{2n}\int_0^{2nx}\frac{\sin a}{\sin\dfrac{a}{2n}}\,da,\quad 0<x\leqq\tfrac{1}{2}\pi.$$

The denominator in the integrand is positive and continually increases in the interval of integration. By considering the successive waves in the graph of $\sin a\operatorname{cosec}\frac{a}{2n}$, the last of which may or may not be completed, it is clear that the integral is positive.

III. *The turning points of* $y=s_n(x)$ *are given by*

$$\begin{cases} x_1=\dfrac{\pi}{2n},\quad x_3=\dfrac{3\pi}{2n},\ \dots,\ x_{2n-1}=\dfrac{2n-1}{2n}\pi\ (\textit{maxima}),\\[2ex] x_2=\dfrac{\pi}{n},\quad x_4=\dfrac{2\pi}{n},\ \dots,\ x_{2(n-1)}=\dfrac{n-1}{n}\pi\ (\textit{minima}).\end{cases}$$

---

*§§ 115–117 are founded on a paper by the author in the *American Journal of Mathematics*, 39 (1917), 185. The computations for Fig. 34 and Fig. 35 were made by Mr. F. G. Brown, B.Sc.

We have

$$y=\int_0^x \frac{\sin 2n\alpha}{\sin\alpha}\,d\alpha, \quad \text{and} \quad \frac{dy}{dx}=\frac{\sin 2nx}{\sin x}.$$

The result follows at once.

IV. *As we proceed from $x=0$ to $x=\frac{1}{2}\pi$, the heights of the maxima continually diminish, and the heights of the minima continually increase, n being kept fixed.*

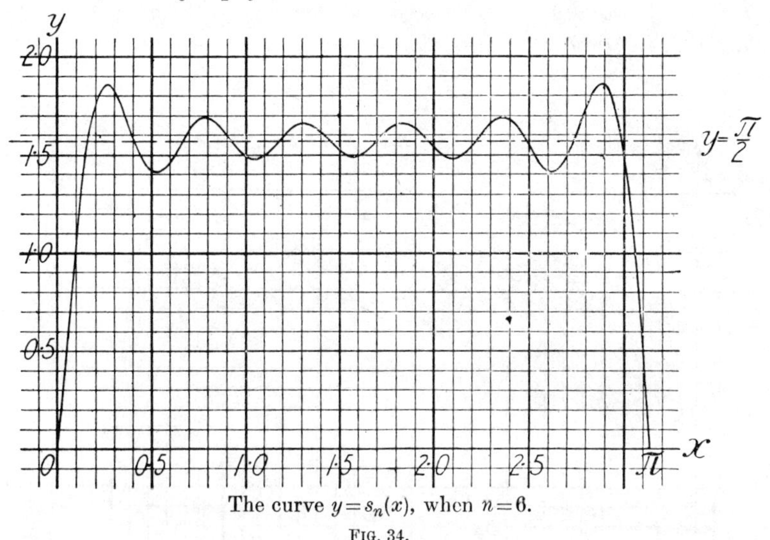

The curve $y=s_n(x)$, when $n=6$.

FIG. 34.

Consider two consecutive maxima in the interval $0<x\leqq\frac{1}{2}\pi$, namely, $s_n\left(\frac{2m-1}{2n}\pi\right)$ and $s_n\left(\frac{2m+1}{2n}\pi\right)$, m being a positive integer less than or equal to $\frac{1}{2}(n-1)$. We have

$$s_n\left(\frac{2m-1}{2n}\pi\right)-s_n\left(\frac{2m+1}{2n}\pi\right)=\frac{1}{2n}\int_{(2m+1)\pi}^{(2m-1)\pi}\frac{\sin\alpha}{\sin\dfrac{\alpha}{2n}}\,d\alpha$$

$$=-\frac{1}{2n}\left\{\int_{(2m-1)\pi}^{2m\pi}\frac{\sin\alpha}{\sin\dfrac{\alpha}{2n}}\,d\alpha+\int_{2m\pi}^{(2m+1)\pi}\frac{\sin\alpha}{\sin\dfrac{\alpha}{2n}}\,d\alpha\right\}.$$

The denominator in both integrands is positive and it continually increases in the interval $(2m-1)\pi\leqq\alpha\leqq(2m+1)\pi$; also the numerator in the first is continually negative and in the second continually positive, the absolute values for elements at equal distances from $(2m-1)\pi$ and $2m\pi$ being the same.

Thus the result follows. Similarly for the minima, we have to examine the sign of

$$s_n\left(\frac{m-1}{n}\pi\right)-s_n\left(\frac{m}{n}\pi\right),$$

where m is a positive integer less than or equal to $\frac{1}{2}n$.

V. *The first maximum to the right of* $x=0$ *is at* $x=\frac{\pi}{2n}$ *and its height continually diminishes as* n *increases. When* n *tends to infinity, its limit is*

$$\int_0^{\pi}\frac{\sin x}{x}\,dx.$$

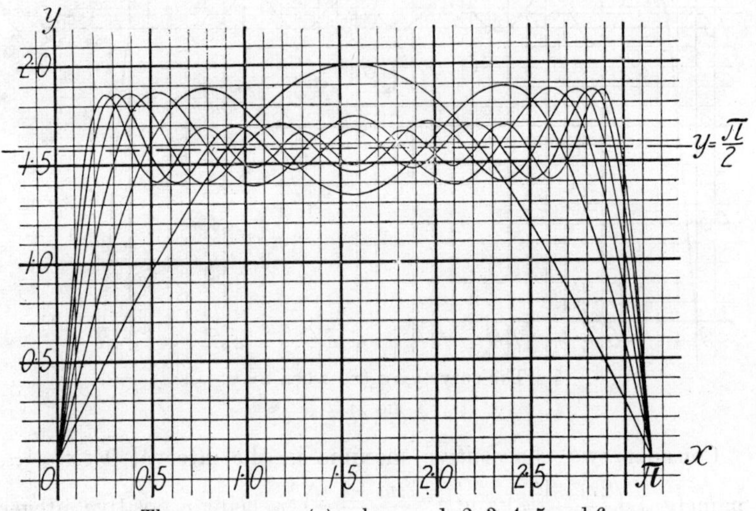

The curves $y=s_n(x)$, when $n=1$, 2, 3, 4, 5 and 6.

FIG. 35.

We have

$$s_n\left(\frac{\pi}{2n}\right)=\int_0^{\frac{\pi}{2n}}\frac{\sin 2n\alpha}{\sin\alpha}\,d\alpha=\frac{1}{2n}\int_0^{\pi}\sin\alpha\operatorname{cosec}\frac{\alpha}{2n}\,d\alpha.$$

Thus

$$s_n\left(\frac{\pi}{2n}\right)-s_{n+1}\left(\frac{\pi}{2n+2}\right)$$

$$=\int_0^{\pi}\sin\alpha\left(\frac{1}{2n}\operatorname{cosec}\frac{\alpha}{2n}-\frac{1}{2n+2}\operatorname{cosec}\frac{\alpha}{2n+2}\right)d\alpha.$$

Since $\alpha/\sin\alpha$ continually increases from 1 to ∞, as α passes from 0 to π, it is clear that in the interval with which we have to deal

$$\frac{1}{2n}\operatorname{cosec}\frac{\alpha}{2n}-\frac{1}{2n+2}\operatorname{cosec}\frac{\alpha}{2n+2}>0.$$

Thus $$s_n\left(\frac{\pi}{2n}\right)-s_{n+1}\left(\frac{\pi}{2(n+1)}\right)>0.$$

But, from (II), $s_n(x)$ is positive when $0<x<\pi$.

It follows that $s_n\left(\frac{\pi}{2n}\right)$ tends to a limit as n tends to infinity.

The value of this limit can be obtained by the method used by Bôcher for the integral $\int_0^x \frac{\sin(n+\frac{1}{2})\alpha}{\sin\frac{1}{2}\alpha}\,d\alpha$.* But it is readily obtained from the definition of the Definite Integral.

For $$s_n\left(\frac{\pi}{2n}\right)=2\left(\frac{\pi}{2n}\right)\left\{\frac{2n}{\pi}\sin\frac{\pi}{2n}+\frac{2n}{3\pi}\sin\frac{3\pi}{2n}+\dots\right.$$
$$\left.+\frac{2n}{(2n-1)\pi}\sin\frac{2n-1}{2n}\pi\right\}$$
$$=2\sum_{\substack{m=1\\2nh=\pi}}^{2n}\left(\frac{\sin mh}{mh}h\right)-\sum_{\substack{m=1\\nh=\pi}}^{n}\left(\frac{\sin mh}{mh}h\right).$$

Therefore
$$\lim_{n\to\infty}s_n\left(\frac{\pi}{2n}\right)=2\int_0^\pi\frac{\sin x}{x}\,dx-\int_0^\pi\frac{\sin x}{x}\,dx=\int_0^\pi\frac{\sin x}{x}\,dx.$$

VI. The result obtained in (V) for the first wave is a special case of the following:

The rth maximum to the right of $x=0$ is at $x_{2r-1}=\frac{2r-1}{2n}\pi$, and its height continually diminishes as n increases, r being kept constant. When n tends to infinity, its limit is
$$\int_0^{(2r-1)\pi}\frac{\sin x}{x}\,dx,$$
which is greater than $\frac{1}{2}\pi$.

The rth minimum to the right of $x=0$ is at $x_{2r}=\frac{r}{n}\pi$, and its height continually increases as n increases, r being kept constant. When n tends to infinity, its limit is $\int_0^{2r\pi}\frac{\sin x}{x}\,dx$, which is less than $\frac{1}{2}\pi$.

To prove these theorems we consider first the integral
$$\int_0^{m\pi}\sin\alpha\left(\frac{1}{2n}\operatorname{cosec}\frac{\alpha}{2n}-\frac{1}{2n+2}\operatorname{cosec}\frac{\alpha}{2n+2}\right)d\alpha,$$
m being a positive integer less than or equal to $2n-1$, so that $0<\frac{\alpha}{2n}<\pi$ in the interval of integration.

---

* *Annals of Math.*, (2), 7 (1906), 124. Also Hobson, *Theory of Functions of a Real Variable*, 2 (2nd ed., 1926), 494.

Then $$F(\alpha)=\frac{1}{2n}\operatorname{cosec}\frac{\alpha}{2n}-\frac{1}{2n+2}\operatorname{cosec}\frac{\alpha}{2n+2}>0$$

in this interval. (Cf. (V).)

Further,

$$F'(\alpha)=\frac{1}{(2n+2)^2}\cos\frac{\alpha}{2n+2}\operatorname{cosec}^2\frac{\alpha}{2n+2}-\frac{1}{(2n)^2}\cos\frac{\alpha}{2n}\operatorname{cosec}^2\frac{\alpha}{2n}$$

$$=\alpha^{-2}\{\phi^2\cos\phi\operatorname{cosec}^2\phi-\psi^2\cos\psi\operatorname{cosec}^2\psi\},$$

where $\phi=\alpha/(2n+2)$ and $\psi=\alpha/2n$.

But $$\frac{d}{d\phi}(\phi^2\cos\phi\operatorname{cosec}^2\phi)$$

$$=-\phi\operatorname{cosec}^3\phi\,[\phi(1\mp\cos\phi)^2\pm 2\cos\phi\,(\phi\mp\sin\phi)].$$

And the right-hand side of the equation will be seen to be negative, choosing the upper signs for $0<\phi<\frac{1}{2}\pi$ and the lower for $\frac{1}{2}\pi<\phi<\pi$.

Therefore $\phi^2\cos\phi\operatorname{cosec}^2\phi$ diminishes as ϕ increases from 0 to π.

It follows, from the expression for $F'(\alpha)$, that $F'(\alpha)>0$, and $F(\alpha)$ increases with α in the interval of integration.

The curve

$$y=\sin x\left(\frac{1}{2n}\operatorname{cosec}\frac{x}{2n}-\frac{1}{2n+2}\operatorname{cosec}\frac{x}{2n+2}\right),\ \ldots,\quad 0<x<m\pi,$$

thus consists of a succession of waves of length π, alternately above and below the axis, and the absolute values of the ordinates at points at the same distance from the beginning of each wave continually increase.

It follows that, when m is equal to 2, 4, ... , $2(n-1)$, the integral

$$\int_0^{m\pi}\sin\alpha\left(\frac{1}{2n}\operatorname{cosec}\frac{\alpha}{2n}-\frac{1}{2n+2}\operatorname{cosec}\frac{\alpha}{2n+2}\right)d\alpha$$

is negative; and, when m is equal to 1, 3, ... , $2n-1$, this integral is positive.

Returning to the maxima and minima, we have, for the rth maximum to the right of $x=0$,

$$s_n(x_{2r-1})-s_{n+1}(x_{2r-1})$$

$$=\int_0^{\frac{2r-1}{2n}\pi}\frac{\sin 2n\alpha}{\sin\alpha}d\alpha-\int_0^{\frac{2r-1}{2(n+1)}\pi}\frac{\sin 2(n+1)\alpha}{\sin\alpha}d\alpha$$

$$=\int_0^{(2r-1)\pi}\sin\alpha\left(\frac{1}{2n}\operatorname{cosec}\frac{\alpha}{2n}-\frac{1}{2n+2}\operatorname{cosec}\frac{\alpha}{2n+2}\right)d\alpha.$$

Therefore, from the above argument, $s_n(x_{2r-1})>s_{n+1}(x_{2r-1})$.

Also for the rth minimum to the right of $x=0$, we have

$$s_n(x_{2r})-s_{n+1}(x_{2r})=\int_0^{2r\pi}\sin a\left(\frac{1}{2n}\operatorname{cosec}\frac{a}{2n}-\frac{1}{2n+2}\operatorname{cosec}\frac{a}{2n+2}\right)da,$$

and $s_n(x_{2r})<s_{n+1}(x_{2r})$.

By an argument similar to that at the close of (V) we have

$$\lim_{n\to\infty} s_n(x_r)=\int_0^{r\pi}\frac{\sin x}{x}\,dx.*$$

It is clear that these limiting values are all greater than $\frac{1}{2}\pi$ for the maxima, and positive and less than $\frac{1}{2}\pi$ for the minima.

The Gibbs Phenomenon for the Series $2(\sin x+\frac{1}{3}\sin 3x+\frac{1}{5}\sin 5x+\ldots)$.

116. From the Theorems I-VI of § 115 all the features of the Gibbs Phenomenon for the series

$$2(\sin x+\tfrac{1}{3}\sin 3x+\tfrac{1}{5}\sin 5x+\ldots)\ldots,\qquad -\pi\leqq x\leqq\pi,$$

follow immediately.

It is obvious that we need only examine the interval $0\leqq x\leqq\pi$, and that a discontinuity occurs at $x=0$.

For large values of n, the curve

$$y=s_n(x),$$

where $s_n(x)=2\left(\sin x+\frac{1}{3}\sin 3x+\ldots+\frac{1}{2n-1}\sin(2n-1)x\right)$, rises at a steep gradient from the origin to its first maximum, which is very near, but above, the point $\left(0,\int_0^{\pi}\frac{\sin x}{x}dx\right)$ (§ 115, V). The curve, then, falls at a steep gradient, without reaching the axis of x (§ 115, II), to its first minimum, which is very near, but below, the point $\left(0,\int_0^{2\pi}\frac{\sin x}{x}dx\right)$ (§ 115, VI). It then oscillates above and below the line $y=\frac{1}{2}\pi$, the heights and (depths) of the waves continually diminishing as we proceed from $x=0$ to $x=\frac{1}{2}\pi$ (§ 115, IV); and from $x=\frac{1}{2}\pi$ to $x=\pi$, the procedure is reversed, the curve in the interval $0\leqq x\leqq\pi$ being symmetrical about $x=\frac{1}{2}\pi$ (§ 115, I).

The highest (or lowest) point of the rth wave to the right of

---

* For the values of $\int_0^{r\pi}\frac{\sin x}{x}dx$, see *Annals of Math.*, (2), **7** (1906), 129.

$x=0$ will, for large values of n, be at a point whose abscissa is $\frac{r\pi}{2n}$ (§ 115, III) and whose ordinate is very nearly

$$\int_0^{r\pi} \frac{\sin x}{x}\, dx \text{ (§ 115, VI).}$$

By increasing n the curve for $0 \leqq x \leqq \pi$ can be brought as close as we please to the lines

$$\left.\begin{array}{r} x=0, \quad 0<y \leqq \int_0^{\pi} \frac{\sin x}{x}\, dx, \\ 0<x<\pi, \quad y=\frac{\pi}{2}, \\ x=\pi, \quad 0<y \leqq \int_0^{\pi} \frac{\sin x}{x}\, dx. \end{array}\right\}$$

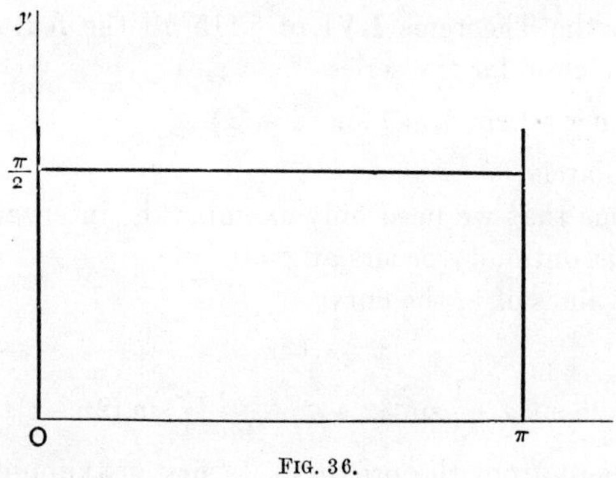

FIG. 36.

We may state these results more definitely as follows :

(i) If ϵ is any positive number, as small as we please, there is a positive integer ν' such that

$$|\tfrac{1}{2}\pi - s_n(x)| < \epsilon, \text{ when } n \geqq \nu', \ \epsilon \leqq x \leqq \tfrac{1}{2}\pi.$$

This follows from the uniform convergence of the Fourier's Series for $f(x)$, as defined in the beginning of this section, in an interval which does not include, either in its interior or at an end, a discontinuity of $f(x)$ (cf. § 107).

(ii) Since the height of the first maximum to the right of $x=0$ tends from above to $\int_0^{\pi} \frac{\sin x}{x}\, dx$ as n tends to infinity, there is a positive integer ν'' such that

$$0 < s_n\left(\frac{\pi}{2n}\right) - \int_0^{\pi} \frac{\sin x}{x}\, dx < \epsilon, \text{ when } n \geqq \nu''.$$

(iii) Let ν''' be the integer next greater than $\frac{\pi}{2\epsilon}$. Then the abscissa of the first maximum to the right of $x=0$, when $n \geqq \nu'''$, is less than ϵ.

It follows from (i), (ii) and (iii) that, if ν is the greatest of the positive integers ν', ν'' and ν''', the curve $y=s_n(x)$, when $n \geqq \nu$, behaves as follows:

It rises at a steep gradient from the origin to its first maximum which is above $\int_0^\pi \frac{\sin x}{x}\,dx$ and within the rectangle

$$0<x<\epsilon, \quad 0<y<\int_0^\pi \frac{\sin x}{x}\,dx+\epsilon.$$

After leaving this rectangle, in which there may be many oscillations about $y=\frac{1}{2}\pi$, it remains within the rectangle

$$\epsilon<x<\pi-\epsilon, \quad \tfrac{1}{2}\pi-\epsilon<y<\tfrac{1}{2}\pi+\epsilon.$$

Finally, it enters the rectangle

$$\pi-\epsilon<x<\pi, \quad 0<y<\int_0^\pi \frac{\sin x}{x}\,dx+\epsilon,$$

and the procedure in the first region is repeated.*

The Gibbs Phenomenon for Fourier's Series in General.

117. 1. Let $f(x)$ be a function with an ordinary discontinuity when $x=a$, which satisfies Dirichlet's Conditions in the interval $-\pi \leqq x \leqq \pi$.

Denote as usual by $f(a+0)$ and $f(a-0)$ the values towards which $f(x)$ tends as x approaches a from the right or left. It will be convenient to consider $f(a+0)$ as greater than $f(a-0)$ in the description of the curve, but this restriction is in no way necessary.

Let $\phi(x-a)=2\sum_1^\infty \frac{1}{(2r-1)} \sin(2r-1)(x-a)$.

Then
$$\left.\begin{aligned} \phi(x-a) &= \tfrac{1}{2}\pi, \quad \text{when } a<x<\pi+a, \\ \phi(x-a) &= -\tfrac{1}{2}\pi, \quad \text{when } -\pi+a<x<a, \\ \phi(+0) &= \tfrac{1}{2}\pi, \quad \phi(-0)=-\tfrac{1}{2}\pi, \\ \phi(0) &= 0 \quad \text{and} \quad \phi(x)=\phi(x+2\pi). \end{aligned}\right\}$$

---

*The cosine series

$$\frac{4}{\pi}\left[\cos x-\frac{1}{3}\cos 3x+\frac{1}{5}\cos 5x+\ldots\right],$$

which represents 0 in the interval $0 \leqq x < \frac{1}{2}\pi$ and $\frac{1}{2}\pi$ in the interval $\frac{1}{2}\pi < x \leqq \pi$. can be treated in the same way as the series discussed in this article.

Now put

$$\psi(x)=f(x)-\tfrac{1}{2}\{f(a+0)+f(a-0)\}-\frac{1}{\pi}\{f(a+0)-f(a-0)\}\phi(x-a),$$

and let $f(x)$, when $x=a$, be defined as $\frac{1}{2}\{f(a+0)+f(a-0)\}$.

Then $\psi(a+0)=\psi(a-0)=\psi(a)=0$, and $\psi(x)$ is continuous at $x=a$.

The following distinct steps in the argument are numbered for the sake of clearness :

(i) Since $\psi(x)$ is continuous at $x=a$ and $\psi(a)=0$, if ϵ is a positive number, as small as we please, a number η exists such that

$$|\psi(x)|<\tfrac{1}{4}\epsilon, \text{ when } |x-a|\leqq\eta.$$

If η is not originally less than ϵ, we can choose this part of η for our interval.

(ii) $\psi(x)$ can be expanded in a Fourier's Series,* this series being uniformly convergent in an interval $\alpha\leqq x\leqq\beta$ contained within an interval which includes neither within it nor as an end-point any other discontinuity of $f(x)$ and $\phi(x-a)$ than $x=a$ (cf. § 108).

Let $s_n(x)$, $\phi_n(x-a)$ and $\psi_n(x)$ be the sums of the terms up to and including those in sin nx and cos nx in the Fourier's Series for $f(x)$, $\phi(x-a)$ and $\psi(x)$. Then ϵ being the positive number of (i), as small as we please, there exists a positive integer ν' such that

$$|\psi_n(x)-\psi(x)|<\tfrac{1}{4}\epsilon, \text{ when } n\leqq\nu',$$

the same ν' serving for every x in $\alpha\leqq x\leqq\beta$.

Also $\quad |\psi_n(x)|\leqq|\psi_n(x)-\psi(x)|+|\psi(x)|<\tfrac{1}{4}\epsilon+\tfrac{1}{4}\epsilon=\tfrac{1}{2}\epsilon,$

in $|x-a|\leqq\eta$, if $\alpha<a-\eta<a<a+\eta<\beta$, and $n\geqq\nu'$.

(iii) Now if n is even, the first maximum in $\phi_n(x-a)$ to the right of $x=a$ is at $a+\frac{\pi}{n}$; and if n is odd, it is at $a+\frac{\pi}{n+1}$. In either case there exists a positive integer ν'' such that the height of the first maximum lies between

$$\int_0^\pi \frac{\sin x}{x}\,dx \text{ and } \int_0^\pi \frac{\sin x}{x}\,dx+\frac{\pi\epsilon}{2\{f(a+0)-f(a-0)\}}, \text{ when } n\geqq\nu''.$$

(iv) This first maximum will have its abscissa between a and $a+\eta$, provided that $\frac{\pi}{n}<\eta$.

---

*If $f(x)$ satisfies Dirichlet's Conditions, it is clear from the definition of $\phi(x-a)$ that $\psi(x)$ does so also.

Let ν''' be the first positive integer which satisfies this inequality.

(v) In the interval $a+\eta \leqq x \leqq \beta$, $s_n(x)$ converges uniformly to $f(x)$. Therefore a positive integer $\nu^{(\text{iv})}$ exists such that, when $n \geqq \nu^{(\text{iv})}$,

$$|f(x)-s_n(x)|<\epsilon,$$

the same $\nu^{(\text{iv})}$ serving for every x in this interval.

Now, from the equation defining $\psi(x)$,

$$s_n(x)=\tfrac{1}{2}(f(a+0)+f(a-0))+\frac{f(a+0)-f(a-0)}{\pi}\phi_n(x-a)+\psi_n(x).$$

It follows from (i)-(v) that if ν is the first positive integer greater than ν', ν'', ν''' and $\nu^{(\text{iv})}$, the curve $y=s_n(x)$, when $n \geqq \nu$, in the interval $a \leqq x \leqq \beta$, behaves as follows:

When $x=a$, it passes through a point whose ordinate is within $\frac{1}{2}\epsilon$ of $\frac{1}{2}(f(a+0)+f(a-0))$, and ascends at a steep gradient to a point within ϵ of

$$\tfrac{1}{2}\{f(a+0)+f(a-0)\}+\frac{f(a+0)-f(a-0)}{\pi}\int_0^\pi \frac{\sin x}{x}\,dx.$$

This may be written

$$f(a+0)-\frac{f(a+0)-f(a-0)}{\pi}\int_\pi^\infty \frac{\sin x}{x}\,dx,$$

and, from Bôcher's table, referred to in § 115, we have

$$\int_\pi^\infty \frac{\sin x}{x}\,dx=-0{\cdot}2811.$$

It then oscillates about $y=f(x)$ till x reaches $a+\eta$, the character of the waves being determined by the function $\phi_n(x-a)$, since the term $\psi_n(x)$ only adds a quantity less than $\frac{1}{2}\epsilon$ to the ordinate.

And on passing beyond $x=a+\eta$, the curve enters, and remains within, the strip of width 2ϵ enclosing $y=f(x)$ from $x=a+\eta$ to $x=\beta$.

On the other side of the point a a similar set of circumstances can be established.

Writing $D=f(a+0)-f(a-0)$, the crest (or hollow) of the first wave to the left and right of $x=a$ tends to a height

$$f(a-0)+\frac{DP_1}{\pi},\ f(a+0)-\frac{DP_1}{\pi},$$

where $P_1=\int_\pi^\infty \frac{\sin x}{x}\,dx$.

117. 2. The argument of the previous section can be at once adapted to the case of summation by Arithmetic Means, and it will be seen that, if we can show that the Gibbs Phenomenon does not occur in the approximation curves for a single example of Fourier's Series with this method of summation, it cannot appear in these curves for any function satisfying the conditions of § 101, at a point of ordinary discontinuity.*

It is, however, easy to show that it does not occur in the curves $y=\sigma_n(x)$ for the Fourier's Series corresponding to the function in § 114. 2.

For this case, we have $f(x)=-\frac{1}{2}\pi$, when $-\pi<x<0$, and $f(x)=\frac{1}{2}\pi$, when $0<x<\pi$.

But from § 101 we know that, when $f(x)$ is bounded and integrable in $(-\pi, \pi)$.

$$\sigma_n(x)=\frac{1}{n}\sum_0^{n-1} s_r=\frac{1}{2n\pi}\int_{-\pi}^{\pi} f(x')\frac{\sin^2\frac{1}{2}n(x'-x)}{\sin^2\frac{1}{2}(x'-x)}\,dx',$$

and
$$\int_{-\pi}^{\pi}\frac{\sin^2\frac{1}{2}n(x'-x)}{\sin^2\frac{1}{2}(x'-x)}\,dx'=2n\pi.$$

Thus, for the function with which we are now concerned,

$$\begin{aligned}|\sigma_n(x)| &\leqq \frac{1}{2n\pi}\int_{-\pi}^{\pi}|f(x')|\frac{\sin^2\frac{1}{2}n(x'-x)}{\sin^2\frac{1}{2}(x'-x)}\,dx' \\ &< \frac{1}{4n}\int_{-\pi}^{\pi}\frac{\sin^2\frac{1}{2}n(x'-x)}{\sin^2\frac{1}{2}(x'-x)}\,dx'. \\ &< \tfrac{1}{2}\pi.\end{aligned}$$

By § 115, II, the sums s_n, when $0<x<\pi$, are all positive.

It follows that $0<\sigma_n(x)<\frac{1}{2}\pi$, when $0<x<\pi$.

But we know by § 101 that $\sigma_n(x)$ converges uniformly to $\frac{1}{2}\pi$ in any interva wholly within $(0, \pi)$.

Thus the Gibbs Phenomenon does not occur when the Fourier's Series

$$2(\sin x+\tfrac{1}{3}\sin 3x+\tfrac{1}{5}\sin 5x+\ldots)$$

is summed by Arithmetic Means.†

In this example, by § 114. 2 (1)

$$\sigma_{2n+1}(x)=\frac{2}{2n+1}\int_0^x\frac{\sum_1^n \sin 2rt}{\sin t}\,dt.$$

It can be shown that, as we proceed from 0 to $\frac{1}{2}\pi$, the ordinates of the maxima continually increase, and that the same holds of the minima.

The last maximum in $0\leqq x\leqq\frac{1}{2}\pi$, has the greatest ordinate for the whole interval, and when $n\to\infty$, this tends to $\frac{1}{2}\pi$ from below.

The curve $y=\sigma_{2n+1}(x)$ for $n=6$ is given in Fig. 37, and it is interesting to compare it with the curve of Fig. 34, which corresponds, with this notation, to $y=s_{2n}(x)$ for $n=6$.

In the case of the ordinary sums the approximation curves $y=s_n(x)$ are brought within the shaded part of Fig. 38 by taking n large enough.

---

*For another proof see § 4 of a paper by Fejér in *Math. Annalen*, **64** (1907), 273.

†It should be noticed that with the notation of this section $s_{2n-1}=2\sum_1^n\frac{\sin(2r-1)x}{2r-1}$, and that $s_{2n-1}=s_{2n}$, since the coefficient of $\sin 2nx$ is zero.

In the case of the arithmetic means the approximation curves $y = \sigma_n(x)$ are brought within the shaded part of Fig. 39 by taking n large enough.

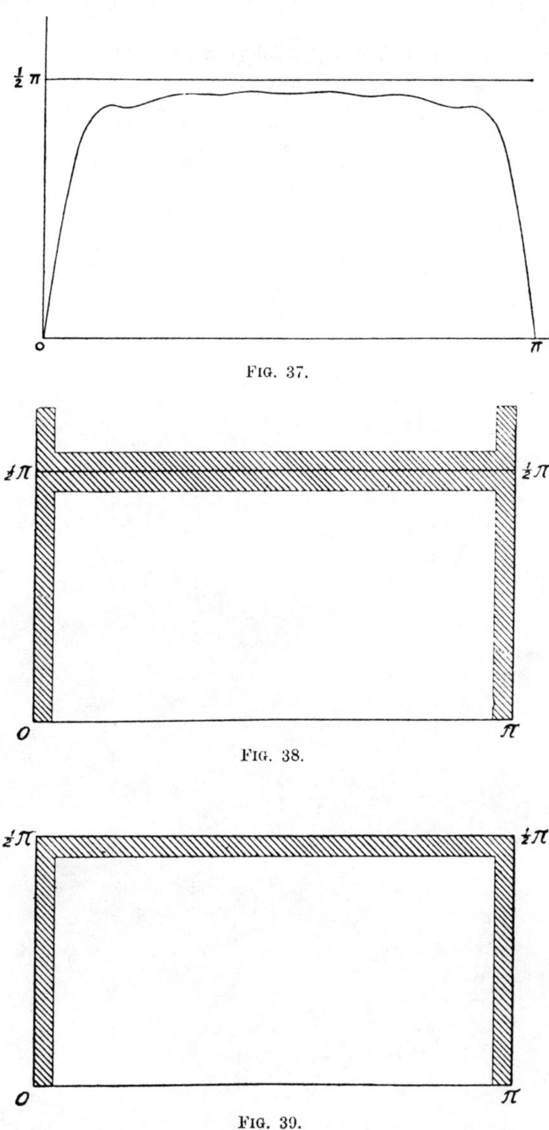

FIG. 37.

FIG. 38.

FIG. 39.

The width of the shaded part in these diagrams may be taken as small as we please.

REFERENCES.

BÔCHER, "Introduction to the Theory of Fourier's Series," *Annals of Math.* (2), **7** (1906).

GIBBS, *Scientific Papers*, Vol. II, p. 258 (London, 1906).

And the other papers named in the text.

CHAPTER X

FOURIER'S INTEGRALS

118. When the arbitrary function $f(x)$ satisfies Dirichlet's Conditions in the interval $(-l, l)$, we have seen in § 98 that the sum of the series

$$\frac{1}{2l}\int_{-l}^{l} f(x')\,dx' + \frac{1}{l}\sum_{1}^{\infty}\int_{-l}^{l} f(x')\cos\frac{n\pi}{l}(x'-x)\,dx' \quad \text{.........(1)}$$

is equal to $\frac{1}{2}[f(x+0)+f(x-0)]$ at every point in $-l<x<l$ where $f(x+0)$ and $f(x-0)$ exist; and that at $x=\pm l$ its sum is $\frac{1}{2}[f(-l+0)+f(l-0)]$, when these limits exist.

Corresponding results were found for the series

$$\frac{1}{l}\int_{0}^{l} f(x')\,dx' + \frac{2}{l}\sum_{1}^{\infty}\cos\frac{n\pi}{l}x\int_{0}^{l} f(x')\cos\frac{n\pi}{l}x'\,dx', \quad \text{........(2)}$$

and

$$\frac{2}{l}\sum_{1}^{\infty}\sin\frac{n\pi}{l}x\int_{0}^{l} f(x')\sin\frac{n\pi}{l}x'\,dx', \quad \text{.................(3)}$$

in the interval $(0, l)$.

Fourier's Integrals are definite integrals which represent the arbitrary function in an unlimited interval. They are suggested, but not established, by the forms these series appear to take as l tends to infinity.

If l is taken large enough, π/l may be made as small as we please, and we may neglect the first term in the series (1), assuming that $\int_{-\infty}^{\infty} f(x)\,dx$ is convergent. Then we may write

$$\frac{1}{l}\sum_{1}^{\infty}\int_{-l}^{l} f(x')\cos\frac{n\pi}{l}(x'-x)\,dx'$$

as

$$\frac{1}{\pi}\left[\Delta a\int_{-l}^{l} f(x')\cos\Delta a\,(x'-x)\,dx' + \Delta a\int_{-l}^{l}\cos 2\Delta a\,(x'-x)\,dx' + \ldots\right],$$

where $\Delta a = \pi/l$.

Assuming that this sum has a limit as $l \to \infty$, an assumption which, of course, would have to be defended if the proof were to be regarded as in any way complete, its value would be

$$\frac{1}{\pi}\int_0^\infty d\alpha \int_{-\infty}^{\infty} f(x') \cos \alpha (x'-x) dx',$$

and it would follow that

$$\frac{1}{\pi}\int_0^\infty d\alpha \int_{-\infty}^{\infty} f(x') \cos \alpha (x'-x) dx' = \tfrac{1}{2}[f(x+0)+f(x-0)], \qquad -\infty < x < \infty,$$

when these limits exist.

In the same way we are led to the Cosine Integral and Sine Integral corresponding to the Cosine Series and the Sine Series:

$$\left.\begin{aligned} \frac{2}{\pi}\int_0^\infty d\alpha \int_0^\infty f(x') \cos \alpha x \cos \alpha x' \, dx' &= \tfrac{1}{2}[f(x+0)+f(x-0)], \\ \frac{2}{\pi}\int_0^\infty d\alpha \int_0^\infty f(x') \sin \alpha x \sin \alpha x' \, dx' &= \tfrac{1}{2}[f(x+0)+f(x-0)], \end{aligned}\right\} 0 < x < \infty,$$

when these limits exist.

It must be remembered that the above argument is not a proof of any of these results. All that it does is to suggest the possibility of representing an arbitrary function $f(x)$, given for all values of x, or for all positive values of x, by these integrals.

We shall now show that this representation is possible, pointing out in our proof the limitation the discussion imposes upon the arbitrary function.*

119. *Let the arbitrary function $f(x)$, defined for all values of x, satisfy Dirichlet's Conditions in any finite interval, and in addition let $\int_{-\infty}^{\infty} f(x)\,dx$ be absolutely convergent.*

Then

$$\frac{1}{\pi}\int_0^\infty d\alpha \int_{-\infty}^{\infty} f(x') \cos \alpha (x'-x) dx' = \tfrac{1}{2}[f(x+0)+f(x-0)],$$

at every point where $f(x+0)$ and $f(x-0)$ exist.

Having fixed upon the value of x for which we wish to evaluate the integral, we can choose a positive number a greater than x such that $f(x')$ is bounded in the interval $a \leqq x' \leqq b$, where b is

---

*See footnote, p. 230.

arbitrary, and $\int_a^\infty |f(x')|dx'$ converges, since with our definition of the infinite integral only a finite number of points of infinite discontinuity were admitted (§ 60).

It follows that $\displaystyle\int_a^\infty f(x')\cos\alpha(x'-x)\,dx'$

converges uniformly for every α, so that this integral is a continuous function of α (§§ 83, 84).

Therefore $\displaystyle\int_0^q d\alpha\int_a^\infty f(x')\cos\alpha(x'-x)\,dx'$ exists.

Also, by § 85,

$$\int_0^q d\alpha\int_a^\infty f(x')\cos\alpha(x'-x)\,dx' = \int_a^\infty dx'\int_0^q f(x')\cos\alpha(x'-x)\,d\alpha$$

$$= \int_a^\infty f(x')\,\frac{\sin q(x'-x)}{x'-x}\,dx'. \quad\text{........(1)}$$

But $x'-x \geqq a-x > 0$ in $x' \geqq a$, since we have chosen $a > x$.

And $\displaystyle\int_a^\infty |f(x')|\,dx'$ converges.

Therefore $\displaystyle\int_a^\infty \frac{|f(x')|}{x'-x}\,dx'$ also converges.

It follows that $\displaystyle\int_a^\infty f(x')\,\frac{\sin q(x'-x)}{x'-x}\,dx'$

converges uniformly for every q.

Thus, to the arbitrary positive number ϵ there corresponds a positive number $A > a$, such that

$$\left|\int_{A'}^\infty f(x')\,\frac{\sin q(x'-x)}{x'-x}\,dx'\right| < \tfrac{1}{2}\epsilon, \text{ when } A' \geqq A > a, \quad\text{........(2)}$$

the same A serving for every value of q.

But we know from § 94 that

$$\lim_{q\to\infty}\int_a^A f(x')\,\frac{\sin q(x'-x)}{x'-x}\,dx'$$

$$= \lim_{q\to\infty}\int_{a-x}^{A-x} f(u+x)\,\frac{\sin qu}{u}\,du$$

$$= 0,$$

since $f(u+x)$ satisfies Dirichlet's Conditions in the interval $(a-x,\ A-x)$, and both these numbers are positive.

Thus we can choose the positive number Q, so that

$$\left|\int_a^A f(x')\frac{\sin q(x'-x)}{x'-x}dx'\right| < \tfrac{1}{2}\epsilon, \text{ when } q \geqq Q. \quad \text{.......... (3)}$$

It follows from (2) and (3) that

$$\left|\int_a^\infty f(x')\frac{\sin q(x'-x)}{x'-x}dx'\right| < \tfrac{1}{2}\epsilon + \tfrac{1}{2}\epsilon = \epsilon,$$

when $q \geqq Q$.

Thus
$$\lim_{q\to\infty}\int_a^\infty f(x')\frac{\sin q(x'-x)}{x'-x}dx' = 0,$$

and from (1),
$$\int_0^\infty d\alpha \int_a^\infty f(x')\cos\alpha(x'-x)\,dx' = 0. \quad \text{..................(4)}$$

But, by § 87,

$$\int_0^q d\alpha \int_x^a f(x')\cos\alpha(x'-x)\,dx' = \int_x^a dx' \int_0^q f(x')\cos\alpha(x'-x)\,d\alpha$$

$$= \int_x^a f(x')\frac{\sin q(x'-x)}{x'-x}dx'$$

$$= \int_0^{a-x} f(u+x)\frac{\sin qu}{u}du.$$

Letting $q\to\infty$, we have

$$\int_0^\infty d\alpha \int_x^a f(x')\cos\alpha(x'-x)\,dx' = \frac{\pi}{2}f(x+0), \quad \text{..............(5)}$$

when $f(x+0)$ exists.

Adding (4) and (5), we have

$$\int_0^\infty d\alpha \int_x^\infty f(x')\cos\alpha(x'-x)\,dx' = \frac{\pi}{2}f(x+0), \quad \text{.............(6)}$$

when $f(x+0)$ exists.

Similarly, under the given conditions,

$$\int_0^\infty d\alpha \int_{-\infty}^x f(x')\cos\alpha(x'-x)\,dx' = \frac{\pi}{2}f(x-0), \quad \text{...........(7)}$$

when $f(x-0)$ exists.

Adding (6) and (7), we obtain Fourier's Integral in the form

$$\frac{1}{\pi}\int_0^\infty d\alpha \int_{-\infty}^\infty f(x')\cos\alpha(x'-x)\,dx' = \tfrac{1}{2}[f(x+0)+f(x-0)]$$

for every point in $-\infty < x < \infty$, where $f(x+0)$ and $f(x-0)$ exist.

120. More general conditions for f(x).* In this section we shall show that Fourier's Integral formula holds if the conditions in any finite interval remain as before, and the absolute convergence of the infinite integral $\int_{-\infty}^{\infty} f(x)dx$ is replaced by the following conditions:

For some positive number and to the right of it, $f(x')$ is bounded and monotonic and $\lim_{x'\to\infty} f(x')=0$; and for some negative number and to the left of it, $f(x')$ is bounded and monotonic and $\lim_{x'\to-\infty} f(x')=0$.

Having fixed upon the value of x for which we wish to evaluate the integral, we can choose a positive number a greater than x, such that $f(x')$ is bounded and monotonic when $x' \geqq a$ and $\lim_{x'\to\infty} f(x')=0$.

Consider the integral $\int_a^{\infty} f(x') \cos \alpha(x'-x)\, dx'$.

By the Second Theorem of Mean Value,

$$\int_{A'}^{A''} f(x') \cos \alpha(x'-x)\, dx' = f(A') \int_{A'}^{\xi} \cos \alpha(x'-x)\, dx' + f(A'') \int_{\xi}^{A''} \cos \alpha(x'-x)\, dx',$$

where $a < A' \leqq \xi \leqq A''$.

Thus

$$\int_{A'}^{A''} f(x') \cos \alpha(x'-x) dx' = \frac{f(A')}{\alpha} \int_{\alpha(A'-x)}^{\alpha(\xi-x)} \cos u\, du + \frac{f(A'')}{\alpha} \int_{\alpha(\xi-x)}^{\alpha(A''-x)} \cos u\, du.$$

Therefore

$$\left| \int_{A'}^{A''} f(x') \cos \alpha(x'-x)\, dx' \right| < \frac{4\,|f(A')|}{q_0} \text{ for } \alpha \geqq q_0 > 0.$$

But we are given that $\lim_{x'\to\infty} f(x')=0$.

It follows that

$$\int_a^{\infty} f(x') \cos \alpha(x'-x)\, dx'$$

is uniformly convergent for $\alpha \geqq q_0 > 0$, and this integral represents a continuous function of α in $\alpha \geqq q_0$.

Also, by § 85,

$$\int_{q_0}^{q} d\alpha \int_a^{\infty} f(x') \cos \alpha(x'-x)\, dx' = \int_a^{\infty} dx' \int_{q_0}^{q} f(x') \cos \alpha(x'-x)\, d\alpha$$

$$= \int_a^{\infty} f(x') \left\{ \frac{\sin q(x'-x)}{x'-x} - \frac{\sin q_0(x'-x)}{x'-x} \right\} dx'. \quad \ldots(1)$$

But $x'-x \geqq a-x > 0$ in the interval $x' \geqq a$, since we have chosen a greater than x.

And $\int_a^{\infty} f(x') \frac{\sin q(x'-x)}{x'-x} dx, \quad \int_a^{\infty} f(x') \frac{\sin q_0(x'-x)}{x'-x} dx'$

both converge.

---

*These extensions are due to Pringsheim. Cf. *Math. Annalen*, **68** (1910), 367, and **71** (1911), 289. Reference should also be made to a paper by W. H. Young in *Proc. Royal Soc. Edinburgh*, **31** (1911), 559.

Therefore, from (1),

$$\int_{q_0}^{q} da \int_a^\infty f(x') \cos a(x'-x)dx'$$
$$= \int_a^\infty f(x') \frac{\sin q(x'-x)}{x'-x} dx' - \int_a^\infty f(x') \frac{\sin q_0(x'-x)}{x'-x} dx'. \quad \text{.....(2)}$$

Now consider the integral

$$\int_a^\infty f(x') \frac{\sin q_0(x'-x)}{x'-x} dx'.$$

From the Second Theorem of Mean Value,

$$\int_{A'}^{A''} f(x') \frac{\sin q_0(x'-x)}{x'-x} dx'$$
$$= f(A') \int_{A'}^{\xi} \frac{\sin q_0(x'-x)}{x'-x} dx' + f(A'') \int_{\xi}^{A''} \frac{\sin q_0(x'-x)}{x'-x} dx', \quad \text{......(3)}$$

where $a < A' \leqq \xi \leqq A''$.

Also $$\int_{A'}^{\xi} \frac{\sin q_0(x'-x)}{x'-x} dx' = \int_{q_0(A'-x)}^{q_0(\xi - x)} \frac{\sin u}{u} du,$$

the limits of the integral being both positive.

Therefore $$\left| \int_{A'}^{\xi} \frac{\sin q_0(x'-x)}{x'-x} dx' \right| < \pi. \quad \text{(Cf. § 91.)}$$

And similarly $$\left| \int_{\xi}^{A''} \frac{\sin q_0(x'-x)}{x'-x} dx' \right| < \pi.$$

Thus, from (3),

$$\left| \int_{A'}^{A''} f(x') \frac{\sin q_0(x'-x)}{x'-x} dx' \right| < 2\pi \, | f(A') |. \quad \text{..................(4)}$$

It follows that $$\int_a^\infty f(x') \frac{\sin q_0(x'-x)}{x'-x} dx'$$

is uniformly convergent for $q_0 \geqq 0$, and by § 84 it is continuous in this range.

Thus $$\lim_{q_0 \to 0} \int_a^\infty f(x') \frac{\sin q_0(x'-x)}{x'-x} dx' = 0, \quad \text{........................(5)}$$

since the integral vanishes when $q_0 = 0$.

Also from (2),

$$\int_0^q da \int_a^\infty f(x') \cos a(x'-x)dx' = \int_a^\infty f(x') \frac{\sin q(x'-x)}{x'-x} dx'. \quad \text{...........(6)}$$

But we have already shown that the integral on the right-hand side of (6) is uniformly convergent for $q \geqq 0$.

Proceeding as in § 119, (2) and (3),* it follows that

$$\lim_{q \to \infty} \int_a^\infty f(x') \frac{\sin q(x'-x)}{x'-x} dx' = 0.$$

Thus, from (6), $$\int_0^\infty da \int_a^\infty f(x') \cos a(x'-x)\, dx' = 0. \quad \text{........................(7)}$$

*Or we might use the Second Theorem of Mean Value as proved in § 58 for the Infinite Integral.

But we know from § 119 that

$$\int_0^\infty da \int_x^a f(x') \cos a(x'-x)dx' = \frac{\pi}{2} f(x+0), \quad \text{.....................(8)}$$

when this limit exists.

Thus
$$\frac{1}{\pi}\int_0^\infty da \int_x^\infty f(x') \cos a(x'-x)dx' = \tfrac{1}{2} f(x+0), \quad \text{.....................(9)}$$

when this limit exists.

Similarly, under the given conditions, we find that

$$\frac{1}{\pi}\int_0^\infty da \int_{-\infty}^x f(x') \cos a(x'-x) = \tfrac{1}{2} f(x-0), \quad \text{.....................(10)}$$

when this limit exists.

Adding (9) and (10), our formula is proved for every point at which $f(x\pm 0)$ exist.

121. Other conditions for f(x). We shall now show that Fourier's Integral formula also holds when the conditions at $\pm\infty$ of the previous section are replaced by the following:

(I) *For some positive number and to the right of it, and for some negative number and to the left of it, $f(x')$ is of the form $g(x') \cos(\lambda x' + \mu)$, where $g(x')$ is bounded and monotonic in these intervals and has the limit zero as $x' \to \pm\infty$.*

Also, (II) $\displaystyle\int^\infty \frac{g(x')}{x'}dx'$ *and* $\displaystyle\int_{-\infty} \frac{g(x')}{x'}dx'$ *converge.*

We have shown in § 120 that when $g(x')$ satisfies the conditions named above in (I), there will be a positive number a greater than x such that

$$\int_0^\infty da \int_a^\infty g(x') \cos a(x'-x)dx' = 0.$$

But, if λ is any positive number,

$$\int_0^\infty da \int_a^\infty g(x') \cos a(x'-x)dx'$$

$$= \int_0^\lambda da \int_a^\infty g(x') \cos a(x'-x)dx' + \int_\lambda^\infty da \int_a^\infty g(x') \cos a(x'-x)dx'$$

$$= \tfrac{1}{2}\int_{-\lambda}^\lambda da \int_a^\infty g(x') \cos a(x'-x)dx' + \int_\lambda^\infty da \int_a^\infty g(x') \cos a(x'-x)dx'$$

$$= \tfrac{1}{2}\int_{-\lambda}^\infty da \int_a^\infty g(x') \cos a(x'-x)dx' + \tfrac{1}{2}\int_\lambda^\infty da \int_a^\infty g(x') \cos a(x'-x)dx'$$

$$= \tfrac{1}{2}\int_0^\infty da \int_a^\infty g(x') \cos (a+\lambda)(x'-x)dx' + \tfrac{1}{2}\int_0^\infty da \int_a^\infty g(x') \cos (a-\lambda)(x'-x)dx'.$$

Therefore

$$\int_0^\infty da \int_a^\infty g(x') \cos a(x'-x)dx'$$

$$= \int_0^\infty da \int_a^\infty g(x') \cos \lambda(x'-x) \cos a(x'-x)dx' = 0. \quad \text{...........(1)}$$

Again we know, from § 120, that

$$\int_a^\infty g(x') \sin \alpha(x'-x)dx'$$

is uniformly convergent for $\alpha \geqq \lambda_0 > 0$.

It follows that $\displaystyle\int_{\lambda_0}^{\lambda_1} d\alpha \int_a^\infty g(x') \sin \alpha(x'-x)dx'$

exists, for $\lambda_1 > \lambda_0 > 0$.

Also

$$\int_{\lambda_0}^{\lambda_1} d\alpha \int_a^\infty g(x') \sin \alpha(x'-x)dx'$$

$$= \int_a^\infty dx' \int_{\lambda_0}^{\lambda_1} g(x') \sin \alpha(x'-x)d\alpha \quad \text{(by § 84)}$$

$$= \int_a^\infty g(x') \left\{ \frac{\cos \lambda_0(x'-x)}{x'-x} - \frac{\cos \lambda_1(x'-x)}{x'-x} \right\} dx'$$

$$= \int_a^\infty g(x') \frac{\cos \lambda_0(x'-x)}{x'-x} dx' - \int_a^\infty g(x') \frac{\cos \lambda_1(x'-x)}{x'-x} dx', \quad \text{...(2)}$$

since both integrals converge.

But we are given that $\displaystyle\int^\infty \frac{g(x')}{x'} dx'$

converges ; and it follows that $\displaystyle\int_a^\infty \frac{g(x')}{x'-x} dx'$

also converges, so that we know that

$$\int_a^\infty g(x') \frac{\cos \lambda_0(x'-x)}{x'-x} dx'$$

is uniformly convergent for $\lambda_0 \geqq 0$, and therefore continuous.

Thus $$\lim_{\lambda_0 \to 0} \int_a^\infty g(x') \frac{\cos \lambda_0(x'-x)}{x'-x} dx' = \int_a^\infty \frac{g(x')}{x'-x} dx'.$$

It follows from (2) that, when $\lambda > 0$,

$$\int_0^\lambda d\alpha \int_a^\infty g(x') \sin \alpha(x'-x)dx' = \int_a^\infty \frac{g(x')}{x'-x} dx' - \int_a^\infty g(x') \frac{\cos \lambda(x'-x)}{x'-x} dx'. \quad \text{...(3)}$$

Also, as before, we find that, with the conditions imposed upon $g(x')$,*

$$\lim_{\lambda \to \infty} \int_a^\infty g(x') \frac{\cos \lambda(x'-x)}{x'-x} dx' = 0.$$

Therefore, from (2) and (3),

$$\int_\lambda^\infty d\alpha \int_a^\infty g(x') \sin \alpha(x'-x)dx' = \int_a^\infty g(x') \frac{\cos \lambda(x'-x)}{x'-x} dx', \quad \text{..........(4)}$$

and $$\int_0^\infty d\alpha \int_a^\infty g(x') \sin \alpha(x'-x)dx' = \int_a^\infty \frac{g(x')}{x'-x} dx'. \quad \text{................(5)}$$

*This can be obtained at once from the Second Theorem of Mean Value, as proved in § 58 for the Infinite Integral; but it is easy to establish the result, as in § 119, without this theorem.

Again, since $$\int_0^{\lambda} da \int_a^{\infty} g(x') \sin \alpha(x'-x)\,dx'$$

exists, we have $$\int_{-\lambda}^{\lambda} da \int_a^{\infty} g(x') \sin \alpha(x'-x)\,dx' = 0. \quad \text{......................(6)}$$

From (6), and the convergence of $\int_{-\lambda}^{\infty} da \int_a^{\infty} g(x') \sin \alpha(x'-x)\,dx'$, it follows that

$$\int_{-\lambda}^{\infty} da \int_a^{\infty} g(x') \sin \alpha(x'-x)\,dx' - \int_{\lambda}^{\infty} da \int_a^{\infty} g(x') \sin \alpha(x'-x)\,dx' = 0.$$

Thus

$$\int_0^{\infty} da \int_a^{\infty} g(x') \sin (\alpha+\lambda)(x'-x)\,dx' - \int_0^{\infty} da \int_a^{\infty} g(x') \sin (\alpha-\lambda)(x'-x)\,dx' = 0.$$

Therefore $$\int_0^{\infty} da \int_a^{\infty} g(x') \sin \lambda(x'-x) \cos \alpha(x'-x)\,dx' = 0. \quad \text{..................(7)}$$

Multiply (1) by $\cos(\lambda x+\mu)$ and (7) by $\sin(\lambda x+\mu)$ and subtract.

It follows that

$$\int_0^{\infty} da \int_a^{\infty} g(x') \cos(\lambda x'+\mu) \cos \alpha\,(x'-x)\,dx' = 0. \quad \text{..................(8)}$$

And in the same way, with the conditions imposed upon $g(x')$, we have

$$\int_0^{\infty} da \int_{-\infty}^{-a'} g(x') \cos(\lambda x'+\mu) \cos \alpha(x'-x)\,dx' = 0. \quad \text{..................(9)}$$

These results, (8) and (9), may be written

$$\left.\begin{aligned} &\int_0^{\infty} da \int_a^{\infty} f(x') \cos \alpha(x'-x)\,dx' = 0, \\ &\int_0^{\infty} da \int_{-\infty}^{-a'} f(x') \cos \alpha\,(x'-x)\,dx' = 0, \end{aligned}\right\} \quad \text{..................(10)}$$

when $f(x') = g(x') \cos(\lambda x'+\mu)$ in (a, ∞) and $(-\infty, -a')$.

But we know that, when $f(x)$ satisfies Dirichlet's Conditions in $(-a', a)$,

$$\int_0^{\infty} da \int_{-a'}^{a} f(x') \cos \alpha(x'-x)\,dx' = \frac{\pi}{2}[f(x+0)+f(x-0)], \quad \text{...........(11)}$$

when these limits exist. [Cf. § 119 (5).]

Adding (10) and (11), we see that Fourier's Integral formula holds, when the arbitrary function satisfies the conditions imposed upon it in this section.

It is clear that the results just established still hold if we replace $\cos(\lambda x'+\mu)$ in (I) by the sum of a number of terms of the type

$$a_n \cos(\lambda_n x'+\mu_n).$$

It can be proved * that the theorem is also valid when this sum is replaced by an infinite series

$$\sum_1^{\infty} a_n \cos(\lambda_n x'+\mu_n),$$

when $\sum_1^{\infty} a_n$ converges absolutely and the constants λ_n, so far arbitrary, tend to infinity with n.

---

*Cf. Pringsheim, *Math. Annalen*, 68 (1910), 399.

122. Fourier's Cosine Integral and Sine Integral. In the case when $f(x)$ is given only for positive values of x, there are two forms of Fourier's Integral which correspond to the Cosine Series and Sine Series respectively.

I. In the first place, consider the result of § 119, when $f(x)$ has the same values for negative values of x as for the corresponding positive values of x: i.e. $f(-x)=f(x)$, $x>0$.

Then
$$\frac{1}{\pi}\int_0^\infty d\alpha \int_{-\infty}^\infty f(x')\cos\alpha(x'-x)\,dx'$$
$$=\frac{1}{\pi}\int_0^\infty d\alpha\int_0^\infty f(x')[\cos\alpha(x'+x)+\cos\alpha(x'-x)]\,dx'$$
$$=\frac{2}{\pi}\int_0^\infty d\alpha\int_0^\infty f(x')\cos\alpha x\cos\alpha x'\,dx'.$$

It follows from § 119 that *when $f(x)$ is defined for positive values of x, and satisfies Dirichlet's Conditions in any finite interval, while $\int_0^\infty f(x)\,dx$ converges absolutely, then*
$$\frac{2}{\pi}\int_0^\infty d\alpha\int_0^\infty f(x')\cos\alpha x\cos\alpha x'\,dx'=\tfrac{1}{2}[f(x+0)+f(x-0)],$$
at every point where $f(x+0)$ and $f(x-0)$ exist, and when $x=0$ the value of the integral is $f(+0)$, if this limit exists.

Also it follows from §§ 120 and 121 that the condition at infinity may be replaced by either of the following:

(i) *For some positive number and to the right of it, $f(x')$ shall be bounded and monotonic and $\lim_{x'\to\infty} f(x')=0$;*

or, (ii) *For some positive number and to the right of it, $f(x')$ shall be of the form $g(x')\cos(\lambda x'+\mu)$, where $g(x')$ is bounded and monotonic and $\lim_{x'\to\infty} g(x')=0$. Also $\int^\infty \frac{g(x')}{x'}\,dx'$ must converge.*

II. In the next place, by taking $f(-x)=-f(x)$, $x>0$, we see that, *when $f(x)$ is defined for positive values of x, and satisfies Dirichlet's Conditions in any finite interval, while $\int_0^\infty f(x)\,dx$ converges absolutely, then,*
$$\frac{2}{\pi}\int_0^\infty d\alpha\int_0^\infty f(x')\sin\alpha x\sin\alpha x'\,dx'=\tfrac{1}{2}[f(x+0)+f(x-0)],$$
at every point where $f(x+0)$ and $f(x-0)$ exist, and when $x=0$ the integral is zero.

Also it follows from §§ 120 and 121 that the condition at infinity may be replaced by one or other of those given under (I).

It should be noticed that, when we express the arbitrary function $f(x)$ by any of Fourier's Integrals, we must first decide for what value of x we wish the value of the integral, and that this value of x must be inserted in the integrand before the integrations indicated are carried out (cf. § 62).

123. Fourier's Integrals. Sommerfeld's Discussion. In many of the problems of Applied Mathematics in the solution of which Fourier's Integrals occur, they appear in a slightly different form, with an exponential factor (e.g. $e^{-\kappa a^2 t}$) added. In these cases we are concerned with the *limiting value as* $t \to 0$ of the integral

$$\phi(t) = \frac{1}{\pi}\int_0^\infty da \int_a^b f(x') \cos a(x'-x) e^{-\kappa a^2 t}\, dx',$$

and, so far as the actual physical problem is concerned, *the value of the integral for* $t=0$ is not required.

It was shown, first of all by Sommerfeld,* that, when the limit on the right-hand side exists,

$$\lim_{t\to 0} \frac{1}{\pi}\int_0^\infty da \int_a^b f(x') \cos a(x'-x) e^{-\kappa a^2 t}\, dx' = \tfrac{1}{2}[f(x+0)+f(x-0)],$$

$$\text{when } a < x < b,$$

$$= \tfrac{1}{2} f(a+0), \quad \text{when } x = a,$$

$$= \tfrac{1}{2} f(b-0), \quad \text{when } x = b,$$

the result holding in the case of any integrable function given in the interval (a, b).

The case when the interval is infinite was also treated by Sommerfeld, but it has been examined in much greater detail by Young.† It will be sufficient in this place to state that, when the arbitrary function satisfies the conditions at infinity imposed in §§ 120-122, Sommerfeld's result still holds for an infinite interval.

However, it should be noticed that we cannot deduce the value of the integral

$$\frac{1}{\pi}\int_0^\infty da \int_a^b f(x') \cos a(x'-x)\, dx'$$

from the above results. This would require the continuity of the function

$$\phi(t) = \frac{1}{\pi}\int_0^\infty da \int_a^b f(x') \cos a(x'-x) e^{-\kappa a^2 t}\, dx'$$

for $t=0$.

We have come across the same point in the discussion of Poisson's treatment of Fourier's Series. [Cf. § 99.]

---

*Sommerfeld, *Die willkürlichen Functionen in der mathematischen Physik*, Diss., Königsberg, 1891.

†W. H. Young, *loc. cit., Proc. Royal Soc. Edinburgh*, **31** (1911).

REFERENCES.

FOURIER, *Théorie analytique de la chaleur*, Ch. IX.

HOBSON, *Theory of Functions of a Real Variable*, 2 (2nd ed., 1926), Ch. IX.

JORDAN, *loc. cit.*, T. II (3e éd.), Ch. V.

LEBESGUE, *Leçons sur les séries trigonométriques* (Paris, 1906), Ch. III.

NEUMANN, *loc. cit.*, Kap. III.

WEBER-RIEMANN, *loc. cit.*, 1 (2 Aufl., 1910), §§ 18-20.

And the papers named in the text.

EXAMPLES ON CHAPTER X.

1. Taking $f(x)$ as 1 in Fourier's Cosine Integral when $0<x<1$, and as zero when $1<x$, show that

$$1=\frac{2}{\pi}\int_0^\infty \frac{\sin a \cos ax}{a}\,da, \quad (0<x<1)$$

$$\frac{1}{2}=\frac{2}{\pi}\int_0^\infty \frac{\sin a \cos ax}{a}\,da, \quad (x=1)$$

$$0=\frac{2}{\pi}\int_0^\infty \frac{\sin a \cos ax}{a}\,dx. \quad (1<x).$$

2. By considering Fourier's Sine Integral for $e^{-\beta x}$ $(\beta>0)$, prove that

$$\int_0^\infty \frac{a \sin ax}{a^2+\beta^2}\,da=\frac{\pi}{2}e^{-\beta x},$$

and in the same way, from the Cosine Integral, prove that

$$\int_0^\infty \frac{\cos ax}{a^2+\beta^2}\,da=\frac{\pi}{2\beta}e^{-\beta x}.$$

3. Show that the expression

$$\frac{2a^2}{\pi}\int_0^\infty \cos\left(\frac{\theta x}{a}\right)\frac{d\theta}{\theta^3}\int_0^\theta v^2 \cos v\,dv$$

is equal to x^2 when $0\leqq x<a$, and to zero when $x>a$.

4. Show that $\quad \dfrac{2}{\pi}\displaystyle\int_0^\infty \sin qx\left\{\frac{h}{q}+\tan a\frac{\sin qb-\sin qa}{q^2}\right\}dq$

is the ordinate of a broken line running parallel to the axis of x from $x=0$ to $x=a$, and from $x=b$ to $x=\infty$, and inclined to the axis of x at an angle a between $x=a$ and $x=b$.

5. Show that $f(x)=\dfrac{1}{\sqrt{|x|}}$ satisfies the conditions of § 120 for Fourier's Integral, and verify independently that

$$\frac{2}{\pi}\int_0^\infty da\int_0^\infty \cos ax \cos ax'\frac{dx'}{\sqrt{x'}}=\frac{1}{\sqrt{x}} \quad \text{when } x>0.$$

6. Show that $f(x)=\dfrac{\sin x}{x}$ satisfies the conditions of § 121 for Fourier's Integral, and verify independently that

$$\frac{1}{\pi}\int_0^\infty da\int_{-\infty}^\infty \sin x' \cos a(x'-x)\frac{dx'}{x'}=\frac{\sin x}{x}.$$

APPENDIX I

PRACTICAL HARMONIC ANALYSIS AND PERIODOGRAM ANALYSIS

1. Let $y = f(x)$ be a given periodic function, with period 2π. We have seen that, for a very general class of functions, we may represent $f(x)$ by its Fourier's Series

$$\left. \begin{array}{l} a_0 + a_1 \cos x + a_2 \cos 2x + \ldots \\ \quad + b_1 \sin x + b_2 \sin 2x + \ldots \end{array} \right\},$$

where $a_0, a_1, a_2, \ldots b_1, b_2, \ldots$ are Fourier's Constants for $f(x)$. We may suppose the range of x to be $0 \leqq x \leqq 2\pi$. If the period is a, instead of 2π, the terms $\begin{smallmatrix}\cos\\ \sin\end{smallmatrix} nx$ are replaced by $\begin{smallmatrix}\cos\\ \sin\end{smallmatrix} 2n\pi x/a$, and the range becomes $0 \leqq x \leqq a$.

However, in many practical applications, y is not known analytically as a function of x, but the relation between the dependent and independent variables is given in the form of a curve obtained by continuous observations. Or again, we may only be given the values of y corresponding to isolated values of x, the observations having been made at definite intervals. In the latter case we may suppose that a continuous curve is drawn through the isolated points in the plane of x, y. And in both cases Fourier's Constants for the function can be obtained by mechanical means. One of the best known machines for the purpose is Kelvin's Harmonic Analyser.*

2. The practical questions referred to above can also be treated by substituting for Fourier's Infinite Series a trigonometrical series with only a limited number of terms.

Suppose the value of the function given at the points

$$0,\ a,\ 2a,\ \ldots\ (m-1)a, \quad \text{where } ma = 2\pi.$$

Denote these points on the interval $(0, 2\pi)$ by

$$x_0,\ x_1,\ x_2,\ \ldots\ x_{m-1},$$

and the corresponding values of y by

$$y_0,\ y_1,\ y_2,\ \cdots\ y_{m-1}.$$

---

*Such mechanical methods are described in the handbook entitled *Modern Instruments and Methods of Calculation*, published by Bell & Sons in connection with the Napier Tercentenary Meeting of 1914.

Let
$$\left.\begin{aligned}s_n(x)=a_0+a_1\cos x+a_2\cos 2x+\ldots+a_n\cos nx\\+b_1\sin x+b_2\sin 2x+\ldots+b_n\sin nx\end{aligned}\right\}.$$

If $2n+1=m$, we can determine these $2n+1$ constants so that

$$s_n(x_r)=y_r,\quad \text{when } r=0,\ 1,\ 2,\ \ldots\ 2n.$$

The $2n+1$ equations giving the values of a_0, a_1, ... a_n, b_1, ... b_n, are as follows:

$$\left.\begin{array}{l}a_0+a_1+\ldots\qquad +a_p+\ldots\qquad +a_n\qquad =y_0\\ \left.\begin{array}{l}a_0+a_1\cos x_1+\ldots+a_p\cos px_1+\ldots+a_n\cos nx_1\\ \quad+b_1\sin x_1+\ldots+b_p\sin px_1+\ldots+b_n\sin nx_1\end{array}\right\}=y_1\\ \ldots\ldots\ldots\ldots\ldots\ldots\ldots\ldots\\ \left.\begin{array}{l}a_0+a_1\cos x_{2n}+\ldots+a_p\cos px_{2n}+\ldots+a_n\cos nx_{2n}\\ \quad+b_1\sin x_{2n}+\ldots+b_p\sin px_{2n}+\ldots+b_n\sin nx_{2n}\end{array}\right\}=y_{2n}\end{array}\right\}.$$

Adding these equations, we see that

$$(2n+1)a_0=\sum_{r=0}^{2n}y_r,$$

since
$$1+\cos pa+\cos 2pa+\ldots+\cos 2npa=0,$$
and
$$\sin pa+\sin 2pa+\ldots+\sin 2npa=0,$$
when
$$(2n+1)a=2\pi.$$

Further, we know that

$$\begin{aligned}&1+\cos pa\cos ra+\cos 2pa\cos 2ra\\&\qquad+\ldots+\cos 2npa\cos 2nra=0,\quad p\neq r,\end{aligned}$$

$$\left.\begin{aligned}&\cos pa\sin ra+\cos 2pa\sin 2ra\\&\qquad+\ldots+\cos 2npa\sin 2nra=0,\end{aligned}\right.\qquad \left\{\begin{aligned}p&=1,\ 2,\ \ldots\ n\\ r&=1,\ 2,\ \ldots\ n\end{aligned}\right\}.$$

And
$$1+\cos^2 pa+\cos^2 2pa+\ldots+\cos^2 2npa=\tfrac{1}{2}(2n+1).$$

It follows that, if we multiply the second equation by $\cos px_1$, the third by $\cos px_2$, etc., and add, we have

$$\tfrac{1}{2}(2n+1)a_p=\sum_{r=0}^{2n}y_r\cos pra.$$

Similarly, we find that

$$\tfrac{1}{2}(2n+1)b_p=\sum_{r=1}^{2n}y_r\sin pra.$$

A trigonometrical series of $(2n+1)$ terms has thus been formed, whose sum takes the required values at the points

$$0,\ a,\ 2a,\ \ldots\ 2na,\quad \text{where } (2n+1)a=2\pi.$$

It will be observed that as $n\to\infty$ the values of a_0, a_1, ... and b_1, b_2, ... reduce to the integral forms for the coefficients, but as remarked in § 90, p. 218, this passage from a finite number of equations to an infinite number requires more careful handling if the proof is to be made rigorous.

3. For purposes of calculation, there are advantages in taking an even

number of equidistant points instead of an odd number. Suppose that to the points

$$0, a, 2a, \dots (2n-1)a, \quad \text{where } na=\pi,$$

we have the corresponding values of y,

$$y_0, y_1, y_2, \dots y_{2n-1}.$$

In this case we can obtain the values of the $2n$ constants in the expression

$$\left.\begin{aligned} a_0 + a_1 \cos x + a_2 \cos 2x + \dots + a_{n-1} \cos (n-1)x + a_n \cos nx \\ + b_1 \sin x + b_2 \sin 2x + \dots + b_{n-1} \sin (n-1)x \end{aligned}\right\},$$

so that the sum shall take the values $y_0, y_1, \dots y_{2n-1}$ at these $2n$ points in $(0, 2\pi)$.

It will be found that

$$\left.\begin{aligned} a_0 &= \frac{1}{2n} \sum_{r=0}^{2n-1} y_r \\ a_p &= \frac{1}{n} \sum_{r=0}^{2n-1} y_r \cos pra, \text{ if } p \neq n \\ a_n &= \frac{1}{2n} \sum_{r=0}^{2n-1} y_r \cos r\pi \\ b_p &= \frac{1}{n} \sum_{r=1}^{2n-1} y_r \sin pra \end{aligned}\right\} a=\pi/n.$$

Runge* gave a convenient scheme for evaluating these constants in the case of 12 equidistant points. This and a similar table devised by Whittaker for the case of 24 equidistant points will be found in Whittaker and Robinson's *Calculus of Operations* (1924), Ch. X.

4. This question may be looked at from another point of view. Suppose we are given the values of y, viz.

$$y_0, y_1, y_2, \dots y_{m-1},$$

corresponding to the points

$$0, a, 2a, \dots (m-1)a, \quad \text{where } ma=2\pi.$$

Denote these values of x, as before, by

$$x_0, x_1, x_2, \dots x_{m-1}.$$

Let $$\left.\begin{aligned} s_n(x) = a_0 + a_1 \cos x + a_2 \cos 2x + \dots + a_n \cos nx \\ + b_1 \sin x + b_2 \sin 2x + \dots + b_n \sin nx \end{aligned}\right\}.$$

For a given value of n, on the understanding that $m > 2n+1$,† the $2n+1$ constants $a_0, a_1, \dots a_n, b_1, \dots b_n$ are to be determined so that $s_n(x)$ shall approximate as closely as possible to $y_0, y_1, \dots y_{m-1}$ at $x_0, x_1, \dots x_{m-1}$.

---

* *Math. Zeitschrift*, 48 (1903) and 52 (1905). Also *Theorie u. Praxis der Reihen* (Leipzig, 1904), 147–164.

† If $m < 2n+1$, we can choose the constants in any number of ways so that $s_n(x)$ shall be equal to $y_0, y_1, \dots y_{m-1}$ at $x_0, x_1, \dots x_{m-1}$, for there are more constants than equations. And if $m = 2n+1$, we can choose the constants in one way so that this condition is satisfied.

The Theory of Least Squares shows that the closest approximation will be obtained by making the function

$$\sum_{r=0}^{m-1} (y_r - s_n(x_r))^2,$$

regarded as a function of $a_0, a_1, \dots a_n, b_1, \dots b_n$, a minimum.

The conditions for a minimum, in this case, are:

$$\left.\begin{aligned} &\sum_{r=0}^{m-1} (y_r - s_n(x_r)) = 0 \\ &\sum_{r=0}^{m-1} (y_r - s_n(x_r)) \cos px_r = 0 \\ &\sum_{r=0}^{m-1} (y_r - s_n(x_r)) \sin px_r = 0 \end{aligned}\right\} p = 1, 2, \dots n.$$

It will be found, as in § 2 above, that these equations lead to the following values for the coefficients:

$$ma_0 = \sum_{r=0}^{m-1} y_r,$$

$$\left.\begin{aligned} \tfrac{1}{2}ma_p &= \sum_{r=0}^{m-1} y_r \cos pra \\ \tfrac{1}{2}mb_p &= \sum_{r=1}^{m-1} y_r \sin pra \end{aligned}\right\} \begin{aligned} &p = 1, 2, \dots n, \\ &m \text{ odd.} \end{aligned}$$

But if m is even, the coefficient a_p (when $p = \frac{1}{2}m$) is given by

$$ma_{\frac{1}{2}m} = \sum_{r=0}^{m-1} y_r \cos r\pi,$$

the others remaining as above.

In some cases, it is sufficient to find the terms up to $\cos x$ and $\sin x$, viz.

$$a_0 + a_1 \cos x + b_1 \sin x.$$

The values of a_0, a_1 and b_1, which will make this expression approximate most closely to

$$y_0, y_1, y_2, \dots y_{m-1}$$

at $\quad 0, a, 2a, \dots (m-1)a, \quad$ when $ma = 2\pi$,

are then given by:

$$ma_0 = \sum_{r=0}^{m-1} y_r,$$

$$\tfrac{1}{2}ma_1 = \sum_{r=0}^{m-1} y_r \cos ra,$$

$$\tfrac{1}{2}mb_1 = \sum_{r=1}^{m-1} y_r \sin ra.$$

Tables for evaluating the coefficients in such cases have been constructed by Turner.*

5. In the preceding sections we have been dealing with a set of observations known to have a definite period. The graph for the observations would repeat

---

* *Tables for Harmonic Analysis*, Oxford University Press, 1913.

itself exactly after the lapse of the period; and the function thus defined could be decomposed into simple undulations by the methods just described.

But when the graph of the observations is not periodic, the function may yet be represented by a sum of periodic terms whose periods are incommensurable with each other. The gravitational attractions of the heavenly bodies, to which the tides are due, are made up of components whose periods are not commensurable. But in the tidal graph of a port the component due to each would be resolvable into simple undulations. A method of extracting these trains of simple waves from the record would allow the schedule of the tidal oscillations at the port to be constructed for the future so far as these components are concerned.

The usual method of extracting from a graph of length L a part that repeats itself periodically in equal lengths λ is to cut up the graph into segments of this length, and superpose them by addition or mechanically. If there are enough segments, the sum thus obtained, divided by the number of the segments, approximates to the periodic part sought; the other oscillations of different periods may be expected to contribute a constant to the sum, namely the sum of the mean part of each.

6. The principle of this method is also used in searching for hidden periodicities in a set of observations taken over a considerable time. Suppose that a period T occurs in these observations and that they are taken at equal intervals, there being n observations in the period T.

Arrange the numbers in rows thus:

$$\begin{array}{cccccc} u_0, & u_1, & u_2, & \cdots & u_{n-2}, & u_{n-1}. \\ u_n, & u_{n+1}, & u_{n+2}, & \cdots & u_{2n-2}, & u_{2n-1}. \\ \cdot & \cdot & \cdot & \cdot & \cdot & \cdot \\ u_{(m-1)n}, & u_{(m-1)n+1}, & u_{(m-1)n+2}, & \cdots & u_{mn-2}, & u_{mn-1}. \end{array}$$

Add the vertical columns, and let the sums be

$$U_0, \quad U_1, \quad U_2, \ldots U_{n-2}, \quad U_{n-1}.$$

In the sequence U_0, U_1, U_2, ... U_{n-1} the component of period T will be multiplied m-fold, and the variable parts of the other components may be expected to disappear, as these will enter with different phases into the horizontal rows, and the rows are supposed to be numerous. The difference between the greatest and least of the numbers U_0, U_1, U_2, ... U_{n-2}, U_{n-1} furnishes a rough indication of the amplitude of the component of period T, if such exists; and the presence of such a period is indicated by this difference being large.

Let y denote the difference between the greatest and least of the numbers U_0, U_1, U_2, ... U_{n-2}, U_{n-1} corresponding to the trial period x. If y is plotted as a function of x, we obtain a "curve of periods." This curve will have peaks at the values of x corresponding to the periodicities which really exist. When the presence of such periods is indicated by the curve, the statistics are then analysed by the methods above described.

This method was devised by Whittaker for the discussion of the periodicities entering into the variation of variable stars.* It is a modification of Schuster's work, applied by him to the discussion of the statistics of sunspots and other cosmical phenomena.† To Schuster, the term "*periodogram analysis*" is due, but the "curve of periods" referred to above is not identical with that finally adopted by Schuster and termed *periodograph* (or *periodogram*).

For numerical examples, and for descriptions of other methods of attacking this problem, reference may be made to Whittaker and Robinson's *Calculus of Observations*, Chapter XIII, already cited, and to Schuster's papers.

**Monthly Notices, R.A.S.* 71 (1911), 686.

See also a paper by Gibb, "The Periodogram Analysis of the Variation of SS Cygni," *ibid.*, 74 (1914), 678.

†The following papers may be mentioned:
Trans. Camb. Phil. Soc., 18 (1900), 108.
Trans. Royal Soc. London, (A), 206 (1906), 69.
Proc. Royal Soc. London (A), 77 (1906), 136.

APPENDIX II

LEBESGUE'S THEORY OF THE DEFINITE INTEGRAL

1. Introductory. In Chapter II we have seen what is meant by the lower and upper bounds of a bounded linear set of points. The limiting points of such a set have been defined, and Weierstrass's Theorem, that an infinite set bounded above and below must have at least one limiting point, has been proved.

Lebesgue's Theory of the Definite Integral depends essentially on the idea of the *measure of a set of points*. This is a number, positive or zero, associated with the set and depending upon it. When the set consists of the points of an interval, open or closed, the measure is to be the same as the length of the interval. And the measure of the set of points, which belong to one or other of two sets without common points, should be the sum of the measures of these two sets.

It is this very subtle and rather difficult idea of the measure of a set that forms the chief obstacle in the way of the introduction of the Lebesgue Integral into Analysis in place of the Riemann Integral. The discussion which follows is confined to bounded linear sets, though one of the advantages of Lebesgue's work is that the extension to two and three dimensions is more or less immediate. Among the alternatives at our disposal the treatment by de la Vallée Poussin in his *Cours d'Analyse Infinitésimale*, 1 (3e éd., 1914), has been adopted. The more compact and direct development in the first and second chapters of his *Intégrales de Lebesgue*, etc. (1916), seems more difficult as an introduction. But much use is made below of the third chapter of that work dealing with the properties of the Lebesgue Integral.

The first step in this extension of the Riemann Theory of Integration was made by Jordan* (1892), when he introduced the idea of the *inner* and *outer content* of a set of points. In 1898 Borel† showed that a much more useful

*Jordan, *Cours d'Analyse*, 1 (2e éd., 1893), p. 28. He uses the terms *aire intérieure* and *aire extérieure*.

†Borel, *Leçons sur la théorie des fonctions* (1e éd., 1898). In the third and last edition (1928), this work is both revised and enlarged.

concept was what he called the measure of the set. The principles that guided Lebesgue in his theory of measure were those of Borel, and his definition and earliest treatment are given in his Paris Thesis—*Intégrale, Longueur, Aire*—published in Ann. di Matematica (**3**), 7, (1902). This was followed by his book, *Leçons sur l'intégration et la recherche des fonctions primitives* (1904), of which a second and greatly enlarged edition has just appeared (1928).

2. We now give such definitions and simple properties of bounded linear sets of points as will be required in the discussion of measure and the Lebesgue Integral.

As before we denote a set by E, and we shall assume that its points lie on the interval $a \leqq x \leqq b$. The points of (a, b), which are not points of E, form the complementary set denoted by CE. Clearly $C(CE) = E$.

A set E is said to be countable (*or enumerable*), *when there is a one-one correspondence between the points of the set and the positive integers.* To every point of E there corresponds a positive integer, and to every positive integer there corresponds a point of E.

The terms of a countable set can be written as

$$u_1, \quad u_2, \quad u_3, \ldots .$$

The set $1, \frac{1}{2}, \frac{1}{3}, \ldots, 1/n, \ldots$ is obviously countable.

The positive rational numbers form a countable set.

For every such number can be expressed as a fraction p/q, where p and q are positive integers without common factor. We arrange these fractions according to the sum $p+q$, beginning in each with the fraction of lowest numerator.

When $p+q=2$, we have 1 only, and this is taken as u_1.

When $p+q=3$, we have $\frac{1}{2}$ and 2, and these are taken as u_2 and u_3.

When $p+q=4$, we have $\frac{1}{3}$ and 3, omitting the number 1 already used: and these are taken as u_4 and u_5: and so on.

A set E is said to be a closed set if it contains its limiting points; *e.g.* the set $1, \frac{1}{2}, \frac{1}{3}, \ldots$ is not closed, but, if the origin is included in the set, it becomes a closed set.

A point P of abscissa x is said to be an interior point of the set, if a neighbourhood of P, $\alpha < x < \beta$, exists all of whose points are points of the set.

A point P is said to be an exterior point of the set E, if it is an interior point of CE: in other words, there must be a neighbourhood of P none of whose points are points of E.

A set E is said to be an open set, if all its points are interior points of the set.

Open sets have the important property that they are composed of a finite, or countably infinite, number of not-overlapping open intervals.

To prove this, take any point P of the set, and let its abscissa be x_0. We can divide all the positive numbers into two classes A and B as follows: a number h is put in the lower class A, if all the points x given by $x_0 \leqq x < x_0 + h$ belong to the set; and it is put in the upper class B, if this is not the case. There are numbers of both classes, and every number in the class A is less than every number in the class B. If μ is the number separating the two classes,

$x_0 + \mu$ is the right-hand end-point of the interval associated with the point x_0. Similarly by taking negative numbers, we obtain the left-hand end-point $x_0 - \lambda$. To all the points of the open interval $x_0 - \lambda < x < x_0 + \mu$ this same interval corresponds. Now consider the set of all such open intervals. They are obviously not overlapping.

Let $\epsilon_1, \epsilon_2, \ldots$ be a monotonic descending sequence of positive numbers, such that $\lim_{n\to\infty} \epsilon_n = 0$. There can only be a finite number of the intervals of the set just described, each of length $\geqq \epsilon_1$. We can arrange these in order from left to right. Again there can only be a finite number of those that remain, each of length $\geqq \epsilon_2$. These we now arrange in order after the first group; and so on.

If the end-points a and b are points of the open set, they are to be the left-hand end-point, and right-hand end-point, respectively, of intervals (a, α), (β, b), open at the ends α and β.

The complement of an open set E with respect to the closed interval $a \leqq x \leqq b$ is a closed set.

For let P be a limiting point of CE. Then P cannot be a point of E, otherwise there would be a neighbourhood of P containing no point of CE. And this is impossible, if P is a limiting point of CE.

Again, *the complement of a closed set E is an open set.*

For let P be a point of CE. There must be a neighbourhood of P without any point of E inside it; otherwise P would be a limiting point of E, and therefore a point of E.

Two sets E_1 and E_2 are said to be equal, if they consist of the same points, and we write $E_1 = E_2$.

A set E_1 is said to be greater than a set E_2, if E_1 consists of the points of E_2 and some other points, and we write $E_1 > E_2$.

And $E_1 < E_2$ means that $E_2 > E_1$.

3. Operations on Sets. The set E is said to be the sum of the sets $E_1, E_2 \ldots E_n$, when it is composed of the points which belong to at least one of these sets, and we write

$$E = E_1 + E_2 + \ldots + E_n = \sum_1^n E_r.$$

The set E is said to be the difference of two sets E_1, E_2, when it consists of the points of E_1 which do not belong to E_2; and we write $E = E_1 - E_2$.

The set of points which are common to all the sets $E_1, E_2, \ldots E_n$ is called the product of these sets, and we write

$$E = E_1 . E_2 . \ldots E_n = \prod_1^n E_r.$$

Multiplication and subtraction are reduced to addition by the use of complementary sets, as it will be seen that

$$CE_1 + CE_2 = C(E_1 E_2), \quad E_1 E_2 = C(CE_1 + CE_2),$$

$$C(E_1 + E_2) = CE_1 . CE_2, \quad C(E_1 - E_2) = CE_1 + E_2, \quad E_1 - E_2 = E_1 . CE_2.$$

The Commutative and Associative Laws of Algebra hold for the addition and multiplication of sets;

e.g.,
$$E_1 + E_2 = E_2 + E_1, \qquad E_1 E_2 = E_2 E_1,$$

$$E_1 + E_2 + E_3 = E_1 + (E_2 + E_3), \quad E_1 E_2 E_3 = E_1(E_2 E_3).$$

Also the Distributive Law applies;

$$E_1(E_2+E_3)=E_1E_2+E_1E_3,$$

and the ordinary algebraical process gives, for example,

$$(E_1+E_2)(E_3+E_4)=E_1E_3+E_2E_3+E_1E_4+E_2E_4.$$

Since in general $(E_1-E_2)+E_2 \neq E_1$, but is actually E_1+E_2, care must be exercised in dealing with subtraction. In practice there is no difficulty, for subtraction is replaced by multiplication and addition, making use of complementary sets. The commutative, associative and distributive laws may then be employed.

The operations above referred to are finite; that is to say, carried out on a finite number of sets. But addition and multiplication can be extended to an infinite number of sets.

The set

$$E=E_1+E_2+\dots \text{ to } \infty = \sum_1^{\infty} E_r$$

is composed of the points which belong to at least one of the infinite number of sets $E_1, E_2, \dots$ to ∞.

The infinite product

$$E=E_1E_2 \dots \text{ to } \infty = \prod_1^{\infty} E_r$$

is made up of the points which are common to all the sets $E_1, E_2, \dots$ to ∞.

The distributive law

$$E_1(E_2+E_3+\dots \text{ to } \infty)=E_1E_2+E_1E_3+\dots \text{ to } \infty$$

applies again here, and it holds also for the case of a finite number of factors, which themselves may be finite or infinite sums.

Further $\quad C(E_1+E_2+E_3+\dots \text{ to } \infty)=CE_1 \,.\, CE_2 \,.\, CE_3 \dots \text{ to } \infty$

and $\quad C(E_1 \,.\, E_2 \,.\, E_3 \dots \text{ to } \infty)=CE_1+CE_2+CE_3+\dots \text{ to } \infty$.

It is clear that the sum of a finite number of countable sets is a countable set, for we can take all the first points of the sets in order, and then place after them all the second points of the sets in their order, and so on.

But it is important to notice that *the sum of a countably infinite number of countable sets is also a countable set.*

To prove this let the sets be

$$\begin{aligned} E_1 &= a_{11}+a_{12}+a_{13}+a_{14}+\dots, \\ E_2 &= a_{21}+a_{22}+a_{23}+\dots, \\ E_3 &= a_{31}+a_{32}+a_{33}+\dots, \\ E_4 &= a_{41}+a_{42}+a_{43}+\dots, \end{aligned}$$

and so on.

Then $\quad \sum_1^{\infty} E_r = a_{11}+a_{21}+a_{12}+a_{31}+a_{22}+a_{13}+\dots,$

where the points in the sum are taken from left to right along the dotted diagonals as in the above scheme.

From this result we can deduce that *an open bounded linear set of points can be broken up into a countably infinite set of not-overlapping closed intervals.**

For we have seen (§ 2) that a bounded open linear set is composed of a finite or countably infinite set of not-overlapping open intervals.

Let (α, β) be one of these intervals.

Divide it into four equal parts, and take the two middle parts, both closed, as intervals Δ_{11}, Δ_{12}. The remaining left-hand quarter and right-hand quarter are then to be divided into two equal parts, and the closed halves lying nearest the centre are to be taken as intervals Δ_{21}, Δ_{22}. The outer parts are again halved, and so on.

We thus replace this open interval $\alpha < x < \beta$ by the countable set of not-overlapping closed intervals $\Delta_{11}, \Delta_{12}, \Delta_{21}, \Delta_{22}, \Delta_{31}, \Delta_{32} \cdots$.

In this way from each of the finite or countably infinite set of open intervals we obtain a countably infinite set of closed intervals, and the result follows.

4. The Measure of a Bounded Linear Set of Points.

I. Let E be a set of points all lying on the interval $a \leqq x \leqq b$, and let all the points of E be enclosed as interior points in a set of intervals $\delta_1, \delta_2, \ldots$, lying in (a, b).

This set of intervals can be replaced by a set of not-overlapping intervals $\Delta_1, \Delta_2, \Delta_3, \ldots$, such that every point of E is an interior point of one of the intervals, or the common end-point of two adjacent intervals. For start with δ_1 and call it Δ_1. Then take δ_2 and suppress the parts, if any, of δ_2 which lie in δ_1. If δ_2 lies altogether outside δ_1, we take it as Δ_2; and also if δ_2 abuts on δ_1 but does not overlap it. If it overlaps, we take for Δ_2 the part of δ_2 outside δ_1, and we have in addition their common end-point. If δ_1 lies wholly within δ_2, we replace δ_2 by the two open intervals outside δ_1 and the two end-points of δ_1.

In this process, at any stage when we take in the interval δ_n, we add to our set of not-overlapping intervals $\Delta_1, \Delta_2, \ldots$, which replaces $\delta_1, \delta_2, \ldots$, a finite number of not-overlapping intervals, and, possibly, the common end-points of adjacent intervals.

Now let $\Delta_1, \Delta_2, \ldots$ *be a finite or countably infinite set of not-overlapping intervals, all in* (a, b), *or with* a *and* b *as left-hand end-point or right-hand end-point, respectively, such that every point of* E *is an interior point of one of the intervals, or the common end-point of two adjacent intervals. Let* $\Sigma\Delta$ *denote the sum of the lengths of the intervals of the set* $\Delta_1, \Delta_2, \ldots$.

The lower bound of $\Sigma\Delta$ *for all such sets of intervals is called the exterior measure of* E *and denoted by* $m_e(E)$.

---

* The word *overlap* is used here in its natural sense; two intervals overlap if they have points in common which are not end-points of either. Thus (0, 1) and ($\frac{1}{2}$, 2) overlap. The closed intervals (0, 1) and (1, 2) in a more exact sense overlap, as there is a point common to both; but in the text this meaning of the term is not used. A pair of such intervals (open or closed) may be said to *abut*.

The *interior measure* of E is defined as $(b-a)-m_e(CE)$, where CE is the set of points of $a \leqq x \leqq b$ which are not points of E.

The interior measure of E is written $m_i(E)$, and is not negative, since $m_e(CE)$ cannot exceed $(b-a)$.

II. *To prove* $m_i(E) \leqq m_e(E)$.

From the above definition, there must be a set of not-overlapping intervals $\alpha_1, \alpha_2, \ldots$ for E, as above, such that

$$m_e(E) \leqq \Sigma\alpha < m_e(E) + \tfrac{1}{2}\epsilon,$$

and similarly a set $\beta_1, \beta_2, \ldots$ for CE, such that

$$m_e(CE) \leqq \Sigma\beta < m_e(CE) + \tfrac{1}{2}\epsilon,$$

ϵ being the usual arbitrary positive number.

The combined set
$$\alpha_1, \beta_1, \alpha_2, \beta_2, \ldots$$
can be replaced as in (i) by a set of not-overlapping intervals $\gamma_1, \gamma_2, \ldots$, with possibly the addition of the common end-points of certain adjacent intervals among the γ's.

This countable set $\gamma_1, \gamma_2, \ldots$ of not-overlapping intervals is such that all the points of $a \leqq x \leqq b$ are either interior points of these intervals, or common end-points of adjacent intervals.

It follows that
$$\sum_1^\infty \gamma_r = b-a.$$

For, when n is any positive integer, $\sum_1^n \gamma_r < (b-a)$.

Thus
$$\lim_{n\to\infty} \sum_1^n \gamma_r \leqq b-a.$$

If possible, let this limit be $b-a-2\epsilon'$.

Form a new set of intervals $\bar{\gamma}_1, \gamma_2, \ldots$ by adding $\dfrac{\epsilon'}{2^{r+1}}$ to $\gamma_r (r=1, 2, \ldots)$ at each end.

Then every point of $a \leqq x \leqq b$ is an *interior* point of at least one of the intervals $\bar{\gamma}_1, \bar{\gamma}_2, \ldots$, and by the Heine-Borel Theorem (§ 31.2, p. 71) this is true for a set made up of a finite number of them.

Hence
$$b-a < \sum_1^\infty \bar{\gamma}_r < \sum_1^\infty \left(\gamma_r + \frac{\epsilon'}{2^r}\right) < (b-a-2\epsilon') + \epsilon',$$
which is impossible.

But it is clear that
$$\sum_1^\infty \gamma_r \leqq \sum_1^\infty \alpha_r + \sum_1^\infty \beta_r.$$

Thus we have
$$b-a < m_e(E) + m_e(CE) + \epsilon,$$
and
$$m_i(E) \leqq m_e(E).$$

III. *If* $E_2 > E_1$, *then* $m_e(E_2) \geqq m_e(E_1)$.

If possible let $m_e E_2 < m_e E_1$. Then there must be a set of intervals $\Delta_1, \Delta_2, \ldots$ for E_2, as in the definition of the exterior measure, such that the sum of their lengths is greater than $m_e(E_2)$ and less than $m_e(E_1)$.

But as E_2 contains E_1, this set of intervals $\Delta_1, \Delta_2, \ldots$ will also serve for E_1 and the sum of their lengths cannot be less than $m_e(E_1)$.

IV. *Further if* $E_2 > E_1$, *then* $m_i(E_2) \geqq m_i(E_1)$.

For
$$CE_1 > CE_2,$$
and
$$m_e(CE_1) \geqq m_e(CE_2).$$
Thus
$$(b-a) - m_e(CE_2) \geqq (b-a) - m_e(CE_1).$$
Therefore
$$m_i(E_2) \geqq m_i(E_1).$$

V. *To prove* $m_e(E_1 + E_2) \leqq m_e E_1 + m_e E_2$.

Let $\alpha_1, \alpha_2, \ldots$ be a set of not-overlapping intervals for E_1, as in the definition of (I), such that

$$m_e(E_1) \leqq \sum_1^\infty \alpha < m_e(E_1) + \tfrac{1}{2}\epsilon,$$

and $\beta_1, \beta_2, \ldots$ a set for E_2, such that

$$m_e(E_2) \leqq \sum_1^\infty \beta_r < m_e(E_2) + \tfrac{1}{2}\epsilon,$$

ϵ being the usual arbitrary positive number.

As before, from the set

$$\alpha_1, \quad \beta_1, \quad \alpha_2, \quad \beta_2, \ldots,$$

we form a not-overlapping set $\gamma_1, \gamma_2, \gamma_3 \ldots$, all in (a, b), such that every point of $(E_1 + E_2)$ is either an interior point of one of the γ's or the common end-point of two adjacent γ's.

Also
$$\sum_1^\infty \gamma_r \leqq \sum_1^\infty \alpha_r + \sum_1^\infty \beta_r,$$
and
$$m_e(E_1 + E_2) \leqq \sum_1^\infty \gamma_r.$$
Thus
$$m_e(E_1 + E_2) < m_e(E_1) + m_e(E_2) + \epsilon.$$
Therefore
$$m_e(E_1 + E_2) \leqq m_e(E_1) + m_e(E_2).$$

VI. *Definition of Measurable Sets.* **If $m_e(E) = m_i(E)$, then E is said to be a measurable set, and its measure m (E) is their common value.**

It is clear that if E is a measurable set, CE is also measurable.

The measure of a finite number of points is obviously zero, since the exterior measure is zero.

Further, *a countably infinite linear set of points is of measure zero.*

Let E be the set of points $x_1, x_2, \ldots$.

Take the arbitrary positive ϵ and enclose x_1 in an interval of length $\frac{\epsilon}{2}$, x_2 in an interval of length $\frac{\epsilon}{2^2}$, and so on.

Thus
$$m_e(E) < \epsilon \sum_1^\infty \frac{1}{2^r} < \epsilon.$$

Therefore the exterior measure of this set is zero, and the measure of the set is also zero.

If E is an interval (α, β) *in* (a, b), *it is measurable and its measure is* $(\beta - \alpha)$.

For we must have $m_e(E) \leqq \beta - \alpha$, since the interval (α, β) is itself a possible interval for the exterior measure with the definition given above.

And the Heine-Borel Theorem, as in (II), shows at once that the sign of inequality is impossible.

Again it is clear that

$$m_e(CE) \leqq (b-a)-(\beta-a).$$

But the sign of inequality can be omitted, for it would give

$$\beta - a < (b-a) - m_e(CE) < m_i(E),$$

and we would have

$$m_e(E) < m_i(E).$$

Similarly, *if E is the sum of a finite number of not-overlapping intervals, it is measurable, and its measure is the sum of the lengths of these intervals.*

In the discussion which follows it will be seen that the sum of a finite, or countably infinite, set of measurable sets is measurable. Thus starting with countable sets of points and countable sets of intervals, we see that open sets are measurable. And as the complement of a measurable set is measurable, closed sets are also measurable. All sets which can be obtained by additions, subtractions and multiplications, finite or infinite, of measurable sets are also measurable.

It is only for measurable sets that this theory of measure is studied. It is still a debatable question whether, and, if so, in what sense, sets which are not measurable do exist.*

5. A Necessary and Sufficient Condition that a Set E be Measurable.†

A necessary and sufficient condition that a linear set E in $a \leqq x \leqq b$ be measurable is that to the arbitrary positive ϵ there shall correspond a set I consisting of a finite number of intervals and two sets e', e'' of exterior measures $< \epsilon$, such that

$$E = I + e' - e''.$$

(i) To prove that this condition is necessary, we note that to the arbitrary ϵ there correspond the sets of not-overlapping intervals $\alpha_1, \alpha_2, \ldots$ for E and $\beta_1, \beta_2, \ldots$ for CE, such that

$$mE \leqq \sum_1^\infty \alpha_r < mE + \tfrac{1}{2}\epsilon,$$

$$mCE \leqq \sum_1^\infty \beta_r < mCE + \tfrac{1}{2}\epsilon,$$

and

$$mE + mCE = b - a.$$

Thus

$$\sum_1^\infty \alpha_r + \sum_1^\infty \beta_r < (b-a) + \epsilon.$$

But, as in § 4, I, from the set

$$\alpha_1, \beta_1, \alpha_2, \beta_2, \ldots \alpha_n, \beta_n,$$

we obtain a set of not-overlapping intervals

$$\gamma_1, \gamma_2, \ldots \gamma_N,$$

such that

$$\sum_1^N \gamma_r = \sum_1^n \alpha_r + \sum_1^n \beta_r - \Sigma(\alpha_r \beta_s),$$

*Cf. Lebesgue, *Leçons sur l'intégration* (2e éd., 1928), footnote, p. 114.

†In this and the following sections, we shall, when convenient, write mE instead of $m(E)$ for the measure of E, when E is measurable, and similarly m_eE for $m_e(E)$, m_iE for $m_i(E)$.

It will be noticed that, when E is measurable, $mE = m_eE = m_iE$.

the last term on the right-hand side comprising the parts common to an α and a β blotted out in forming the γ's.

As n tends to ∞, so does N and we know that $\sum_1^\infty \gamma_r = (b-a)$.

Thus the sum of the lengths of the finite or countably infinite set of not-overlapping intervals blotted out as above is less than ϵ.

Now take n so large that $\sum_{n+1}^\infty \alpha_r < \epsilon$, and denote the points of $\alpha_1, \alpha_2, \ldots \alpha_n$ by S_n and the points of $\alpha_{n+1}, \alpha_{n+2}, \ldots$ to ∞ by R_n.

Then $$E = S_n + R_n E - S_n CE.$$

Also $R_n E < R_n$ and $S_n CE <$ the points common to all the α's and β's.

Thus $m_e(R_n E) < \epsilon$ and $m_e(S_n CE) < \sum (\alpha\beta) < \epsilon$.

(ii) If the condition is satisfied, E is measurable.

We have $$E = I + e' - e'',$$
where I consists of a finite set of intervals and e', e'' are two sets of exterior measure $< \epsilon$, which we suppose have no common points.

Thus $$E < I + e',$$
and $$m_e E \leqq m_e(I + e') < mI + \epsilon.$$

Also $$E > I - e''.$$

Thus $$CE < C(I - e'') = CI + e''.$$

Therefore $$m_e CE \leqq m_e(CI + e'') < mCI + \epsilon.$$

It follows that $$m_e E + m_e CE < mI + mCI + 2\epsilon \doteq (b-a) + 2\epsilon.$$

And $$m_e E < m_i E + 2\epsilon.$$

Thus $$m_e E \leqq m_i E.$$

But $$m_e E \geqq m_i E.$$

Hence $$m_e E = m_i E.$$

6. I. *If E_1 and E_2 are measurable, so is $E_1 + E_2$.*

We have to show that $E_1 + E_2$ can be broken up as in the theorem of § 5.

Given the arbitrary positive ϵ, we have
$$E_1 = I_1 + e_1' - e_1'',$$
$$E_2 = I_2 + e_2' - e_2'',$$
where I_1, I_2 are sets of a finite number of intervals and $m_e e_1' < \frac{1}{2}\epsilon$, etc.

Thus $$E_1 + E_2 = I_1 + I_2 + (e_1' + e_2') - e'',$$
where e'' is contained in $e_1'' + e_2''$.

Also $$m_e(e_1' + e_2') \leqq m_e e_1' + m_e e_2' < \epsilon,$$
and $$m_e e'' \leqq m_e(e_1'' + e_2'') \leqq m_e e_1'' + m_e e_2'' < \epsilon.$$

Thus $E_1 + E_2$ is measurable.

II. *If E_1 and E_2 are measurable, so is $E_1 - E_2$.*

We know that $$C(E_1 - E_2) = CE_1 + E_2.$$

Thus $C(E_1 - E_2)$ is measurable, and $E_1 - E_2$ also.

III. *If E_1 and E_2 are measurable, so is $E_1 E_2$.*

We know that $C(E_1 E_2) = CE_1 + CE_2$, and the result follows.

7. I. *If* E_1 *and* E_2 *are measurable and without common points, then*

$$m(E_1+E_2)=mE_1+mE_2.$$

With the same notation as in § 6, we have

$$E_1=I_1+e_1'-e_1'',$$
$$E_2=I_2+e_2'-e_2'',$$

where $m_e e_1'<\frac{1}{2}\epsilon$, etc.

Thus $$E_1+E_2<I_1+I_2+e_1'+e_2',$$
and $$m(E_1+E_2)=m_e(E_1+E_2)<m(I_1+I_2)+\epsilon.$$
Also $$E_1+E_2>I_1+I_2-e'', \quad \text{where} \quad e''<e_1''+e_2'',$$
and $$C(E_1+E_2)<C(I_1+I_2-e'')=C(I_1+I_2)+e''.$$
Therefore $$mC(E_1+E_2)=m_eC(E_1+E_2)<mC(I_1+I_2)+\epsilon,$$
and $$m(E_1+E_2)>m(I_1+I_2)-\epsilon.$$
Thus $$|m(E_1+E_2)-m(I_1+I_2)|<\epsilon.$$

But for the finite sets of intervals it is clear that

$$m(I_1+I_2)=mI_1+mI_2-mI_1I_2.$$

Also since $$E_1<I_1+e_1', \quad mE_1<mI_1+\tfrac{1}{2}\epsilon.$$
And since $$E_1>I_1-e_1'',$$
$$CE_1<C(I_1-e_1'')=CI_1+e_1'',$$
we have as before $$mE_1>mI_1-\tfrac{1}{2}\epsilon.$$
Thus $$|mE_1-mI_1|<\tfrac{1}{2}\epsilon.$$
Similarly $$|mE_2-mI_2|<\tfrac{1}{2}\epsilon.$$
Again $$I_1<E_1+e_1'' \quad \text{and} \quad I_2<E_2+e_2''.$$
It follows that $$I_1I_2<E_1E_2+E_1e_2''+e_1''(E_2+e_2'')$$
$$<e_1''+e_2'', \quad \text{since} \quad E_1E_2=0.$$
Therefore $$m(I_1I_2)\leqq m_e(e_1''+e_2'')\leqq m_e e_1''+m_e e_2''<\epsilon.$$
But $|m(E_1+E_2)-mE_1-mE_2|$
$$\leqq|m(E_1+E_2)-m(I_1+I_2)|+|m(I_1+I_2)-mI_1-mI_2|$$
$$+|mI_1-mE_1|+|mI_2-mE_2|.$$
Thus $$|m(E_1+E_2)-mE_1-mE_2|<3\epsilon.$$
It follows that $$m(E_1+E_2)=mE_1+mE_2.$$

This theorem is a special case of the following:

II. *If* E_1, E_2 *are measurable, then*

$$m(E_1+E_2)+mE_1E_2=mE_1+mE_2.$$

For let $$E_1=E_1E_2+e_1 \quad \text{and} \quad E_2=E_1E_2+e_2.$$

Then e_1, e_2 and E_1E_2 are measurable and without common points, and

$$E_1+E_2=e_1+e_2+E_1E_2.$$

But by (I) $$mE_1=mE_1E_2+me_1,$$
$$mE_2=mE_1E_2+me_2,$$
$$m(E_1+E_2)=m(e_1+e_2+E_1E_2)=me_1+me_2+mE_1E_2.$$

Therefore $$m(E_1+E_2)+mE_1E_2=mE_1+mE_2.$$

III. *If E_1 and E_2 are measurable and $E_1 > E_2$, then*

$$m(E_1-E_2)=mE_1-mE_2.$$

We have $$(E_1-E_2)+E_2=E_1,$$
and (E_1-E_2), E_2 have no common point.

Thus from (I.) the result follows.

8. I. *If E_1, E_2, ... are measurable and without common points, then $E=\sum_1^\infty E_r$ is measurable, and $mE=\sum_1^\infty mE_r$.*

Since $E_1+E_2+\ldots+E_n$ is measurable and contained in (a, b), we have, for any positive integer (by § 7),

$$m(\sum_1^n E_r)=\sum_1^n mE_r < (b-a).$$

Thus $$\sum_1^\infty mE_r \leqq (b-a).$$

Taking the arbitrary positive ϵ, there is a positive integer ν such that

$$mE_{n+1}+mE_{n+2}+\ldots<\epsilon, \quad \text{when} \quad n \geqq \nu.$$

Now for E_{n+r} there is a set of not-overlapping intervals $\alpha_{n+r,1}$, $\alpha_{n+r,2}$, ... such that

$$mE_{n+r} \leqq \sum_{s=1}^\infty \alpha_{n+r,s} < mE_{n+r}+\frac{\epsilon}{2^r}, \quad r=1, 2, \ldots .$$

Also the countably infinite set of intervals

$$\sum_{s=1}^\infty \alpha_{n+1,s}, \quad \sum_{s=1}^\infty \alpha_{n+2,s}, \ldots$$

are such that

$$m_e(E_{n+1}+E_{n+2}+\ldots) < \sum_{r=1}^\infty (\sum_{s=1}^\infty \alpha_{n+r,s})$$
$$< \sum_{r=1}^\infty \left(mE_{n+r}+\frac{\epsilon}{2^r}\right)$$
$$< \epsilon+\epsilon.$$

Now let $$S_n=\sum_1^n E_r \quad \text{and} \quad R_n=\sum_{n+1}^\infty E_r.$$

Then $$m_eE=m_e(S_n+R_n) \leqq m_eS_n+m_eR_n$$
$$< mS_n+2\epsilon.$$

But $$E > S_n.$$
Therefore $$m_iE \geqq mS_n.$$
Thus $$m_eE < m_iE+2\epsilon.$$
It follows as before that $$m_eE=m_iE.$$

Again $$\sum_1^n mE_r=mS_n \leqq mE < mS_n+2\epsilon, \quad \text{when } n \geqq \nu.$$

Thus $$0 \leqq mE-mS_n < 2\epsilon, \quad \text{when } n \geqq \nu,$$

and $$mE=\lim_{n\to\infty} \sum_1^n mE_r=\sum_1^\infty mE_r.$$

II. *If the measurable sets E_1, E_2, ... have common points, then $E=\sum_1^\infty E_r$ is measurable.*

For any integer n, we have

$$E_1+E_2+\dots+E_n$$
$$=E_1+(E_2-E_1)+(E_3-E_1-E_2)+\dots+(E_n-E_1-E_2\dots-E_{n-1}),$$

and the different sets on the right-hand side have no common points.

Thus $$E_1+E_2+\dots \text{ to } \infty=\mathcal{E}_1+\mathcal{E}_2+\dots \text{ to } \infty,$$

where $$\mathcal{E}_1=E_1, \quad \text{and} \quad \mathcal{E}_r=E_r-E_1-E_2\dots-E_{r-1}, \quad r\geqq 2,$$

and $\sum_1^\infty \mathcal{E}_r$ is measurable.

9. I. *Let $E, E_1, E_2, \dots$ be all measurable. Then $E-E_1-E_2\dots$ is measurable.*

This follows from § 8 and § 6, II.

II. *If in addition $E_1, E_2, \dots$ are all contained in E, and have no common points,*

$$m(E-\sum_1^\infty E_r)=mE-\sum_1^\infty mE_r.$$

This follows from § 7, III and § 8, I.

III. *If $E_1, E_2, \dots$ are all measurable, then $E=E_1 . E_2 . E_3 \dots$ is also measurable.*

We know that $$C(E_1 . E_2)=CE_1+CE_2,$$
$$C(E_1 . E_2 . E_3)=CE_1+CE_2+CE_3, \text{ and so on.}$$

Thus $$C(E_1 . E_2 . E_3 \dots)=CE_1+CE_2+CE_3+\dots,$$

and the result follows at once.

IV. *If $E_1, E_2, \dots$ are all measurable and $E_1<E_2<E_3<\dots$, then*

$$E=E_1+E_2+E_3+\dots$$

is measurable, and $mE=\lim_{n\to\infty}(mE_n)$.

In this case

$$E=E_1+(E_2-E_1)+(E_3-E_2)+\dots,$$

and $$mE=\lim_{n\to\infty}\sum_1^n m(E_r-E_{r-1}) \text{ by § 8, I}$$
$$=\lim_{n\to\infty}\sum_1^n (mE_r-mE_{r-1}) \text{ by § 7, III}$$
$$=\lim_{n\to\infty}(mE_n).$$

V. *If $E_1, E_2, \dots$ are all measurable and $E_1>E_2>E_3>\dots$, then*

$$E=E_1 . E_2 . E_3 \dots$$

is measurable and $mE=\lim_{n\to\infty}(mE_n)$.

We have as in (III),

$$m(CE)=m(CE_1+CE_2+\dots)$$
$$=\lim_{n\to\infty} m(CE_n) \text{ by (IV).}$$

And $$mE=(b-a)-m(CE)=(b-a)-\lim_{n\to\infty}(mCE_n)=\lim_{n\to\infty}(mE_n).$$

10. *A necessary and sufficient condition that the bounded function $f(x)$ be integrable in (a, b) according to Riemann's definition is that its points of discontinuity form a set of measure zero.*

We prove first that this condition is sufficient.

Take the arbitrary positive ϵ, and $k < \dfrac{\epsilon}{2(b-a)}$.

Let G_k be the set of points of discontinuity at which the oscillation* $\geqq k$

Then the measure of this set is zero since it is a part of a set of measure zero and it is a closed set.†

Thus the points of G_k can be enclosed as interior points in a finite set of not-overlapping intervals, the sum of their lengths being $\leqq \dfrac{\epsilon}{2(M-m)}$, where M and m are the bounds of $f(x)$ in (a, b).‡

The complement of this set of intervals is a finite number of closed not-overlapping intervals, the oscillation at every point of these being $< k$.

Each of these can be divided up into a finite number of partial intervals, such that the oscillation in each of these partial intervals $< k$. (Cf. § 31. 1, footnote, p. 71.)

Thus we have a mode of division of (a, b) for which

$$S - s < k(b-a) + \frac{\epsilon}{2(M-m)} \times (M-m) < \epsilon,$$

and $f(x)$ is integrable in (a, b), according to Riemann's definition.

We now prove that the condition is necessary.

Let $k_1 > k_2 > \dots$ and $\lim k_n = 0$, and G_r be the set of points for which the oscillation $\geqq k_r$.

Let the measure of this closed set be $C > 0$, and take $\epsilon = \frac{1}{4} k_r C$.

Since $f(x)$ is integrable (R), there is a mode of division of (a, b)

$$a = x_0, x_1 \dots, \quad x_{n-1}, x_n = b,$$

such that $S - s < \epsilon$.

If a point of G_r is inside one of these intervals (x_{s-1}, x_s), then the oscillation in $(x_{s-1}, x_s) \geqq k_r$.

If it coincides with the common end-point of (x_{s-1}, x_s), (x_s, x_{s+1}), the oscillation in at least one of the two must be $\frac{1}{2} k_r$.

---

*Cf. § 29. 4, p. 66.

†For let $P(x_0)$ be a limiting point of G_k and not a point of G_k. Then the oscillation at P is equal to $k_1 < k$, and there is a neighbourhood $|x - x_0| \leqq \eta$ in which the oscillation $< k$. But P is a limiting point of G_k, so there is a point P' of G_k inside $(x_0 - \eta, x_0 + \eta)$, and the oscillation at $P' \geqq k$. Therefore there is a neighbourhood of P' inside $(x_0 - \eta, x_0 + \eta)$ for which the oscillation $\geqq k$, which is impossible.

‡G_k is a closed set of measure zero and CG_k is thus an open set of measure $(b-a)$. CG_k can therefore be broken up into a countably infinite set of not-overlapping closed intervals

$$\Delta_1, \Delta_2, \dots \quad \text{and} \quad \sum_1^{\infty} \Delta_r = (b-a).$$

We can choose the positive integer n so that

$$(b-a) - \sum_1^n \Delta_r < \frac{\epsilon}{2(M-m)}.$$

The points of G_k are interior points of the finite set of not-overlapping intervals left, when $\Delta_1, \Delta_2, \dots \Delta_n$ are removed from (a, b).

Thus these two adjacent partial intervals give to $(S-s)$ a contribution $\geqq \frac{1}{4}k_r$ multiplied by the sum of the lengths of these intervals.

Hence all the intervals of $a=x_0, x_1 \ldots, x_{n-1}, x_n=b$, which have a point of G_r inside them or at an end contribute to $S-s$ an amount $\geqq \frac{1}{4}k_r$ (their sum).

But the sum of the lengths of these intervals $\geqq C$.

Therefore, for this mode of division,

$$S-s \geqq \tfrac{1}{4}k_r C > \epsilon,$$

which is impossible.

Thus $$C=0 \quad \text{and} \quad mG_r=0.$$

But the points of discontinuity are given by the sum

$$E=G_1+G_2+\ldots \text{ to } \infty,$$

and $$G_1 \leqq G_2 \leqq G_3 \ldots .$$

Therefore (by § 9, IV), $$m(E)=\lim_{n\to\infty} m(G_n)=0.$$

11. Measurable Functions.

Let E be a bounded measurable set of points on the axis of x. The function $f(x)$, defined at the points of E, is said to be measurable in E, if the set of points of E for which $f(x)>A$ is measurable, for every constant A.

We denote the set of points of E for which $f(x)>A$ by $E[f(x)>A]$, and similarly $E[f(x)\geqq A]$ denotes the set of points of E for which $f(x)\geqq A$.

We shall now show that if $f(x)$ is measurable in E as defined above, the sets

$$E[f(x)\geqq A], \quad E[f(x)<A], \quad E[f(x)\leqq A]$$

are also measurable.

We are given that $E[f(x)>A]$ is measurable.

Then its complement with respect to E is also measurable, that is

$$E[f(x)\leqq A]$$

is measurable.

Again if E_n is the set of points of E for which $f(x)>A-\dfrac{1}{n}$, we know that E_n is measurable.

And the infinite product

$$E[f(x)\leqq A]\,.\,E_1\,.\,E_2 \ldots$$

is measurable (§ 9, III).

Therefore $E[f(x)=A]$ is measurable.

It follows by addition of $E[f(x)>A]$, that $E[f(x)\geqq A]$ is measurable, and by subtraction from $E[f(x)\leqq A]$, we see that $E[f(x)<A]$ is measurable.

It is clear too that, if A and B are any constants,

$$E[A<f(x)<B], \quad E[A\leqq f(x)<B], \quad E[A<f(x)\leqq B]$$

and $$E[A\leqq f(x)\leqq B]$$

are all measurable.

12. Operations on Measurable Functions.

I. *If $f(x)$ is measurable, so are $a+f(x)$, $af(x)$ and $|f(x)|$, where a is a constant.*

We know that $E[f(x)>A-a]$ is measurable.

Thus $E[f(x)+a>A]$ is measurable.

The others follow from the fact that $E[f(x) > A/a]$ is measurable, and that

$$E[f(x) > A] + E[f(x) < -A]$$

is measurable.

II. *If $f_1(x)$ and $f_2(x)$ are finite and measurable, then $E[f_1(x) > f_2(x)]$ is measurable.*

$E[f_1(x) > f_2(x)]$ is the sum of the countably infinite measurable sets

$$E[f_1(x) > r] \,.\, E[f_2(x) < r],$$

where r is any rational number.

III. *If $f_1(x)$ and $f_2(x)$ are finite and measurable, their sum, difference and product are measurable.*

We know $\quad E[f_1(x) \pm f_2(x) > A] = E[f_1(x) > A \mp f_2(x)].$

Thus the sum and difference of $f_1(x), f_2(x)$ are measurable by (II).

Also if $f(x)$ is measurable, so is $(f(x))^2$, for $E[(f(x))^2 > A]$ is equal to

$$E[f(x) > \sqrt{A}] + E[f(x) < -\sqrt{A}].$$

Thus $(f_1(x) + f_2(x))^2$ and $(f_1(x) - f_2(x))^2$ are measurable, and the result follows.

IV. *Let $f_1(x), f_2(x), \ldots$ be an infinite monotonic sequence of measurable functions. Then, for every x in E, $\lim_{n\to\infty} f_n(x)$ exists (finite or infinite) and this limit is measurable.*

For example, for every x in E let $f_1(x) < f_2(x) < f_3(x) < \cdots$.

Then $E[\lim_{n\to\infty} f_n(x) > A]$ is the sum of $E[f_1(x) > A]$, $E[f_2(x) > A]$, ..., and this sum is measurable (§ 8).

V. *Let $f_1(x), f_2(x), \ldots$ be an infinite sequence of measurable functions. Then $\overline{\lim}_{n\to\infty} f_n(x)$ and $\underline{\lim}_{n\to\infty} f_n(x)$ exist (finite or infinite), and these limits are measurable.*

Suppose $\phi_1(x)$ to be the upper bound of $f_1(x), f_2(x), f_3(x) \ldots$ and $\phi_2(x)$ to be the upper bound of $f_2(x), f_3(x), \ldots$ and so on.

Then $$\phi_1(x) \geqq \phi_2(x) \geqq \phi_3(x) \ldots .$$

Also $$\overline{\lim_{n\to\infty}} f_n(x) = \lim_{n\to\infty} \phi_n(x),$$

and this is measurable by (IV).

VI. *Let $f_1(x), f_2(x), \ldots$ be an infinite sequence of measurable functions, and $\lim_{n\to\infty} f_n(x)$ exist (finite or infinite).*

Then this limit is a measurable function.

This follows from (V), since $\lim_{n\to\infty} f_n(x) = \overline{\lim_{n\to\infty}} f_n(x) = \underline{\lim_{n\to\infty}} f_n(x)$.

We note that every monotonic function is measurable in an interval; and, in particular, that $f(x) =$ constant and $f(x) = x$ are so. Applying the above results we see that every polynomial is measurable: and, as a continuous function, by a theorem due to Weierstrass,* is the limit of a sequence of polynomials,

---

*Cf. Hobson, *Theory of Functions of a Real Variable*, 2 (2nd. ed., 1926), 228.

we see that every continuous function is measurable. Then discontinuous functions, which are the limits of sequences of continuous functions, are also measurable; and so to more complicated classes of measurable functions.

13. The Lebesgue Integral. Let $f(x)$ be a bounded and measurable function for the measurable set E contained in (a, b).*

Let A and B be the lower and upper bounds of $f(x)$ in E.

Divide the interval (A, B) on the axis of y into n partial intervals

$$(A, l_1), \quad (l_1, l_2), \ldots (l_{n-1}, B).$$

Denoting A by l_0 and B by l_n we thus have the mode of division of (A, B),

$$A = l_0, \ l_1, \ldots l_{n-1}, \ l_n = B.$$

Let e_r be the set of points of E for which $l_{r-1} \leqq f(x) < l_r \ldots r = 1, 2 \ldots (n-1)$, and e_n the set of points of E for which $l_{n-1} \leqq f(x) \leqq l_n$.

Then $e_1, e_2 \ldots e_n$ are measurable sets without common points.

Form the sums S and s, where

$$S = \sum_1^n l_r m(e_r) \quad \text{and} \quad s = \sum_1^n l_{r-1} m(e_r),$$

$m(e_r)$ being the measure of the set e_r.

Then
$$S \geqq Am(E) \quad \text{and} \quad s \leqq Bm(E).$$

Thus the sums S and s for all possible modes of division of (A, B) have a lower bound J and an upper bound I, respectively; and for the same mode of division $S \geqq s$.

We shall now show that
$$I \leqq J.$$

Let some or all of the intervals (l_{r-1}, l_r) in $l_0, l_1 \ldots l_{n-1}, l_n$ be divided into smaller intervals, and

$$l_0, \ \lambda_1, \ \lambda_2, \ldots \lambda_{k-1}, \ l_1, \ldots$$

be the new mode of division of (A, B) thus obtained.

This mode of division is said to be *consecutive* to the former.

Let its sums, as above, be Σ, σ.

Compare, for example, the parts of S and Σ which come from (l_0, l_1).

From Σ we have

$$\lambda_1 m(e_{11}) + \lambda_2 m(e_{12}) \ldots + l_1 m(e_{1k}),$$

where $e_{11}, e_{12}, \ldots e_{1k}$ are the sets of points of E for which

$$(l_0 \leqq f(x) < \lambda_1), \ldots (\lambda_{k-1} \leqq f(x) < l_1).$$

And
$$e_1 = e_{11} + e_{12} \ldots + e_{1k},$$

the sets on the right-hand having no common points.

Thus
$$\lambda_1 m(e_{11}) + \lambda_2 m(e_{12}) \ldots + l_1 m(e_{1k}) \leqq l_1 m(e_1).$$

It follows that $\Sigma \leqq S$; and similarly we have $\sigma \geqq s$.

Now take *any* two modes of division of (A, B),

$$A = l_0, \ l_1, \ldots \ l_{m-1}, \ l_m = B, \text{ with sums } S \text{ and } s, \quad \ldots\ldots\ldots\ldots\ldots\ldots (1)$$

$$A = l_0, \ l_1', \ldots \ l'_{n-1}, \ l_n = B, \text{ with sums } S' \text{ and } s'. \quad \ldots\ldots\ldots\ldots\ldots\ldots (2)$$

---

*In this section, so far as possible, the notation corresponds to that of §§ 39-41, pp. 91-4, and the argument proceeds on exactly the same lines.

On superposing (1) and (2) we obtain a third mode of division (3) consecutive to both (1) and (2).

Let its sums be Σ and σ.

Then $$S \geqq \Sigma \quad \text{and} \quad \sigma \geqq s'.$$

But $$\Sigma \geqq \sigma.$$

Thus $$S \geqq s',$$

and the sum S arising from any mode of division is not less than the sum s arising from the same or any other mode of division of (A, B).

It follows at once that $$I \leqq J.$$

For suppose $I > J$. Since J is the lower bound of the sums S, there must be a mode of division giving a sum S to the left of the middle point of JI, and since I is the upper bound of the sums s, there must be a mode of division giving a sum s to the right of this middle point. This is impossible as a sum S cannot be less than any sum s.

Now let ϵ be an arbitrary positive number, as small as we please. Take a mode of division
$$A = l_0, l_1, l_2, \dots l_{n-1}, l_n = B,$$
in which all the partial intervals are less than ϵ.

For this mode of division it is clear that
$$0 \leqq S - s = (l_1 - l_0)m(e_1) + (l_2 - l_1)m(e_2) + \dots + (l_n - l_{n-1})m(e_n) < \epsilon m(E).$$

Therefore we must have $I = J$, *and the sums S, s tend to the common value of I and J as the number of points of division of (A, B) tends to infinity in such a way that the largest of these partial intervals tends to zero.*

This number I is called the Lebesgue Integral of the bounded and measurable function $f(x)$ in the measurable set E, and we write
$$I = \int_E f(x)dx.$$

If $f(x) = C$ in E, where C is a constant, we define $\int_E f(x)dx$ as $Cm(E)$.

If E consists of all the points of an interval (a, b), we use the ordinary notation $\int_a^b f(x)dx$, and the integral is now called the Lebesgue Integral of $f(x)$ between the limits a and b.

Sometimes it is convenient to make clear that the integral is taken in Lebesgue's sense by placing a capital L before the ordinary symbol. In such a case the Riemann Integral would be written as $(R)\int_a^b f(x)dx$ and the Lebesgue Integral as $(L)\int_a^b f(x)dx$.

Again, if the bounded function $f(x)$ is integrable with Lebesgue's definition for the interval (a, b), we say that it is integrable (L) ; and, if it is integrable with Riemann's definition, we say that it is integrable (R).

We shall see below that for bounded functions, if $f(x)$ is integrable (R), it is also integrable (L), and the two integrals are equal: but that $f(x)$ may be integrable (L) and not integrable (R).

Just as in the case of the Riemann Integral, we define the Lebesgue Integral $\int_a^b f(x)dx$ for $a > b$, by the equation $\int_a^b f(x)dx = -\int_b^a f(x)dx$.

14. Properties of the Lebesgue Integral of a Bounded and Measurable Function.

I. $\int_E Cf(x)dx = C\int_E f(x)dx$, *where C is a constant.*

This follows at once from the definition.

II. *If $f(x)$ is measurable in E and $f(x) \geqq C$, then* $\int_E f(x)dx \geqq Cm(E)$.

Since we have for any sum S for $f(x)$, $S \geqq Cm(E)$.

Therefore the limit of the sums $S \geqq Cm(E)$.

A similar result holds for $f(x) \leqq C$.

Thus it is clear that, if A, B are the lower and upper bounds of $f(x)$ in E, then

$$Am(E) \leqq \int_E f(x)dx \leqq Bm(E).$$

III. *Let $f(x)$ be measurable in E, and let $E_1 + E_2 = E$, where E_1 and E_2 are measurable and without common points.**

Then
$$\int_E f(x)dx = \int_{E_1} f(x)dx + \int_{E_2} f(x)dx.$$

Let the bounds of $f(x)$ be A, B in E, α_1, β_1 in E_1, and α_2, β_2 in E_2, respectively.

Then A is the smaller of α_1 and α_2, while B is the larger of β_1 and β_2.

Consider any mode of division of (A, B),

$$(1) \quad A = l_0, l_1, \ldots l_{n-1}, l_n = B.$$

If two of these points do not coincide with the smaller of the β's and the larger of the α's, by introducing these points we have a consecutive mode of division and S is not increased, s not decreased.

Thus the sum S for (1) $\geqq$ a sum S for E_1 + a sum S for E_2

$$\geqq \int_{E_1} f(x)dx + \int_{E_2} f(x)dx.$$

Therefore
$$\int_E f(x)dx \geqq \int_{E_1} f(x)dx + \int_{E_2} f(x)dx.$$

Similarly from the sum s,

$$\int_E f(x)dx \leqq \int_{E_1} f(x)dx + \int_{E_2} f(x)dx.$$

Hence
$$\int_E f(x)dx = \int_{E_1} f(x)dx + \int_{E_2} f(x)dx.$$

IV. The theorem of (III) can be at once extended to the sum of n measurable sets with no common points, two by two.

We now prove that it holds also for a countably infinite number of measurable sets.

---

*It is clear that if $f(x)$ is measurable in E, it is measurable in E_1 and E_2, for $E_1[f(x) > C]$ is the product of $E[f(x) > C]$ and E_1.

Let $f(x)$ be measurable in E and $E = \sum_1^\infty E_r$, where $E_1, E_2, \ldots$ are all measurable and without common points, two by two.

Then
$$\int_E f(x)dx = \sum_1^\infty \int_{E_r} f(x)dx.$$

Let
$$E = \sum_1^n E_r + R_n.$$

Then we know that $f(x)$ is measurable in R_n and that $m(R_n) \to 0$ as $n \to \infty$ (§ 8).

But
$$\int_E f(x)dx = \sum_1^n \int_{E_r} f(x)dx + \int_{R_n} f(x)dx.$$

Thus
$$\left| \int_E f(x)dx - \sum_1^n \int_{E_r} f(x)dx \right| \leqq Mm(R_n), \text{ by (II), above.}$$

when $|M|$ is the upper bound of $|f(x)|$ in E.

Therefore
$$\int_E f(x)dx = \sum_1^\infty \int_{E_r} f(x)dx.$$

V. *If $f(x)$ and $g(x)$ are measurable in E and $f(x) \geqq g(x)$, then*
$$\int_E f(x)dx \geqq \int_E g(x)dx.$$

Let (A, B) be the bounds of $g(x)$ in E and $A = l_0, l_1 \ldots l_{n-1}, l_n = B$ any mode of division of (A, B).

Let $e_1, e_2, \ldots e_n$ be the sets of points as in § 13 for $g(x)$; *i.e.* e_r is the set of points of E for which $l_{r-1} \leqq g(x) < l_r$, when $r = 1, 2, \ldots (n-1)$, and e_n is the set of points for which $l_{n-1} \leqq g(x) \leqq l_n$.

Then
$$\int_E f(x)dx = \sum_1^n \int_{e_r} f(x)dx \geqq \sum_1^n l_{r-1} m(e_r),$$

since $g(x) \geqq l_{r-1}$ in e_r and therefore $f(x) \geqq l_{r-1}$.

Thus
$$\int_E f(x)dx \geqq \text{any sum } s \text{ for } g(x).$$

It follows that
$$\int_E f(x)dx \geqq \int_E g(x)dx.$$

VI. *Let $f(x)$ and $g(x)$ be measurable in E. Then*
$$\int_E (f(x) + g(x))dx = \int_E f(x)dx + \int_E g(x)dx.$$

(i) Let $g(x) = C$ in E, and let A, B be the lower and upper bounds of $f(x)$ as in § 13.

Let $A = l_0, l_1, \ldots l_{n-1}, l_n = B$ be a mode of division of (A, B) with sums S and s for $f(x)$.

Then $f(x) + C$ has $A + C$, $B + C$ for its bounds and $l_0 + C$, $l_1 + C$, $\ldots$ $l_n + C$ is a mode of division of the interval.

If S' and s' are the sums for $f(x) + C$ for this mode of division, we have
$$S' = S + Cm(E).$$

On proceeding to the limit, this gives
$$\int_E (f(x) + C)dx = \int_E f(x)dx + Cm(E) = \int_E f(x)dx + \int_E C\,dx.$$

(ii) With the same notation as in (i) for $f(x)$, let

$$A=l_0, \quad l_1 \dots l_{n-1}, \quad l_n=B$$

be a mode of division of (A, B), and let e_r be the set of points of E for which $l_{r-1} \leqq f(x) < l_r$ when $r=1, 2, \dots (n-1)$, and e_n the set for which $l_{n-1} \leqq f(x) \leqq l_n$.

Then

$$\begin{aligned}\int_E (f(x)+g(x))dx &= \sum_1^n \int_{e_r} (f(x)+g(x))dx \\ &\geqq \sum_1^n \int_{e_r} (l_{r-1}+g(x))dx \\ &\geqq \sum_1^n l_{r-1} m(e_r) + \sum_1^n \int_{e_r} g(x)dx \\ &\geqq s + \sum_1^n \int_{e_r} g(x)dx,\end{aligned}$$

where S, s are the sums for $f(x)$ for this mode of division.

It follows that

$$\int_E (f(x)+g(x))dx \geqq \int_E f(x)dx + \int_E g(x)dx.$$

Again

$$\int_E (f(x)+g(x))dx \leqq \sum_1^n \int_{e_r} (l_r+g(x))dx,$$

and from this we see that

$$\int_E (f(x)+g(x))dx \leqq \int_E f(x)dx + \int_E g(x)dx.$$

Thus we have

$$\int_E (f(x)+g(x))dx = \int_E f(x)dx + \int_E g(x)dx.$$

It is clear that we also have

$$\int_E (f(x)-g(x))dx = \int_E f(x)dx - \int_E g(x)dx.$$

VII. *Let $f(x)$ be measurable in E. Then*

$$\left| \int_E f(x)dx \right| \leqq \int_E \left| f(x) \right| dx.$$

This follows at once from (V), since

$$-|f(x)| \leqq f(x) \leqq |f(x)|.$$

VIII. *Let $f(x)$ be measurable in E and $g(x)$ be bounded and equal to $f(x)$ at all points of E other than points of a component E_1 of E whose measure is zero.*

Then

$$\int_E f(x)dx = \int_E g(x)dx.$$

Since $g(x)$ is also measurable in E,

$$\begin{aligned}\int_E f(x)dx - \int_E g(x)dx &= \int_E (f(x)-g(x))dx \\ &= \int_{E_1} (f(x)-g(x))dx \\ &= 0, \text{ by (II), since } m(E_1)=0.\end{aligned}$$

15. The theorem of this section is of great importance in the application of the Lebesgue Integral.

I. *Let* $f_1(x), f_2(x), \ldots$ *be a sequence of functions which are measurable in* E *and not negative.*

Also let $f_n(x)$ *be uniformly bounded* in* E, *and* $\lim_{n\to\infty} f_n(x)=0$, *for every point* x *of* E.

Then
$$\lim_{n\to\infty}\int_E f_n(x)dx=0.$$

Take the arbitrary positive ϵ.

Let E_1 be the points of E for which $f_1(x), f_2(x) \ldots$ are all $<\dfrac{\epsilon}{2m(E)}$;

E_2 the points for which $f_1(x) \geqq \dfrac{\epsilon}{2m(E)}$, and $f_2(x), f_3(x), \ldots$ all $<\dfrac{\epsilon}{2m(E)}$;

E_3 the points for which $f_2(x) \geqq \dfrac{\epsilon}{2m(E)}$ and $f_3(x), f_4(x), \ldots$ all $<\dfrac{\epsilon}{2m(E)}$;

and so on.

Then $E_1, E_2, \ldots$ are all measurable, by § 9, III, and no two of them have common points.

Every point of $\sum_1^\infty E_r$ is a point of E: and, since $\lim_{n\to\infty} f_n(x)=0$, every point of E is a point of $\sum_1^\infty E_r$†.

Now we are given that $f_n(x)$ is uniformly bounded in E, and that it is not negative.

Therefore there is a positive number K such that $0 \leqq f_n(x) < K$, the same K serving for every x in E, and every positive integer n.

But
$$E=\sum_1^\infty E_r, \quad \text{and} \quad m(E)=\sum_1^\infty m(E_r).$$

Therefore we can choose the positive integer N, so that
$$\sum_{N+1}^\infty m(E_r)<\frac{\epsilon}{2K}.$$

But when $n \geqq N$, $f_n(x)<\dfrac{\epsilon}{2m(E)}$, for every point in $E_1, E_2, \ldots E_N$.

Thus
$$\sum_{r=1}^{N}\int_{E_r} f_n(x)dx \leqq \frac{\epsilon}{2m(E)}\sum_1^N m(E_r)$$
$$<\tfrac{1}{2}\epsilon, \quad \text{when } n \geqq N.$$

---

*See footnote § 67.2, II, p. 149.

†Let x be any point of E.

Then we know that $\lim_{n\to\infty} f_n(x)=0$, and there is a positive integer ν such that
$$0 \leqq f_n(x) < \frac{\epsilon}{2m(E)}, \quad \text{when } n \geqq \nu.$$

The smallest integer ν which will satisfy this inequality for the point in question is supposed taken.

Then this point x is a point of E_ν.

And
$$\sum_{r=N+1}^{\infty}\int_{E_r} f_n(x)dx \leqq K \sum_{N+1}^{\infty} m(E_r).$$
$$< \tfrac{1}{2}\epsilon, \text{ for every } n.$$

But
$$\int_E f_n(x)dx = \sum_{r=1}^{\infty}\int_{E_r} f_n(x)dx.$$

Therefore
$$0 \leqq \int_E f_n(x)dx < \tfrac{1}{2}\epsilon + \tfrac{1}{2}\epsilon, \text{ when } n \geqq N.$$

Thus
$$\lim_{n\to\infty}\int_E f_n(x)dx = 0.$$

We can now prove the following more general theorem :

II. *Let* $f_1(x), f_2(x), \ldots$ *be a sequence of functions measurable in* E.

Also let $f_n(x)$ *be uniformly bounded in* E, *and* $\lim_{n\to\infty} f_n(x) = f(x)$ *for every point* x *of* E.*

Then
$$\lim_{n\to\infty}\int_E f_n(x)dx = \int_E f(x)dx.$$

Since
$$|f_n(x)| < K,$$
the same constant K serving for every x in E and every n, it follows that
$$|f(x)| \leqq K \quad\text{and}\quad |f(x) - f_n(x)| \leqq 2K.$$

Let
$$F_n(x) = |f(x) - f_n(x)|.$$

Then $F_1(x)$, $F_2(x)$, ... are uniformly bounded and measurable in E, and not negative.

Also
$$\lim_{n\to\infty} F_n(x) = 0.$$

Therefore, by (I),
$$\lim_{n\to\infty}\int_E |f(x) - f_n(x)|dx = 0.$$

It follows that
$$\lim_{n\to\infty}\int_E (f(x) - f_n(x))dx = 0.$$

Thus
$$\lim_{n\to\infty}\int_E f_n(x)dx = \int_E f(x)dx.$$

The theorem just proved makes the question of the possibility of term by term integration of an infinite series much simpler to answer when Lebesgue Integrals are used.

Let the functions $u_1(x)$, $u_2(x)$, ... be given in an interval (a, b), and the series $\sum_1^\infty u_n(x)$ converge to $f(x)$ in that interval.

If $u_1(x)$, $u_2(x)$, ... are measurable functions, we know that $f(x)$ is also measurable.

If, in addition, we are told that $s_n(x)$ is uniformly bounded, this theorem establishes that
$$\int_a^b f(x)dx = \lim_{n\to\infty}\int_a^b s_n(x)dx,$$
and term by term integration is possible.

* We know (§ 12, VI) that $f(x)$ is measurable and bounded, and therefore integrable in E.

In the case of the Riemann Integral, we have to add the condition that the sum of the series be integrable.*

16. *If $f(x)$ is bounded and integrable in (a, b) according to Riemann's definition of the integral, then its Lebesgue Integral also exists and it is the same as its Riemann Integral.*

Let E be the set of points in $a \leqq x \leqq b$ for which $f(x) > A$, any constant.

Then if x is one of these points and $f(x)$ is continuous there, x is an interior point of E, with the usual convention as to the ends $x=a$ and $x=b$.

And if $f(x)$ is discontinuous at this point, it is a point of a set of zero measure. (Cf. § 10, above.)

Thus E consists of an open set and a set of zero measure.

Therefore E is measurable, as it is the sum of two measurable sets.

Hence if $f(x)$ is integrable according to Riemann's definition, it is measurable in (a, b) and its Lebesgue Integral exists, since $f(x)$ is also bounded.

Now let $a=x_0, x_1, x_2, \dots x_{n-1}, x_n=b$ be a mode of division of (a, b).

The sum s, with the notation of Riemann's Integral, is given by

$$s=m_1(x_1-x_0)+m_2(x_2-x_1)+\dots+m_n(x_n-x_{n-1}).$$

Also $f(x) \geqq m_r$ in (x_{r-1}, x_r).

Therefore $$(L)\int_{x_{r-1}}^{x_r} f(x)dx \geqq m_r(x_r-x_{r-1}),$$

where L denotes that this is the Lebesgue Integral for this interval.

Hence $$s \leqq (L)\int_a^b f(x)dx.$$

Similarly $$S \geqq (L)\int_a^b f(x)dx.$$

Since the sums S and s tend to their common value, the Riemann Integral $\int_a^b f(x)dx$, it follows that for a bounded function integrable (R) the two integrals are the same.

It is easy to give examples of bounded functions integrable (L) and not integrable (R).

Let $f(x)=1$ for every irrational value of x in $0 \leqq x \leqq 1$ and $f(x)=0$ for every rational value.

Then its Lebesgue Integral $\int_0^1 f(x)dx=1$. (Cf. § 14, VIII.)

But this function is not integrable (R).

17. The Lebesgue Integral for an Unbounded Function. In §§ 13-16, $f(x)$ has been supposed bounded in the bounded and measurable set E. The definition of the Lebesgue Integral is now modified, so that it will include a class of unbounded functions.

Take first the case *when* $f(x) \geqq 0$ *in* E, and define an auxiliary function $f_n(x)$ as follows:

$$f_n(x)=f(x), \text{ at all points of } E, \text{ where } f(x) \leqq n$$
$$=n, \text{ at all points of } E, \text{ where } f(x) > n.$$

*Cf. footnote on p. 161.

The number n is any assigned positive number.

Then $f_n(x)$, for every n, is bounded and measurable in E.

Also $\int_E f_n(x)dx$ exists, and is a monotonic increasing function of n.

Thus $\int_E f_n(x)dx$ either converges to a definite limit, or it tends to $+\infty$ as $n \to \infty$.

When the limit exists and is finite, the integral $\int_E f(x)dx$ is defined by this limit.

When $f(x) \leqq 0$ in E, then the integral $\int_E f(x)dx$ is defined as $-\int_E |f(x)|dx$, when this integral exists and is finite.

Again when $f(x)$ does not keep the same sign in E, write

$$2f_1(x) = |f(x)| + f(x),$$
$$2f_2(x) = |f(x)| - f(x),$$

so that $f_1(x) = f(x)$, at all points of E where $f(x) \geqq 0$, and it vanishes at all other points of E.

Similarly $f_2(x) = -f(x)$, at all points of E where $f(x) \leqq 0$, and it vanishes at all other points of E.

When $\int_E f_1(x)dx$ and $\int_E f_2(x)dx$ exist and are finite, the integral $\int_E f(x)dx$ is defined by the equation

$$\int_E f(x)dx = \int_E f_1(x)dx - \int_E f_2(x)dx.$$

Also *when a measurable function $f(x)$ is such that $\int_E f(x)dx$ exists as a finite number, $f(x)$ is said to be* **summable** *in E.*

A measurable function is always summable in E, if it is bounded, but not necessarily so, if it is unbounded.

When $f(x)$ is summable in E, it is also said to be integrable (L) in E, and the integral $\int_E f(x)dx$ is called the Lebesgue Integral of $f(x)$ in E.

18. Properties of the Lebesgue Integral $\int_E f(x)dx$, when $f(x)$ is not bounded.

I. It is obvious that for a summable function

$$\int_E Cf(x)dx = C\int_E f(x)dx,$$

and that if $f(x) \geqq g(x) \geqq 0$ in E and $f(x)$ is summable, then $g(x)$ is also summable and

$$\int_E f(x)dx \geqq \int_E g(x)dx.$$

II. *Let $f(x)$ be summable in E and let $E = \sum_1^\infty E_r$, where E_1, E_2, ... are measurable and without common points.*

Then
$$\int_E f(x)dx = \sum_1^\infty \int_{E_r} f(x)dx.$$

It is only necessary to prove this for the case when $f(x) \geqq 0$ in E.

Define $f_n(x)$ as in § 17, and let the number of sets E_1, E_2, ... be finite, say k.

Then
$$\int_E f_n(x)dx = \sum_{r=1}^{k} \int_{E_r} f_n(x)dx.$$

Proceeding to the limit, we have
$$\int_E f(x)dx = \sum_{r=1}^{k} \int_{E_r} f(x)dx.$$

Again, when the number of sets E_1, E_2, ... is infinite, we know from § 14, IV that
$$\int_E f_n(x)dx = \sum_{r=1}^{\infty} \int_{E_r} f_n(x)dx \geqq \sum_{r=1}^{s} \int_{E_r} f_n(x)dx.$$

Thus, letting $n \to \infty$, we have
$$\int_E f(x)dx \geqq \sum_{r=1}^{s} \int_{E_r} f(x)dx.$$

And letting $s \to \infty$, we have
$$\int_E f(x)dx \geqq \sum_{1}^{\infty} \int_{E_r} f(x)dx.$$

But $\int_E f_n(x)dx \leqq \sum_{r=1}^{\infty} \int_{E_r} f(x)dx$, since $\int_{E_r} f_n(x)dx \leqq \int_{E_r} f(x)dx.$

Thus
$$\int_E f(x)dx \leqq \sum_{1}^{\infty} \int_{E_r} f(x)dx.$$

Therefore
$$\int_E f(x)dx = \sum_{1}^{\infty} \int_{E_r} f(x)dx.$$

III. *Let $f(x)$ and $g(x)$ be summable in E.*

Then
$$\int_E (f(x) \pm g(x))dx = \int_E f(x)dx \pm \int_E g(x)dx.$$

It will be sufficient to take the sum of the two functions.

Let $F(x) = f(x) + g(x)$ and define $F_n(x)$, $f_n(x)$, $g_n(x)$ as before:

e.g. $F_n(x) = F(x)$ at all points of E where $F(x) \leqq n$,

$= n$, at all points of E where $F(x) > n$.

(i) But, when $f(x)$ and $g(x)$ are both not-negative in E, it is easy to verify that
$$F_{2n}(x) \geqq f_n(x) + g_n(x) \geqq F_n(x).$$

Also $f_n(x)$, $g_n(x)$, $F_n(x)$ and $F_{2n}(x)$ are bounded and measurable.

Therefore
$$\int_E F_{2n}(x)dx \geqq \int_E f_n(x)dx + \int_E g_n(x)dx \geqq \int_E F_n(x)dx.$$

Letting $n \to \infty$, we have
$$\int_E F(x)dx = \int_E f(x)dx + \int_E g(x)dx.$$

(ii) And the same result holds for the case when $f(x)$ and $g(x)$ are both zero or negative in E.

(iii) Now let $f(x) \geqq 0$, $g(x) \leqq 0$, and $F(x) = f(x) + g(x) \geqq 0$ in E.

Then
$$F(x) + |g(x)| = f(x),$$

and
$$\int_E f(x)dx = \int_E F(x)dx + \int_E |g(x)|dx = \int_E F(x)dx - \int_E g(x)dx.$$

(iv) But when $f(x)$ and $g(x)$ do not each keep the same sign in E, we can break up E into a certain number of measurable sets without common points two by two, to which we can apply the results just found.

For example, with the notation of § 11, the points which belong to $E[F(x)\geqq 0]$ will come from the following products:

$$\begin{array}{lll} E[F(x)\geqq 0]. & E[f(x)>0]. & E[g(x)>0], \\ E[F(x)\geqq 0]. & E[f(x)>0]. & E[g(x)<0], \\ E[F(x)\geqq 0]. & E[f(x)<0]. & E[g(x)>0], \\ E[F(x)\geqq 0]. & E[f(x)=0]. & E[g(x)>0], \\ E[F(x)\geqq 0]. & E[f(x)=0]. & E[g(x)=0], \\ E[F(x)\geqq 0]. & E[f(x)>0]. & E[g(x)=0], \end{array}$$

and similarly for the points which belong to $E[F(x)<0]$.

Denote these sets by $E_1, E_2 \dots E_k$.

Thus
$$\int_E F(x)dx = \sum_{r=1}^{k}\int_{E_r} F(x)dx,$$
$$= \sum_1^k\left(\int_{E_r} f(x)dx + \int_{E_r} g(x)dx\right)$$
$$= \int_E f(x)dx + \int_E g(x)dx.$$

IV. With the notation of § 17, we have
$$|f(x)| = f_1(x) + f_2(x).$$

It follows from (III) *that, if $f(x)$ is summable, $|f(x)|$ is also summable, and*
$$\int_E |f(x)|dx = \int_E f_1(x)dx + \int_E f_2(x)dx.$$

Thus the Lebesgue Integral of an unbounded function is an absolutely convergent integral.

The theorem of § 15 applies also to the case of unbounded functions, with some alteration in the conditions as there given. Term by term integration is permissible in this case also, but for the discussion of this question and a fuller treatment of the Lebesgue Integral reference must be made to other works.*

19. Fourier's Series, using Lebesgue Integrals. Before discussing the convergence of Fourier's Series for $f(x)$, when the coefficients are Lebesgue Integrals, we must prove the Riemann-Lebesgue Theorem (Cf. § 105, p. 271) for the case when $f(x)$ is summable in $(-\pi, \pi)$, and in doing so we require the following approximation theorem :—

If $f(x)$ is summable in the interval (a, b) and ϵ is an arbitrary positive number, there is a continuous function $\phi(x)$, such that
$$\int_a^b |f(x) - \phi(x)| < \epsilon.$$

*Cf., for example, Hobson, *Theory of Functions of a Real Variable*, **1** (3rd ed., 1927), Ch. VII, and **2** (2nd ed. 1926), Ch. V.

This important approximation theorem is obtained by proceeding from simple measurable functions to the general summable function.*

(i) Let $\quad f(x)=1$ in the interval (α, β), where $a<\alpha<\beta<b$

$\qquad\qquad =0$, in the rest of (a, b).

Take $\alpha-\epsilon'$ and $\beta+\epsilon'$ on the interval (a, b) between $x=a$, $x=\alpha$, and $x=\beta$, $x=b$ respectively, where $0<\epsilon'<\epsilon$, and join these points to the points $x=\alpha$, $y=1$, and $x=\beta$, $y=1$ respectively.

Then if $\phi(x)=0$ in $a<x<\alpha-\epsilon'$ and $\beta+\epsilon'<x<b$, and $\phi(x)$ is equal to the ordinate of this broken line in $\alpha-\epsilon'\leqq x<\beta+\epsilon'$, we have a continuous function satisfying

$$\int_a^b |f(x)-\phi(x)|\,dx<\epsilon.$$

(ii) Let $f(x)=1$ in the finite set of not-overlapping intervals $\Delta_1, \Delta_2, \dots \Delta_n$, all in (a, b), and $f(x)=0$ elsewhere in (a, b).

Also let $\phi_r(x)$ be the continuous function obtained in (i) such that, when $f_r(x)=1$ in Δ_r and zero elsewhere in (a, b), we have

$$\int_a^b |f_r(x)-\phi_r(x)|\,dx<\frac{\epsilon}{n}.$$

Take
$$\phi(x)=\sum_1^n \phi_r(x).$$

Then we have

$$\int_a^b |f(x)-\phi(x)|\,dx\leqq\sum_1^n\int_a^b |f_r(x)-\phi_r(x)|dx<\epsilon.$$

(iii) Let $f(x)=1$ in a measurable set E in (a, b) and zero elsewhere in (a, b).

Then a set of not-overlapping intervals $\Delta_1, \Delta_2, \dots$, all in (a, b), exists such that the points of E are interior points of these intervals or end-points of two adjacent intervals, and

$$m(E)\leqq\sum_1^\infty m(\Delta_r)<m(E)+\tfrac{1}{4}\epsilon.$$

Also there is a positive integer N such that

$$m(E)\leqq\sum_1^N m(\Delta_r)<m(E)+\tfrac{1}{4}\epsilon,$$

and
$$\sum_{N+1}^\infty m(\Delta_r)<\tfrac{1}{4}\epsilon.$$

Let $f_N(x)=1$ in $\Delta_1, \Delta_2, \dots \Delta_N$ and zero elsewhere in (a, b).

Then by (ii) we can find a continuous function $\phi(x)$ such that

$$\int_a^b |f_N(x)-\phi(x)|<\tfrac{1}{4}\epsilon.$$

Let E_r be the points of E in $\Delta_r (r=1, 2, \dots)$.

Then
$$E=\sum_1^\infty E_r.$$

*Cf. Hobson, *loc. cit.*, 1 (3rd ed., 1927), 632.

Also $\displaystyle\int_a^b |f(x)-f_N(x)|\,dx=\sum_1^\infty\int_{\Delta_r}|f(x)-f_N(x)|\,dx$

$$=\sum_1^N\int_{\Delta_r}|f(x)-f_N(x)|\,dx+\sum_{N+1}^\infty\int_{\Delta_r}|f(x)-f_N(x)|\,dx.$$

But $\displaystyle\int_{\Delta_r}|f(x)-f_N(x)|\,dx=m(\Delta_r)-m(E_r)$, when $r=1, 2, \dots N$.

And $\displaystyle\sum_{N+1}^\infty\int_{\Delta_r}|f(x)-f_N(x)|\,dx\leqq\sum_{N+1}^\infty m(\Delta_r)$, since $f_N(x)=0$ in $\displaystyle\sum_{N+1}^\infty\Delta_r$.

It follows that

$$\int_a^b |f(x)-f_N(x)|\,dx\leqq\left[\sum_1^N m(\Delta_r)-m(E)\right]+\left[m(E)-\sum_1^N m(E_r)\right]+\sum_{N+1}^\infty m(\Delta_r)$$

$$<\tfrac{1}{4}\epsilon+\tfrac{1}{4}\epsilon+\tfrac{1}{4}\epsilon,$$

since $\displaystyle m(E)-\sum_1^N m(E_r)=\sum_{N+1}^\infty m(E_r)\leqq\sum_{N+1}^\infty m(\Delta_r)<\tfrac{1}{4}\epsilon.$

Thus we have

$$\int_a^b |f(x)-\phi(x)|\,dx<\epsilon.$$

(iv) Let $E_1, E_2, \dots E_n$ be measurable sets in (a, b), no two of them having common points.

Let $f_r(x)=1$ in E_r, and zero elsewhere in (a, b), and $f(x)=\sum_1^n c_r f_r(x)$, where $c_1, c_2, \dots c_n$ are constants.

We find a continuous function $\phi_r(x)$ as in (iii), such that

$$\int_a^b |f_r(x)-\phi_r(x)|\,dx<\frac{\epsilon}{|c_1|+|c_2|\dots+|c_n|}.$$

Take $$\phi(x)=\sum_1^n c_r\phi_r(x).$$

Then $$\int_a^b |f(x)-\phi(x)|\,dx\leqq\sum_1^n|c_r|\int_a^b|f_r(x)-\phi_r(x)|\,dx<\epsilon.$$

(v) Now let $f(x)$ be bounded and measurable in (a, b).

With the notation of § 13, A, B are its lower and upper bounds in (a, b), and a mode of division

$$A=l_0, l_1\dots, l_{n-1}, l_n=B$$

is taken, the largest of its partial intervals being η.

Also e_r is defined as in that section.

A function $F(x)$ is defined for $a\leqq x\leqq b$ as being equal to l_{r-1} in e_r

$$(r=1, 2, \dots n).$$

A continuous function $\phi(x)$ is obtained for $F(x)$ as in (iv), such that

$$\int_a^b |F(x)-\phi(x)|\,dx<\tfrac{1}{2}\epsilon.$$

Then $\int_a^b |f(x)-\phi(x)|dx \leqq \int_a^b |f(x)-F(x)|dx + \int_a^b |F(x)-\phi(x)|dx$

$$\leqq \sum_{r=1}^{\nu} \int_{c_r} |f(x)-F(x)|dx + \int_a^b |F(x)-\phi(x)|dx$$

$$< \eta(b-a) + \tfrac{1}{2}\epsilon$$

$$< \epsilon, \quad \text{when } \eta = \frac{\epsilon}{2(b-a)}.$$

(vi) Let $f(x)$ be summable in (a, b) and unbounded.

With the notation of § 17, we have

$$\int_a^b f(x)dx = \int_a^b f_1(x)dx - \int_a^b f_2(x)dx.$$

Let
$$f_{1\cdot n}(x) = f_1(x), \quad \text{when } f_1(x) \leqq n$$
$$= n, \quad \text{when } f_1(x) > n,$$

and similarly for $f_{2\cdot n}(x)$.

Choose N, so that

$$\int_a^b f_1(x)dx - \int_a^b f_{1,N}(x)dx < \tfrac{1}{4}\epsilon, \quad \text{and} \quad \int_a^b f_2(x) - \int_a^b f_{2,N}(x)dx < \tfrac{1}{4}\epsilon.$$

Then obtain continuous functions $\phi_1(x)$ for $f_{1,N}(x)$ and $\phi_2(x)$ for $f_{2,N}(x)$, such that

$$\int_a^b |f_{1,N}(x) - \phi_1(x)|dx < \tfrac{1}{4}\epsilon, \quad \text{and} \quad \int_a^b |f_{2,N}(x) - \phi_2(x)|dx < \tfrac{1}{4}\epsilon,$$

and let
$$\phi(x) = \phi_1(x) - \phi_2(x).$$

But $\int_a^b |f(x)-\phi(x)|dx \leqq \int_a^b |f_1(x) - f_{1,N}(x)|dx + \int_a^b |f_{1,N}(x) - \phi_1(x)|dx$

$$+ \int_a^b |f_2(x) - f_{2,N}(x)|dx + \int_a^b |f_{2,N}(x) - \phi_2(x)|dx.$$

Therefore
$$\int_a^b |f(x)-\phi(x)|dx < \epsilon.$$

20. The Riemann-Lebesgue Theorem.

If $f(x)$ is summable in $(-\pi, \pi)$ then $\lim_{n\to\infty} \int_{-\pi}^{\pi} f(x) \begin{matrix}\sin\\ \cos\end{matrix} nx\, dx = 0.$

Defining $f(x)$ outside $(-\pi, \pi)$ by the equation $f(x \pm 2\pi) = f(x)$, we have

$$\int_{-\pi}^{\pi} f(x) \sin nx\, dx = \int_{-\pi+\frac{\pi}{n}}^{\pi+\frac{\pi}{n}} f(x) \sin nx\, dx$$

$$= -\int_{-\pi}^{\pi} f\left(x + \frac{\pi}{n}\right) \sin nx\, dx.$$

Thus
$$2\int_{-\pi}^{\pi} f(x) \sin nx\, dx = \int_{-\pi}^{\pi} \left(f(x) - f\left(x+\frac{\pi}{n}\right)\right) \sin nx\, dx$$

and
$$2\left|\int_{-\pi}^{\pi} f(x) \sin nx\, dx\right| \leqq \int_{-\pi}^{\pi} \left|f\left(x+\frac{\pi}{n}\right) - f(x)\right| dx.$$

Let (α, β) be an interval enclosing $(-\pi, \pi)$.

Then there is a continuous function $\phi(x)$, such that

$$\int_\alpha^\beta |f(x)-\phi(x)|dx<\tfrac{1}{3}\epsilon,$$

when ϵ is the usual arbitrary positive number.

Since $\phi(x)$ is continuous in (α, β), given ϵ there is an η such that

$$|\phi(x+h)-\phi(x)|<\frac{\epsilon}{6\pi},$$

when $|h|\leqq\eta$, the same η serving for every x in (α, β).

Take ν so large that $\dfrac{\pi}{\nu}<\eta$ and $\left(\pi+\dfrac{\pi}{\nu}\right)<\beta$.

Then
$$\int_{-\pi}^{\pi}\left|f\left(x+\frac{\pi}{n}\right)-f(x)\right|dx\leqq\int_{-\pi}^{\pi}\left|f\left(x+\frac{\pi}{n}\right)-\phi\left(x+\frac{\pi}{n}\right)\right|dx$$
$$+\int_{-\pi}^{\pi}\left|\phi\left(x+\frac{\pi}{n}\right)-\phi(x)\right|dx+\int_{-\pi}^{\pi}|\phi(x)-f(x)|\,dx$$
$$<\tfrac{1}{3}\epsilon+\tfrac{1}{3}\epsilon+\tfrac{1}{3}\epsilon$$
$$<\epsilon,\ \text{when } n\geqq\nu.$$

Therefore
$$\lim_{n\to\infty}\int_{-\pi}^{\pi}f(x)\sin nx\,dx=0,$$

and the proof applies equally to $\displaystyle\int_{-\pi}^{\pi}f(x)\cos nx\,dx$.

Thus the Fourier Constants tend to zero as $n\to\infty$, when $f(x)$ is summable in $(-\pi, \pi)$.

Corollary. *If $f(x)$ is summable in any interval (a, b), then*

$$\lim_{n\to\infty}\int_a^b f(x)\,{\sin \atop \cos}\,nx\,dx=0.$$

If (a, b) lies in $(-\pi, \pi)$, this result follows at once from the theorem just proved, for we may put $f(x)=0$ in the remainder of the interval $(-\pi, \pi)$.

If (a, b) extends beyond $(-\pi, \pi)$, we apply, as above, the theorem to the intervals $((m-1)\pi, (m+1)\pi)$ in which it lies, m being a positive integer.

21. We now apply these theorems to the discussion of the Fourier's Series corresponding to the arbitrary function $f(x)$. We replace the conditions attached to $f(x)$ in § 105 by the condition that $f(x)$ shall be summable (cf. § 17 above) in $(-\pi, \pi)$.

As before, let $s_n(x)$ be the sum of the Fourier's Series for $f(x)$ up to the terms in $\sin nx$ and $\cos nx$.

Then, if x_0 is a point in the interval $(-\pi, \pi)$, we have

$$s_n(x_0)=\frac{1}{\pi}\int_0^{\frac{1}{2}\pi}[f(x_0+2\alpha)+f(x_0-2\alpha)]\frac{\sin(2n+1)\alpha}{\sin\alpha}d\alpha,$$
$$=\frac{1}{\pi}\left(\int_0^\eta+\int_\eta^{\frac{1}{2}\pi}\right)[f(x_0+2\alpha)+f(x_0-2\alpha)]\frac{\sin(2n+1)\alpha}{\sin\alpha}d\alpha.$$

The second integral vanishes, by the Riemann-Lebesgue Theorem, and it follows that:

I. *The behaviour of the Fourier's Series corresponding to $f(x)$, as to conver-*

gence, divergence, or oscillation at a point x_0, depends only on the values of $f(x)$ in the neighbourhood of x_0.

And similarly:

II. *The behaviour of the Fourier's Series corresponding to $f(x)$ in an interval (a, b), where $-\pi < a < b < \pi$, as to convergence, divergence, or oscillation, depends only on the values of $f(x)$ in $(a-\delta, b+\delta)$, where δ is an arbitrarily small positive number.*

Now let x_0 be a point in $(-\pi, \pi)$ for which $\frac{1}{2}\lim_{h\to 0}[f(x_0+h)+f(x_0-h)]$ exists.

We may give $f(x)$ at the point x_0 the value of this limit.

Then
$$s_n(x_0)-f(x_0)=\frac{1}{\pi}\int_0^{\frac{1}{2}\pi}\phi(a)\frac{\sin(2n+1)a}{\sin a}\,da,$$
where
$$\phi(a)=f(x_0+2a)+f(x_0-2a)-2f(x_0).$$

Therefore
$$s_n(x_0)-f(x_0)=\frac{1}{\pi}\int_0^{\frac{1}{2}\pi}\phi(a)\frac{\sin(2n+1)a}{a}\,da$$
$$+\frac{1}{\pi}\int_0^{\frac{1}{2}\pi}\phi(a)\left(\frac{1}{\sin a}-\frac{1}{a}\right)\sin(2n+1)a\,da.$$

The second integral vanishes when $n\to\infty$, by the Riemann-Lebesgue Theorem, as $\phi(a)\left(\frac{1}{\sin a}-\frac{1}{a}\right)$ is summable in $(0, \frac{1}{2}\pi)$.

And the first integral also vanishes when $n\to\infty$, provided that $\phi(a)/a$ is summable in $(0, \frac{1}{2}\pi)$.

Thus we again have Dini's Condition that:

III. *A sufficient condition for the convergence of the Fourier's Series corresponding to the function $f(x)$, summable in $(-\pi, \pi)$, to $f(x_0)$ at a point x_0 in $(-\pi, \pi)$, where $\frac{1}{2}\lim_{h\to 0}[f(x_0+h)+f(x_0-h)]$ exists and is equal to $f(x_0)$ is that*

$\left\{\frac{f(x_0+2a)+f(x_0-2a)-2f(x_0)}{a}\right\}$ *is summable in some interval $(0, \eta)$.*

This condition is satisfied *when $f(x)$ is summable in $(-\pi, \pi)$ and at the point x_0 satisfies Lipschitz's Condition; namely that positive numbers C and k exist such that*
$$|f(x_0+t)-f(x_0)|<C|t|^k,$$
when $|t|\leqq$ some fixed positive number.

We can also show that:

IV. *When $f(x)$ is summable in $(-\pi, \pi)$, and is of bounded variation in the neighbourhood of a point x_0, the Fourier's Series converges there to*
$$\tfrac{1}{2}[f(x_0+0)+f(x_0-0)].$$

Let $f(x)$ be of bounded variation in the interval $(x_0-2\eta, x_0+2\eta)$.

Then
$$\int_0^{\frac{1}{2}\pi}\phi(a)\frac{\sin(2n+1)a}{a}\,da=\int_0^{\eta}\phi(a)\frac{\sin(2n+1)a}{a}\,da+\int_{\eta}^{\frac{1}{2}\pi}\phi(a)\frac{\sin(2n+1)a}{a}\,da.$$

To the second integral we can apply the Riemann-Lebesgue Theorem; and in the first $\phi(a)$ is of bounded variation; so the Lebesgue Integral and the Riemann integral are the same. (Cf. § 16 above.)

But the Riemann Integral has zero for its limit (cf. § 92), and the required result follows.

22. To show the full bearing of the Lebesgue Integral on the Theory of Fourier's Series it would be necessary to discuss at much greater length the properties of that integral.

We have only touched upon these properties in the case of the integrals $\int_E f(x)dx$ and $\int_a^b f(x)dx$, where $f(x)$ is summable in the bounded and measurable set E, or in (a, b).

The properties of the Indefinite Lebesgue Integral $\int_a^x f(x)dx$ have not been dealt with at all. Some of them may be mentioned here without proof. In several of the works named in the list of books at the end of this section a full discussion of the Lebesgue Integral (and other associated integrals) will be found.

This integral $F(x)=\int_a^x f(x)dx$ is continuous and of bounded variation in the interval (a, b), when $f(x)$ is summable, whether bounded or not, in (a, b).

Also $F'(x)$ exists and is equal to $f(x)$ almost everywhere in (a, b), and certainly at all points of continuity of $f(x)$.

And, further, if $f(x)$ is a function which has at every point of (a, b) a differential coefficient $f'(x)$, bounded in that interval, then $f'(x)$ is integrable (L) in the interval (a, x) and its integral differs from $f(x)$ by a constant only.

In the case of the Riemann Integral this last theorem is subject also to the condition that $f'(x)$ shall be integrable (R).

REFERENCES.

The literature dealing with the Theory of Sets of Points and the Lebesgue Integral and other associated integrals is extensive.

Article II C 9—entitled "Neuere Untersuchungen über Funktionen reeller Veränderlichen"—in the *Enc. d. math. Wiss.*, Bd. II, Tl. III, 2, will be found useful. It is divided into three parts:

II C 9 *a*, "Die Punktmengen," by Zoretti and Rosenthal;

II C 9 *b*, "Integration und Differentiation," by Montel and Rosenthal; and II C 9 *c*, "Funktionenfolgen," by Fréchet and Rosenthal.

And the article II C 10 in the same volume—"Neuere Untersuchungen über trigonometrische Reihen"—by Hilb and Riesz covers the work in trigonometrical series up to about 1922.

In Pascal's *Repertorium der höheren Mathematik* (2 Aufl.) Bd. I, Tl. III, there are useful chapters: "Neuere Theorie der reellen Funktionen" (XX), by Kamke, and "Trigonometrische Reihen" (XXV), by Plessner. Both bring their survey up to about the date of publication (1929).

The following books are either devoted wholly to the subject, or contain useful chapters bearing upon it:

Borel, *Leçons sur la théorie des fonctions* (3e éd., Paris, 1928); *Leçons sur les fonctions de variables réelles et les développements en séries de polynomes* (2e éd., Paris, 1928).

De la Vallée Poussin, *Cours d'Analyse*, 1 (3e éd., Paris, 1914); *Intégrales de Lebesgue* (Paris, 1916).

Carathéodory, *Vorlesungen über reelle Funktionen* (2 Aufl., Leipzig, 1927).

Hahn, *Theorie der reellen Funktionen*, 1 (Berlin, 1921).

Hausdorff, *Mengenlehre* (2 Aufl., Leipzig, 1927).

Hobson, *Theory of Functions of a Real Variable*, 1 (3rd ed., 1927), and 2 (2nd ed., 1926).

Kamke, *Das Lebesguesche Integral* (Leipzig, 1925).

Lebesgue, *Leçons sur l'intégration et la recherche des fonctions primitives* (2e éd., Paris, 1928); *Leçons sur les séries trigonométriques* (Paris, 1906).

Pierpont, *Theory of Functions of Real Variables*, 1 (1905) and 2 (1912).

Scheslinger u. Plessner, *Lebesguesche Integrale und Fouriersche Reihen* (Leipzig, 1926).

Schönflies, *Die Entwickelung der Lehre der Punktmannigfaltigkeiten*, Tl. I (*Jahresber. D. Math. Ver., Leipzig*, 8 (1900)); Tl. II (*ibid.* Ergänzungsband, 1908).

Schönflies u. Hahn, *Entwicklung der Mengenlehre und ihre Anwendungen, Erste Hälfte: Allgemeine Theorie der unendlichen Mengen und Theorie der Punktmengen* (Leipzig, 1913).

Tonelli, *Serie trigonometriche* (Bologna, 1928).

Young, W. H. and G. C., *The Theory of Sets of Points* (1908).

Young, L. C., *The Theory of Integration* (1927), being No. 21 in the series of Cambridge Tracts in Mathematics and Mathematical Physics.

INDEX OF PROPER NAMES

The numbers refer to pages.

GENERAL INDEX

The numbers refer to pages.

PRINTED IN GREAT BRITAIN BY ROBERT MACLEHOSE AND CO. LTD. THE UNIVERSITY PRESS, GLASGOW.

A CATALOGUE OF SELECTED DOVER BOOKS IN ALL FIELDS OF INTEREST

A CATALOGUE OF SELECTED DOVER BOOKS IN ALL FIELDS OF INTEREST

DESIGN BY ACCIDENT; A BOOK OF "ACCIDENTAL EFFECTS" FOR ARTISTS AND DESIGNERS, James F. O'Brien. Create your own unique, striking, imaginative effects by "controlled accident" interaction of materials: paints and lacquers, oil and water based paints, splatter, crackling materials, shatter, similar items. Everything you do will be different; first book on this limitless art, so useful to both fine artist and commercial artist. Full instructions. 192 plates showing "accidents," 8 in color. viii + 215pp. 8⅜ x 11¼. 21942-9 Paperbound $3.50

THE BOOK OF SIGNS, Rudolf Koch. Famed German type designer draws 493 beautiful symbols: religious, mystical, alchemical, imperial, property marks, runes, etc. Remarkable fusion of traditional and modern. Good for suggestions of timelessness, smartness, modernity. Text. vi + 104pp. 6⅛ x 9¼. 20162-7 Paperbound $1.25

HISTORY OF INDIAN AND INDONESIAN ART, Ananda K. Coomaraswamy. An unabridged republication of one of the finest books by a great scholar in Eastern art. Rich in descriptive material, history, social backgrounds; Sunga reliefs, Rajput paintings, Gupta temples, Burmese frescoes, textiles, jewelry, sculpture, etc. 400 photos. viii + 423pp. 6⅜ x 9¾. 21436-2 Paperbound $5.00

PRIMITIVE ART, Franz Boas. America's foremost anthropologist surveys textiles, ceramics, woodcarving, basketry, metalwork, etc.; patterns, technology, creation of symbols, style origins. All areas of world, but very full on Northwest Coast Indians. More than 350 illustrations of baskets, boxes, totem poles, weapons, etc. 378 pp. 20025-6 Paperbound $3.00

THE GENTLEMAN AND CABINET MAKER'S DIRECTOR, Thomas Chippendale. Full reprint (third edition, 1762) of most influential furniture book of all time, by master cabinetmaker. 200 plates, illustrating chairs, sofas, mirrors, tables, cabinets, plus 24 photographs of surviving pieces. Biographical introduction by N. Bienenstock. vi + 249pp. 9⅞ x 12¾. 21601-2 Paperbound $4.00

AMERICAN ANTIQUE FURNITURE, Edgar G. Miller, Jr. The basic coverage of all American furniture before 1840. Individual chapters cover type of furniture—clocks, tables, sideboards, etc.—chronologically, with inexhaustible wealth of data. More than 2100 photographs, all identified, commented on. Essential to all early American collectors. Introduction by H. E. Keyes. vi + 1106pp. 7⅞ x 10¾. 21599-7, 21600-4 Two volumes, Paperbound $11.00

PENNSYLVANIA DUTCH AMERICAN FOLK ART, Henry J. Kauffman. 279 photos, 28 drawings of tulipware, Fraktur script, painted tinware, toys, flowered furniture, quilts, samplers, hex signs, house interiors, etc. Full descriptive text. Excellent for tourist, rewarding for designer, collector. Map. 146pp. 7⅞ x 10¾. 21205-X Paperbound $2.50

EARLY NEW ENGLAND GRAVESTONE RUBBINGS, Edmund V. Gillon, Jr. 43 photographs, 226 carefully reproduced rubbings show heavily symbolic, sometimes macabre early gravestones, up to early 19th century. Remarkable early American primitive art, occasionally strikingly beautiful; always powerful. Text. xxvi + 207pp. 8⅜ x 11¼. 21380-3 Paperbound $3.50

ALPHABETS AND ORNAMENTS, Ernst Lehner. Well-known pictorial source for decorative alphabets, script examples, cartouches, frames, decorative title pages, calligraphic initials, borders, similar material. 14th to 19th century, mostly European. Useful in almost any graphic arts designing, varied styles. 750 illustrations. 256pp. 7 x 10. 21905-4 Paperbound $4.00

PAINTING: A CREATIVE APPROACH, Norman Colquhoun. For the beginner simple guide provides an instructive approach to painting: major stumbling blocks for beginner; overcoming them, technical points; paints and pigments; oil painting; watercolor and other media and color. New section on "plastic" paints. Glossary. Formerly *Paint Your Own Pictures*. 221pp. 22000-1 Paperbound $1.75

THE ENJOYMENT AND USE OF COLOR, Walter Sargent. Explanation of the relations between colors themselves and between colors in nature and art, including hundreds of little-known facts about color values, intensities, effects of high and low illumination, complementary colors. Many practical hints for painters, references to great masters. 7 color plates, 29 illustrations. x + 274pp.
20944-X Paperbound $2.75

THE NOTEBOOKS OF LEONARDO DA VINCI, compiled and edited by Jean Paul Richter. 1566 extracts from original manuscripts reveal the full range of Leonardo's versatile genius: all his writings on painting, sculpture, architecture, anatomy, astronomy, geography, topography, physiology, mining, music, etc., in both Italian and English, with 186 plates of manuscript pages and more than 500 additional drawings. Includes studies for the Last Supper, the lost Sforza monument, and other works. Total of xlvii + 866pp. 7⅞ x 10¾.
22572-0, 22573-9 Two volumes, Paperbound $10.00

MONTGOMERY WARD CATALOGUE OF 1895. Tea gowns, yards of flannel and pillow-case lace, stereoscopes, books of gospel hymns, the New Improved Singer Sewing Machine, side saddles, milk skimmers, straight-edged razors, high-button shoes, spittoons, and on and on . . . listing some 25,000 items, practically all illustrated. Essential to the shoppers of the 1890's, it is our truest record of the spirit of the period. Unaltered reprint of Issue No. 57, Spring and Summer 1895. Introduction by Boris Emmet. Innumerable illustrations. xiii + 624pp. 8½ x 11⅝.
22377-9 Paperbound $6.95

THE CRYSTAL PALACE EXHIBITION ILLUSTRATED CATALOGUE (LONDON, 1851). One of the wonders of the modern world—the Crystal Palace Exhibition in which all the nations of the civilized world exhibited their achievements in the arts and sciences—presented in an equally important illustrated catalogue. More than 1700 items pictured with accompanying text—ceramics, textiles, cast-iron work, carpets, pianos, sleds, razors, wall-papers, billiard tables, beehives, silverware and hundreds of other artifacts—represent the focal point of Victorian culture in the Western World. Probably the largest collection of Victorian decorative art ever assembled—indispensable for antiquarians and designers. Unabridged republication of the Art-Journal Catalogue of the Great Exhibition of 1851, with all terminal essays. New introduction by John Gloag, F.S.A. xxxiv + 426pp. 9 x 12.
22503-8 Paperbound $5.00

A HISTORY OF COSTUME, Carl Köhler. Definitive history, based on surviving pieces of clothing primarily, and paintings, statues, etc. secondarily. Highly readable text, supplemented by 594 illustrations of costumes of the ancient Mediterranean peoples, Greece and Rome, the Teutonic prehistoric period; costumes of the Middle Ages, Renaissance, Baroque, 18th and 19th centuries. Clear, measured patterns are provided for many clothing articles. Approach is practical throughout. Enlarged by Emma von Sichart. 464pp. 21030-8 Paperbound $3.50

ORIENTAL RUGS, ANTIQUE AND MODERN, Walter A. Hawley. A complete and authoritative treatise on the Oriental rug—where they are made, by whom and how, designs and symbols, characteristics in detail of the six major groups, how to distinguish them and how to buy them. Detailed technical data is provided on periods, weaves, warps, wefts, textures, sides, ends and knots, although no technical background is required for an understanding. 11 color plates, 80 halftones, 4 maps. vi + 320pp. 6⅛ x 9⅛. 22366-3 Paperbound $5.00

TEN BOOKS ON ARCHITECTURE, Vitruvius. By any standards the most important book on architecture ever written. Early Roman discussion of aesthetics of building, construction methods, orders, sites, and every other aspect of architecture has inspired, instructed architecture for about 2,000 years. Stands behind Palladio, Michelangelo, Bramante, Wren, countless others. Definitive Morris H. Morgan translation. 68 illustrations. xii + 331pp. 20645-9 Paperbound $3.00

THE FOUR BOOKS OF ARCHITECTURE, Andrea Palladio. Translated into every major Western European language in the two centuries following its publication in 1570, this has been one of the most influential books in the history of architecture. Complete reprint of the 1738 Isaac Ware edition. New introduction by Adolf Placzek, Columbia Univ. 216 plates. xxii + 110pp. of text. 9½ x 12¾. 21308-0 Clothbound $12.50

STICKS AND STONES: A STUDY OF AMERICAN ARCHITECTURE AND CIVILIZATION, Lewis Mumford.One of the great classics of American cultural history. American architecture from the medieval-inspired earliest forms to the early 20th century; evolution of structure and style, and reciprocal influences on environment. 21 photographic illustrations. 238pp. 20202-X Paperbound $2.00

THE AMERICAN BUILDER'S COMPANION, Asher Benjamin. The most widely used early 19th century architectural style and source book, for colonial up into Greek Revival periods. Extensive development of geometry of carpentering, construction of sashes, frames, doors, stairs; plans and elevations of domestic and other buildings. Hundreds of thousands of houses were built according to this book, now invaluable to historians, architects, restorers, etc. 1827 edition. 59 plates. 114pp. 7⅞ x 10¾. 22236-5 Paperbound $3.50

DUTCH HOUSES IN THE HUDSON VALLEY BEFORE 1776, Helen Wilkinson Reynolds. The standard survey of the Dutch colonial house and outbuildings, with constructional features, decoration, and local history associated with individual homesteads. Introduction by Franklin D. Roosevelt. Map. 150 illustrations. 469pp. 6⅝ x 9¼. 21469-9 Paperbound $5.00

THE ARCHITECTURE OF COUNTRY HOUSES, Andrew J. Downing. Together with Vaux's *Villas and Cottages* this is the basic book for Hudson River Gothic architecture of the middle Victorian period. Full, sound discussions of general aspects of housing, architecture, style, decoration, furnishing, together with scores of detailed house plans, illustrations of specific buildings, accompanied by full text. Perhaps the most influential single American architectural book. 1850 edition. Introduction by J. Stewart Johnson. 321 figures, 34 architectural designs. xvi + 560pp.
22003-6 Paperbound $4.00

LOST EXAMPLES OF COLONIAL ARCHITECTURE, John Mead Howells. Full-page photographs of buildings that have disappeared or been so altered as to be denatured, including many designed by major early American architects. 245 plates. xvii + 248pp. 7⅞ x 10¾. 21143-6 Paperbound $3.50

DOMESTIC ARCHITECTURE OF THE AMERICAN COLONIES AND OF THE EARLY REPUBLIC, Fiske Kimball. Foremost architect and restorer of Williamsburg and Monticello covers nearly 200 homes between 1620-1825. Architectural details, construction, style features, special fixtures, floor plans, etc. Generally considered finest work in its area. 219 illustrations of houses, doorways, windows, capital mantels. xx + 314pp. 7⅞ x 10¾. 21743-4 Paperbound $4.00

EARLY AMERICAN ROOMS: 1650-1858, edited by Russell Hawes Kettell. Tour of 12 rooms, each representative of a different era in American history and each furnished, decorated, designed and occupied in the style of the era. 72 plans and elevations, 8-page color section, etc., show fabrics, wall papers, arrangements, etc. Full descriptive text. xvii + 200pp. of text. 8⅜ x 11¼.
21633-0 Paperbound $5.00

THE FITZWILLIAM VIRGINAL BOOK, edited by J. Fuller Maitland and W. B. Squire. Full modern printing of famous early 17th-century ms. volume of 300 works by Morley, Byrd, Bull, Gibbons, etc. For piano or other modern keyboard instrument; easy to read format. xxxvi + 938pp. 8⅜ x 11.
21068-5, 21069-3 Two volumes, Paperbound $10.00

KEYBOARD MUSIC, Johann Sebastian Bach. Bach Gesellschaft edition. A rich selection of Bach's masterpieces for the harpsichord: the six English Suites, six French Suites, the six Partitas (Clavierübung part I), the Goldberg Variations (Clavierübung part IV), the fifteen Two-Part Inventions and the fifteen Three-Part Sinfonias. Clearly reproduced on large sheets with ample margins; eminently playable. vi + 312pp. 8⅛ x 11. 22360-4 Paperbound $5.00

THE MUSIC OF BACH: AN INTRODUCTION, Charles Sanford Terry. A fine, nontechnical introduction to Bach's music, both instrumental and vocal. Covers organ music, chamber music, passion music, other types. Analyzes themes, developments, innovations. x + 114pp. 21075-8 Paperbound $1.50

BEETHOVEN AND HIS NINE SYMPHONIES, Sir George Grove. Noted British musicologist provides best history, analysis, commentary on symphonies. Very thorough, rigorously accurate; necessary to both advanced student and amateur music lover. 436 musical passages. vii + 407 pp. 20334-4 Paperbound $2.75

JOHANN SEBASTIAN BACH, Philipp Spitta. One of the great classics of musicology, this definitive analysis of Bach's music (and life) has never been surpassed. Lucid, nontechnical analyses of hundreds of pieces (30 pages devoted to St. Matthew Passion, 26 to B Minor Mass). Also includes major analysis of 18th-century music. 450 musical examples. 40-page musical supplement. Total of xx + 1799pp.
(EUK) 22278-0, 22279-9 Two volumes, Clothbound $17.50

MOZART AND HIS PIANO CONCERTOS, Cuthbert Girdlestone. The only full-length study of an important area of Mozart's creativity. Provides detailed analyses of all 23 concertos, traces inspirational sources. 417 musical examples. Second edition. 509pp. 21271-8 Paperbound $3.50

THE PERFECT WAGNERITE: A COMMENTARY ON THE NIBLUNG'S RING, George Bernard Shaw. Brilliant and still relevant criticism in remarkable essays on Wagner's Ring cycle, Shaw's ideas on political and social ideology behind the plots, role of Leitmotifs, vocal requisites, etc. Prefaces. xxi + 136pp.
(USO) 21707-8 Paperbound $1.50

DON GIOVANNI, W. A. Mozart. Complete libretto, modern English translation; biographies of composer and librettist; accounts of early performances and critical reaction. Lavishly illustrated. All the material you need to understand and appreciate this great work. Dover Opera Guide and Libretto Series; translated and introduced by Ellen Bleiler. 92 illustrations. 209pp.
21134-7 Paperbound $2.00

BASIC ELECTRICITY, U. S. Bureau of Naval Personel. Originally a training course, best non-technical coverage of basic theory of electricity and its applications. Fundamental concepts, batteries, circuits, conductors and wiring techniques, AC and DC, inductance and capacitance, generators, motors, transformers, magnetic amplifiers, synchros, servomechanisms, etc. Also covers blue-prints, electrical diagrams, etc. Many questions, with answers. 349 illustrations. x + 448pp. 6½ x 9¼.
20973-3 Paperbound $3.50

REPRODUCTION OF SOUND, Edgar Villchur. Thorough coverage for laymen of high fidelity systems, reproducing systems in general, needles, amplifiers, preamps, loudspeakers, feedback, explaining physical background. "A rare talent for making technicalities vividly comprehensible," R. Darrell, *High Fidelity*. 69 figures. iv + 92pp. 21515-6 Paperbound $1.25

HEAR ME TALKIN' TO YA: THE STORY OF JAZZ AS TOLD BY THE MEN WHO MADE IT, Nat Shapiro and Nat Hentoff. Louis Armstrong, Fats Waller, Jo Jones, Clarence Williams, Billy Holiday, Duke Ellington, Jelly Roll Morton and dozens of other jazz greats tell how it was in Chicago's South Side, New Orleans, depression Harlem and the modern West Coast as jazz was born and grew. xvi + 429pp.
21726-4 Paperbound $3.00

FABLES OF AESOP, translated by Sir Roger L'Estrange. A reproduction of the very rare 1931 Paris edition; a selection of the most interesting fables, together with 50 imaginative drawings by Alexander Calder. v + 128pp. 6½x9¼.
21780-9 Paperbound $1.50

AGAINST THE GRAIN (A REBOURS), Joris K. Huysmans. Filled with weird images, evidences of a bizarre imagination, exotic experiments with hallucinatory drugs, rich tastes and smells and the diversions of its sybarite hero Duc Jean des Esseintes, this classic novel pushed 19th-century literary decadence to its limits. Full unabridged edition. Do not confuse this with abridged editions generally sold. Introduction by Havelock Ellis. xlix + 206pp. 22190-3 Paperbound $2.00

VARIORUM SHAKESPEARE: HAMLET. Edited by Horace H. Furness; a landmark of American scholarship. Exhaustive footnotes and appendices treat all doubtful words and phrases, as well as suggested critical emendations throughout the play's history. First volume contains editor's own text, collated with all Quartos and Folios. Second volume contains full first Quarto, translations of Shakespeare's sources (Belleforest, and Saxo Grammaticus), Der Bestrafte Brudermord, and many essays on critical and historical points of interest by major authorities of past and present. Includes details of staging and costuming over the years. By far the best edition available for serious students of Shakespeare. Total of xx + 905pp.
21004-9, 21005-7, 2 volumes, Paperbound $7.00

A LIFE OF WILLIAM SHAKESPEARE, Sir Sidney Lee. This is the standard life of Shakespeare, summarizing everything known about Shakespeare and his plays. Incredibly rich in material, broad in coverage, clear and judicious, it has served thousands as the best introduction to Shakespeare. 1931 edition. 9 plates. xxix + 792pp. (USO) 21967-4 Paperbound $3.75

MASTERS OF THE DRAMA, John Gassner. Most comprehensive history of the drama in print, covering every tradition from Greeks to modern Europe and America, including India, Far East, etc. Covers more than 800 dramatists, 2000 plays, with biographical material, plot summaries, theatre history, criticism, etc. "Best of its kind in English," *New Republic*. 77 illustrations. xxii + 890pp.
20100-7 Clothbound $8.50

THE EVOLUTION OF THE ENGLISH LANGUAGE, George McKnight. The growth of English, from the 14th century to the present. Unusual, non-technical account presents basic information in very interesting form: sound shifts, change in grammar and syntax, vocabulary growth, similar topics. Abundantly illustrated with quotations. Formerly *Modern English in the Making*. xii + 590pp.
21932-1 Paperbound $3.50

AN ETYMOLOGICAL DICTIONARY OF MODERN ENGLISH, Ernest Weekley. Fullest, richest work of its sort, by foremost British lexicographer. Detailed word histories, including many colloquial and archaic words; extensive quotations. Do not confuse this with the Concise Etymological Dictionary, which is much abridged. Total of xxvii + 830pp. 6½ x 9¼.
21873-2, 21874-0 Two volumes, Paperbound $7.90

FLATLAND: A ROMANCE OF MANY DIMENSIONS, E. A. Abbott. Classic of science-fiction explores ramifications of life in a two-dimensional world, and what happens when a three-dimensional being intrudes. Amusing reading, but also useful as introduction to thought about hyperspace. Introduction by Banesh Hoffmann. 16 illustrations. xx + 103pp. 20001-9 Paperbound $1.00

POEMS OF ANNE BRADSTREET, edited with an introduction by Robert Hutchinson. A new selection of poems by America's first poet and perhaps the first significant woman poet in the English language. 48 poems display her development in works of considerable variety—love poems, domestic poems, religious meditations, formal elegies, "quaternions," etc. Notes, bibliography. viii + 222pp.
22160-1 Paperbound $2.50

THREE GOTHIC NOVELS: THE CASTLE OF OTRANTO BY HORACE WALPOLE; VATHEK BY WILLIAM BECKFORD; THE VAMPYRE BY JOHN POLIDORI, WITH FRAGMENT OF A NOVEL BY LORD BYRON, edited by E. F. Bleiler. The first Gothic novel, by Walpole; the finest Oriental tale in English, by Beckford; powerful Romantic supernatural story in versions by Polidori and Byron. All extremely important in history of literature; all still exciting, packed with supernatural thrills, ghosts, haunted castles, magic, etc. xl + 291pp.
21232-7 Paperbound $2.50

THE BEST TALES OF HOFFMANN, E. T. A. Hoffmann. 10 of Hoffmann's most important stories, in modern re-editings of standard translations: Nutcracker and the King of Mice, Signor Formica, Automata, The Sandman, Rath Krespel, The Golden Flowerpot, Master Martin the Cooper, The Mines of Falun, The King's Betrothed, A New Year's Eve Adventure. 7 illustrations by Hoffmann. Edited by E. F. Bleiler. xxxix + 419pp. 21793-0 Paperbound $3.00

GHOST AND HORROR STORIES OF AMBROSE BIERCE, Ambrose Bierce. 23 strikingly modern stories of the horrors latent in the human mind: The Eyes of the Panther, The Damned Thing, An Occurrence at Owl Creek Bridge, An Inhabitant of Carcosa, etc., plus the dream-essay, Visions of the Night. Edited by E. F. Bleiler. xxii + 199pp. 20767-6 Paperbound $1.50

BEST GHOST STORIES OF J. S. LEFANU, J. Sheridan LeFanu. Finest stories by Victorian master often considered greatest supernatural writer of all. Carmilla, Green Tea, The Haunted Baronet, The Familiar, and 12 others. Most never before available in the U. S. A. Edited by E. F. Bleiler. 8 illustrations from Victorian publications. xvii + 467pp. 20415-4 Paperbound $3.00

MATHEMATICAL FOUNDATIONS OF INFORMATION THEORY, A. I. Khinchin. Comprehensive introduction to work of Shannon, McMillan, Feinstein and Khinchin, placing these investigations on a rigorous mathematical basis. Covers entropy concept in probability theory, uniqueness theorem, Shannon's inequality, ergodic sources, the E property, martingale concept, noise, Feinstein's fundamental lemma, Shanon's first and second theorems. Translated by R. A. Silverman and M. D. Friedman. iii + 120pp. 60434-9 Paperbound $2.00

SEVEN SCIENCE FICTION NOVELS, H. G. Wells. The standard collection of the great novels. Complete, unabridged. *First Men in the Moon, Island of Dr. Moreau, War of the Worlds, Food of the Gods, Invisible Man, Time Machine, In the Days of the Comet.* Not only science fiction fans, but every educated person owes it to himself to read these novels. 1015pp. (USO) 20264-X Clothbound $6.00

LAST AND FIRST MEN AND STAR MAKER, TWO SCIENCE FICTION NOVELS, Olaf Stapledon. Greatest future histories in science fiction. In the first, human intelligence is the "hero," through strange paths of evolution, interplanetary invasions, incredible technologies, near extinctions and reemergences. Star Maker describes the quest of a band of star rovers for intelligence itself, through time and space: weird inhuman civilizations, crustacean minds, symbiotic worlds, etc. Complete, unabridged. v + 438pp. (USO) 21962-3 Paperbound $2.50

THREE PROPHETIC NOVELS, H. G. WELLS. Stages of a consistently planned future for mankind. *When the Sleeper Wakes,* and *A Story of the Days to Come,* anticipate *Brave New World* and *1984,* in the 21st Century; *The Time Machine,* only complete version in print, shows farther future and the end of mankind. All show Wells's greatest gifts as storyteller and novelist. Edited by E. F. Bleiler. x + 335pp. (USO) 20605-X Paperbound $2.50

THE DEVIL'S DICTIONARY, Ambrose Bierce. America's own Oscar Wilde—Ambrose Bierce—offers his barbed iconoclastic wisdom in over 1,000 definitions hailed by H. L. Mencken as "some of the most gorgeous witticisms in the English language." 145pp. 20487-1 Paperbound $1.25

MAX AND MORITZ, Wilhelm Busch. Great children's classic, father of comic strip, of two bad boys, Max and Moritz. Also Ker and Plunk (Plisch und Plumm), Cat and Mouse, Deceitful Henry, Ice-Peter, The Boy and the Pipe, and five other pieces. Original German, with English translation. Edited by H. Arthur Klein; translations by various hands and H. Arthur Klein. vi + 216pp. 20181-3 Paperbound $2.00

PIGS IS PIGS AND OTHER FAVORITES, Ellis Parker Butler. The title story is one of the best humor short stories, as Mike Flannery obfuscates biology and English. Also included, That Pup of Murchison's, The Great American Pie Company, and Perkins of Portland. 14 illustrations. v + 109pp. 21532-6 Paperbound $1.25

THE PETERKIN PAPERS, Lucretia P. Hale. It takes genius to be as stupidly mad as the Peterkins, as they decide to become wise, celebrate the "Fourth," keep a cow, and otherwise strain the resources of the Lady from Philadelphia. Basic book of American humor. 153 illustrations. 219pp. 20794-3 Paperbound $1.50

PERRAULT'S FAIRY TALES, translated by A. E. Johnson and S. R. Littlewood, with 34 full-page illustrations by Gustave Doré. All the original Perrault stories—Cinderella, Sleeping Beauty, Bluebeard, Little Red Riding Hood, Puss in Boots, Tom Thumb, etc.—with their witty verse morals and the magnificent illustrations of Doré. One of the five or six great books of European fairy tales. viii + 117pp. 8⅛ x 11. 22311-6 Paperbound $2.00

OLD HUNGARIAN FAIRY TALES, Baroness Orczy. Favorites translated and adapted by author of the *Scarlet Pimpernel.* Eight fairy tales include "The Suitors of Princess Fire-Fly," "The Twin Hunchbacks," "Mr. Cuttlefish's Love Story," and "The Enchanted Cat." This little volume of magic and adventure will captivate children as it has for generations. 90 drawings by Montagu Barstow. 96pp. 22293-4 Paperbound $1.95

THE RED FAIRY BOOK, Andrew Lang. Lang's color fairy books have long been children's favorites. This volume includes Rapunzel, Jack and the Bean-stalk and 35 other stories, familiar and unfamiliar. 4 plates, 93 illustrations x + 367pp.
21673-X Paperbound $2.50

THE BLUE FAIRY BOOK, Andrew Lang. Lang's tales come from all countries and all times. Here are 37 tales from Grimm, the Arabian Nights, Greek Mythology, and other fascinating sources. 8 plates, 130 illustrations. xi + 390pp.
21437-0 Paperbound $2.50

HOUSEHOLD STORIES BY THE BROTHERS GRIMM. Classic English-language edition of the well-known tales — Rumpelstiltskin, Snow White, Hansel and Gretel, The Twelve Brothers, Faithful John, Rapunzel, Tom Thumb (52 stories in all). Translated into simple, straightforward English by Lucy Crane. Ornamented with headpieces, vignettes, elaborate decorative initials and a dozen full-page illustrations by Walter Crane. x + 269pp. 21080-4 Paperbound **$2.00**

THE MERRY ADVENTURES OF ROBIN HOOD, Howard Pyle. The finest modern versions of the traditional ballads and tales about the great English outlaw. Howard Pyle's complete prose version, with every word, every illustration of the first edition. Do not confuse this facsimile of the original (1883) with modern editions that change text or illustrations. 23 plates plus many page decorations. xxii + 296pp.
22043-5 Paperbound $2.50

THE STORY OF KING ARTHUR AND HIS KNIGHTS, Howard Pyle. The finest children's version of the life of King Arthur; brilliantly retold by Pyle, with 48 of his most imaginative illustrations. xviii + 313pp. 6⅛ x 9¼.
21445-1 Paperbound $2.50

THE WONDERFUL WIZARD OF OZ, L. Frank Baum. America's finest children's book in facsimile of first edition with all Denslow illustrations in full color. The edition a child should have. Introduction by Martin Gardner. 23 color plates, scores of drawings. iv + 267pp. 20691-2 Paperbound $2.50

THE MARVELOUS LAND OF OZ, L. Frank Baum. The second Oz book, every bit as imaginative as the Wizard. The hero is a boy named Tip, but the Scarecrow and the Tin Woodman are back, as is the Oz magic. 16 color plates, 120 drawings by John R. Neill. 287pp. 20692-0 Paperbound $2.50

THE MAGICAL MONARCH OF MO, L. Frank Baum. Remarkable adventures in a land even stranger than Oz. The best of Baum's books not in the Oz series. 15 color plates and dozens of drawings by Frank Verbeck. xviii + 237pp.
21892-9 Paperbound $2.25

THE BAD CHILD'S BOOK OF BEASTS, MORE BEASTS FOR WORSE CHILDREN, A MORAL ALPHABET, Hilaire Belloc. Three complete humor classics in one volume. Be kind to the frog, and do not call him names . . . and 28 other whimsical animals. Familiar favorites and some not so well known. Illustrated by Basil Blackwell. 156pp. (USO) 20749-8 Paperbound $1.50

EAST O' THE SUN AND WEST O' THE MOON, George W. Dasent. Considered the best of all translations of these Norwegian folk tales, this collection has been enjoyed by generations of children (and folklorists too). Includes True and Untrue, Why the Sea is Salt, East O' the Sun and West O' the Moon, Why the Bear is Stumpy-Tailed, Boots and the Troll, The Cock and the Hen, Rich Peter the Pedlar, and 52 more. The only edition with all 59 tales. 77 illustrations by Erik Werenskiold and Theodor Kittelsen. xv + 418pp. 22521-6 Paperbound $3.50

GOOPS AND HOW TO BE THEM, Gelett Burgess. Classic of tongue-in-cheek humor, masquerading as etiquette book. 87 verses, twice as many cartoons, show mischievous Goops as they demonstrate to children virtues of table manners, neatness, courtesy, etc. Favorite for generations. viii + 88pp. 6½ x 9¼.
22233-0 Paperbound $1.25

ALICE'S ADVENTURES UNDER GROUND, Lewis Carroll. The first version, quite different from the final *Alice in Wonderland,* printed out by Carroll himself with his own illustrations. Complete facsimile of the "million dollar" manuscript Carroll gave to Alice Liddell in 1864. Introduction by Martin Gardner. viii + 96pp. Title and dedication pages in color. 21482-6 Paperbound $1.25

THE BROWNIES, THEIR BOOK, Palmer Cox. Small as mice, cunning as foxes, exuberant and full of mischief, the Brownies go to the zoo, toy shop, seashore, circus, etc., in 24 verse adventures and 266 illustrations. Long a favorite, since their first appearance in St. Nicholas Magazine. xi + 144pp. 6⅝ x 9¼.
21265-3 Paperbound $1.75

SONGS OF CHILDHOOD, Walter De La Mare. Published (under the pseudonym Walter Ramal) when De La Mare was only 29, this charming collection has long been a favorite children's book. A facsimile of the first edition in paper, the 47 poems capture the simplicity of the nursery rhyme and the ballad, including such lyrics as I Met Eve, Tartary, The Silver Penny. vii + 106pp. (USO) 21972-0 Paperbound $1.25

THE COMPLETE NONSENSE OF EDWARD LEAR, Edward Lear. The finest 19th-century humorist-cartoonist in full: all nonsense limericks, zany alphabets, Owl and Pussycat, songs, nonsense botany, and more than 500 illustrations by Lear himself. Edited by Holbrook Jackson. xxix + 287pp. (USO) 20167-8 Paperbound $2.00

BILLY WHISKERS: THE AUTOBIOGRAPHY OF A GOAT, Frances Trego Montgomery. A favorite of children since the early 20th century, here are the escapades of that rambunctious, irresistible and mischievous goat—Billy Whiskers. Much in the spirit of *Peck's Bad Boy,* this is a book that children never tire of reading or hearing. All the original familiar illustrations by W. H. Fry are included: 6 color plates, 18 black and white drawings. 159pp. 22345-0 Paperbound $2.00

MOTHER GOOSE MELODIES. Faithful republication of the fabulously rare Munroe and Francis "copyright 1833" Boston edition—the most important Mother Goose collection, usually referred to as the "original." Familiar rhymes plus many rare ones, with wonderful old woodcut illustrations. Edited by E. F. Bleiler. 128pp. 4½ x 6⅜. 22577-1 Paperbound $1.00

TWO LITTLE SAVAGES; BEING THE ADVENTURES OF TWO BOYS WHO LIVED AS INDIANS AND WHAT THEY LEARNED, Ernest Thompson Seton. Great classic of nature and boyhood provides a vast range of woodlore in most palatable form, a genuinely entertaining story. Two farm boys build a teepee in woods and live in it for a month, working out Indian solutions to living problems, star lore, birds and animals, plants, etc. 293 illustrations. vii + 286pp.
20985-7 Paperbound $2.50

PETER PIPER'S PRACTICAL PRINCIPLES OF PLAIN & PERFECT PRONUNCIATION. Alliterative jingles and tongue-twisters of surprising charm, that made their first appearance in America about 1830. Republished in full with the spirited woodcut illustrations from this earliest American edition. 32pp. 4½ x 6⅜.
22560-7 Paperbound $1.00

SCIENCE EXPERIMENTS AND AMUSEMENTS FOR CHILDREN, Charles Vivian. 73 easy experiments, requiring only materials found at home or easily available, such as candles, coins, steel wool, etc.; illustrate basic phenomena like vacuum, simple chemical reaction, etc. All safe. Modern, well-planned. Formerly *Science Games for Children.* 102 photos, numerous drawings. 96pp. 6⅛ x 9¼.
21856-2 Paperbound $1.25

AN INTRODUCTION TO CHESS MOVES AND TACTICS SIMPLY EXPLAINED, Leonard Barden. Informal intermediate introduction, quite strong in explaining reasons for moves. Covers basic material, tactics, important openings, traps, positional play in middle game, end game. Attempts to isolate patterns and recurrent configurations. Formerly *Chess.* 58 figures. 102pp. (USO) 21210-6 Paperbound $1.25

LASKER'S MANUAL OF CHESS, Dr. Emanuel Lasker. Lasker was not only one of the five great World Champions, he was also one of the ablest expositors, theorists, and analysts. In many ways, his Manual, permeated with his philosophy of battle, filled with keen insights, is one of the greatest works ever written on chess. Filled with analyzed games by the great players. A single-volume library that will profit almost any chess player, beginner or master. 308 diagrams. xli x 349pp.
20640-8 Paperbound $2.75

THE MASTER BOOK OF MATHEMATICAL RECREATIONS, Fred Schuh. In opinion of many the finest work ever prepared on mathematical puzzles, stunts, recreations; exhaustively thorough explanations of mathematics involved, analysis of effects, citation of puzzles and games. Mathematics involved is elementary. Translated bv F. Göbel. 194 figures. xxiv + 430pp. 22134-2 Paperbound $3.50

MATHEMATICS, MAGIC AND MYSTERY, Martin Gardner. Puzzle editor for Scientific American explains mathematics behind various mystifying tricks: card tricks, stage "mind reading," coin and match tricks, counting out games, geometric dissections, etc. Probability sets, theory of numbers clearly explained. Also provides more than 400 tricks, guaranteed to work, that you can do. 135 illustrations. xii + 176pp.
20335-2 Paperbound $1.75

$a + ar + ar^2 + \ldots ar^n = S_n$

$rS_n = ar + ar^2 + ar^3 + \ldots + ar^n$

$rS_n = S_n - a$

$rS_n + a = a + ar + ar^2 + \ldots ar^n + \ldots$

$S_n + \frac{a}{r} = \frac{a}{r} + a + ar^1 + \ldots + ar^n + \ldots$

$S_n + \frac{a}{r} - \frac{a}{r}$

$S_n = 1 + x + x^2 + \ldots + x^n$

$\frac{S_n - 1}{x} = S_n \qquad S_n - 1 = S_n x$

$S_n - 1 - S_n x = S_n\left(1 - \frac{1}{S_n} - x\right) = 0$

B&T 3-800